Tolley's
Fire Safety Management Handbook

2nd Edit

Whilst every care has been taken to ensure the accuracy of the contents of this work, no responsibility for loss occasioned to any person acting or refraining from action as a result of any statement in it can be accepted by the authors or publisher.

Tolley's Fire Safety Management Handbook

2nd Edition

By Ron Miller FIFireE, QFSM
of Fire Safety Consultants Ltd

Members of the LexisNexis Group worldwide

United Kingdom	LexisNexis UK, a Division of Reed Elsevier (UK) Ltd, 2 Addiscombe Road, Croydon CR9 5AF
Argentina	LexisNexis Argentina, BUENOS AIRES
Australia	LexisNexis Butterworths, CHATSWOOD, New South Wales
Austria	LexisNexis Verlag ARD Orac GmbH & Co KG, VIENNA
Canada	LexisNexis Butterworths, MARKHAM, Ontario
Chile	LexisNexis Chile Ltda, SANTIAGO DE CHILE
Czech Republic	Nakladatelství Orac sro, PRAGUE
France	Editions du Juris-Classeur SA, PARIS
Germany	LexisNexis Deutschland GmbH, FRANKFURT and MUNSTER
Hong Kong	LexisNexis Butterworths, HONG KONG
Hungary	HVG-Orac, BUDAPEST
India	LexisNexis Butterworths, NEW DELHI
Ireland	Butterworths (Ireland) Ltd, DUBLIN
Italy	Giuffrè Editore, MILAN
Malaysia	Malayan Law Journal Sdn Bhd, KUALA LUMPUR
New Zealand	LexisNexis Butterworths, WELLINGTON
Poland	Wydawnictwo Prawnicze LexisNexis, WARSAW
Singapore	LexisNexis Butterworths, SINGAPORE
South Africa	LexisNexis Butterworths, Durban
Switzerland	Stämpfli Verlag AG, BERNE
USA	LexisNexis, DAYTON, Ohio

First published in 2000
© Reed Elsevier (UK) Ltd 2003

All rights reserved. No part of this publication may be reproduced in any material form (including photocopying or storing it in any medium by electronic means and whether or not transiently or incidentally to some other use of this publication) without the written permission of the copyright owner except in accordance with the provisions of the Copyright, Designs and Patents Act 1988 or under the terms of a licence issued by the Copyright Licensing Agency Ltd, 90 Tottenham Court Road, London, England W1T 4LP. Applications for the copyright owner's written permission to reproduce any part of this publication should be addressed to the publisher.

Warning: The doing of an unauthorised act in relation to a copyright work may result in both a civil claim for damages and criminal prosecution.

Crown copyright material is reproduced with the permission of the Controller of HMSO and the Queen's Printer for Scotland. Any European material in this work which has been reproduced from EUR-lex, the official European Communities legislation website, is European Communities copyright.

A CIP Catalogue record for this book is available from the British Library.

ISBN 0 754 520935

Typeset by Letterpart Ltd, Reigate, Surrey
Printed and bound in Great Britain by Antony Rowe Ltd, Chippenham, Wilts

Visit LexisNexis UK at www.lexisnexis.co.uk

Foreword

Every year fire statistics show the extent of the tragedies that can be caused by fire. I am sure we all appreciate that it will not be possible to completely eliminate the incidence of fire. However, by the use of sensible fire safety management the frequency of fire in the workplace can be significantly reduced. The preparation of an emergency plan, indicating the fire routine that should be adopted, detailed written instructions for all key members of staff and periodic fire instruction, will all assist this process. Fire risk assessment is the key to ensure that the appropriate fire safety standard is achieved.

This book covers all the main acts, ie the *Fire Precautions Act 1971* and the *Fire Precautions (Workplace) Regulations 1997* and *1999*. Fire safety officers and local authority officers, together with relatively inexperienced members of an employer's staff who may have suddenly become responsible for fire safety, will be able to benefit from explanations covering the subject being written in clear and concise terms. This edition outlines what legislation requires and offers suggested solutions. However experienced the user may be, the various 'checklists' will be of great assistance.

This book will be a useful addition to anyone involved in fire safety management.

Gerald D Clarkson

CBE, QFSM, BA (Hons), Hon FIFireE, FRSH

Former Chief Officer, London Fire Brigade

Chief Executive, London Fire and Civil Defence Authority

Preface

Since the publication of the first edition of this handbook, the tragic events of the 9 September 2001 (9/11) have occurred. The overwhelming loss of life in the 'Twin Towers', New York City has concentrated the minds of most people with regard to the effects of fire when a terrorist attack of this kind occurs. The emergency plans that relate to individual premises must be kept under continual review to ensure that they are up to date.

It must be remembered that the fire safety arrangements that are provided must be maintained at all times. When fire safety equipment has to be tested in accordance with a specific British Standard this is, to all intents and purposes, a requirement. The aspect of adequate testing will be raised at relevant points later in this handbook.

Irrespective of the particular and special circumstances relating to the 'Twin Towers', fire can be a devastating event for any business, whether the business is a corner shop or a large manufacturing plant. Not only does fire destroy property, it also destroys lives. According to statistics published by the Office of the Deputy Prime Minister, during 2000, 25 people died and nearly 1,700 people were injured in fires in commercial or similar premises.

From a statistical viewpoint more than 70 per cent of businesses involved in a major fire either do not re-open or subsequently fail within three years of the fire.

Most businesses are well insured and all losses, which result from a fire, will be covered. Theoretically this is correct, but what a business is actually doing in taking out fire insurance is transferring the majority of the risk to the insurance company. The management of many commercial enterprises fail to appreciate that although the more obvious aspects of fire loss are covered, such as the cost of rebuilding and the replacement of stock and equipment, many of the less obvious losses are not covered.

A fire insurance policy may include an element of cover for a 'consequential loss', however, it is extremely difficult to quantify and anticipate all the potential losses and business interruption problems, which result from fire.

Some examples of the less obvious 'knock-on' effects of a fire are:
- damaged reputation;
- financial insecurity;
- loss of key employees;

Preface

- loss of market share and loss of customers; and
- loss of key suppliers.

No amount of insurance will provide cover for all of these potential losses.

The cost of fire claims on commercial and domestic properties topped £1 billion in 2001 for the first time in ten years, and are now running at record levels according to figures released on 19 June 2002 by the Association of British Insurers (ABI).

With the cost of commercial property fire claims in 2001 rising by 30 per cent to £679 million in 2000, the ABI is warning businesses that greater emphasis on fire prevention is needed if they are to minimise the impact of the current rises in commercial insurance premiums.

John Parker, the Association of British Insurers, Head of General Insurance, said on 19 June 2002:

> 'The current level of fire losses must be reduced. The insured losses are bad enough, but they are only part of a wider disturbing picture. Fires are estimated to cost the UK economy nearly £7 billion a year. Arson in particular remains a major problem.
>
> 'Businesses must face the fact that they need to put more effort and resources into reducing the risk of fire. Effective risk management is key, especially as there has been an increase in large-scale fires. All businesses need to ensure they have comprehensive risk management programmes to identify the fire dangers they face, and implement measures to minimise the risks. Costly disruption, lost productivity, risk of business failure and higher operating costs, including more expensive insurance premiums, all follow from the failure to treat the risk of fire seriously.
>
> 'Since the events of September 11th, 2001 insurers have been faced with rising costs in obtaining reinsurance cover. These losses exacerbate the current trend in increasing commercial premiums being faced by many businesses.'

This Fire Safety Management Handbook has been produced to assist in providing a sensible approach to fire safety and the primary aims of this handbook are to:

(i) provide owners, managers and employees with sound advice on how to prevent fires from occurring within their businesses; and

(ii) help ensure that if the worst happens, and a business does suffer the devastating effects of a fire, their business is in the 30 per cent of businesses, which continue to trade.

Who will find this handbook helpful?

The handbook should be particularly helpful to owners and managers of small to medium sized enterprises who have a legal responsibility to ensure the health and safety of their employees. It will also be helpful to those individuals, in larger organisations, when they are given the 'day-to-day' fire safety responsibility within their firm, possibly in addition to their normal role within the organisation. This additional role is quite a commonplace occurrence within smaller organisations, where the appointment of a full-time fire safety manager cannot be justified. Nevertheless the organisation has a legal responsibility to discharge its health and safety responsibilities.

This book offers practical advice on the fire safety and fire risk management issues, which all organisations are likely to encounter at some time during their normal 'day-to-day' business activities.

As far as practicable, the material has been deliberately laid out in a readable format designed to provide clear information on particular problems avoiding, wherever possible, the use of legal or technical jargon. However, if further research is necessary, it seeks to highlight other publications, ie Codes of Practice and Standards, websites and organisations that will provide further information.

Who should manage fire safety?

In many workplaces the management of fire safety is not given sufficient priority and importance within the overall management of that organisation. This is perhaps because owners and managers consider that fire safety is a 'non-productive' area of their business and one where they cannot see a return on their investment. There is also perhaps an element of the 'it will never happen to us' psychology.

This handbook advocates sound fire prevention, fire safety and fire risk management. These matters should all be seen as vital and integrated parts of the overall management of any successful business.

Are there any cost benefits?

It is virtually impossible to carry out a conventional cost benefit analysis on a business's investment in preventing fires. The problem with fire prevention is that when it is done effectively, there will be a cost. The only benefit or measure of how successful such investment has been is that the business never suffers a fire. An employer will never know how many fires have actually been prevented.

Preface

From a different perspective, if the business continues to prosper and employees have job security, then it is arguable that these are the cost benefits.

Who needs to do it?

Firstly, if an employer employs at least one person, or is the owner or occupier of premises, which require a fire certificate under the *Fire Precautions Act 1971*, there is a legal responsibility to manage fire safety. In **CHAPTER 2: LEGISLATION AND REGULATIONS (FIRE PRECAUTIONS ACT 1971)** details the fire safety responsibilities under the *Fire Precautions (Workplace) Regulations 1997 (SI 1997 No 1840)* (as amended) and the *Fire Precautions Act 1971*.

Secondly, an employer is bound by the general or 'catch-all' provision of the *Health and Safety at Work etc Act 1974, section 2(1)*, which states that:

> 'It shall be the duty of every employer to ensure, so far as is reasonably practicable, the health, safety and welfare at work of all his employees.'

Can an employer do it?

The short answer is 'yes', if there is sufficient time available to do it effectively. Much will depend on the complexity and nature of the business, the size of the premises and the number of employees. Obviously the larger the business the more time consuming the task becomes. This handbook aims to provide the necessary information needed to carry out the task 'in-house', if an employer or owner has the relevant expertise.

Where can an employer get help?

A good starting point, where information on fire safety is needed, is the local fire service. However, with the increasing trend in fire safety legislation for 'self-compliance', coupled with the financial constraints that have been imposed upon fire authorities, the amount of assistance provided by the fire service may be limited.

The trend for 'self-compliance', has been spearheaded by the *Fire Precautions (Workplace) Regulations 1997 (SI 1997 No 1840)* which came into force on 1 December 1997, as a result of an EC initiative which led to the drafting of various EC Directives. These Directives were designed to introduce a common approach to the health and safety of workers across all member states within the European Union. The 1997 Regulations were subsequently amended by the *Fire Precautions (Workplace) (Amendment) Regulations 1999 (SI 1999 No 1877)*, which came in to effect on 1 December 1999. These Regulations are modelled on a typically European approach.

Premises are no longer policed in the same way by the enforcing authority. Employers are now required to find out for themselves how to comply with legislation.

As a consequence of this trend, many companies now 'buy-in' fire safety advice and expertise, as and when needed, from one of the independent fire safety consultancy firms.

If or when a company uses fire safety consultants, it is wise to ensure that their inspectors hold nationally recognised fire engineering qualifications and have adequate and appropriate public indemnity insurance cover. It may also be prudent to obtain references from other companies and organisations, for whom these consultants may have worked. Reputable consultancy firms will be only too pleased to supply you with the information required.

A list of fire safety consultants, many of whom operate on a nationwide basis, can be obtained from the Institution of Fire Engineers, 148 New Walk, Leicester, LE1 7QB. (tel: 0116 255 3654).

E-mail: General Enquiries: www.info@ife.org.uk
Ron Miller
April 2003

Contents

1	**Causes of fire**	
	Main causes of fire	1
	Arson	1
	Construction and refurbishment	7
2	**Legislation and Regulations (Fire Precautions Act 1971)**	
	Introduction	8
	What is a fire certificate?	14
	What does a fire certificate contain?	20
	Fire authority's power of inspection	24
	Improvement Notice	28
	Which premises are exempt from needing a fire certificate?	30
	Fire authority inspectors' powers of inspection	36
	Falsification	38
	Offences by bodies corporate	39
	Defence of 'due diligence'	40
	What are 'special premises'?	40
	Publications providing further guidance on fire safety and Fire Precautions Act 1971	44
3	**Legislation and Regulations (The Fire Precautions (Workplace) Regulations 1997 and 1999)**	
	Introduction	46
	European approach	47
	What do the Regulations cover?	50
	What to do after completing a fire risk assessment	51
	Co-operation and co-ordination	52
	Enforcement	54
	Serious cases: offence	54
	Serious cases: Prohibition Notices	55
	Enforcement Notices	57
	Enforcement Notices: rights of appeal	60
	Enforcement Notice: offence	61
	Where the Regulations do not apply	62
	Other major legislation relating to fire safety in the workplace	63
4	**Legislation and Regulations (Disability Discrimination Act 1995)**	
	Disability Discrimination Act 1995	70
	How does 'Part III – Access to Goods and Services' apply?	71

What are auxiliary aids and services? 75

5 Legislation and Regulations (British and European Standards)
Introduction 79
What are they? 80
Relationship with International and European Standards 81
Contacting the British Standards Institution 82

6 Legislation and Regulations (Health and Safety)
Introduction 83
Recent trends in United Kingdom health and safety law 84
High profile cases and future trends 88
Risk assessments 90
Recent changes in legislation 92
Amendments to existing Regulations 98
Criminal law 99
Levels of duty 103
'Six pack' 104
Other important Regulations 105
Management of health and safety at work 110
Application of the 'management cycle' 114

7 How to carry out a fire risk assessment
Introduction 120
'Five steps' approach 124
Home Office guidance 127
Fire risk categories 129
General principle for escape routes 130
Means of fighting fire 135
Maintenance and testing of fire safety equipment 138

8 Managing fire safety (awareness, prevention and fire safety policy)
Fire awareness and prevention 145
Formulating a fire safety policy for the workplace 145
Developing a 'fire-safe' environment 146
Who should be involved in the process? 146
Fire marshals 147
Raising staff awareness 147
Planning for emergencies: suggested fire safety checklists 148
What is risk management? 153
Policy for the disabled, the public and visitors 154
Protecting computer records 154
Bomb alerts and threats 155
Plan of action 155
Bomb call checklist 156
How the plan should work 157
Home Office guidance 159

	Bombs: protecting people and property – A Guide for Small Business	159
	Bombs – some useful information	160
	Taking precautions	163
	What to do if a telephone bomb threat is received	166

9 Managing fire safety (fire safety procedures, training and fire manual)

Introduction	169
Legal responsibilities	170
Fire safety training policy	172
Fire Training Manual	175
Damage control/salvage	186
Where to obtain assistance with training	187

10 Managing fire safety (prevention of fire and liaison with authorities)

Introduction	188
Statistics	189
Legal obligation	189
Main causes of fire	189
A proactive approach	190
The 'triangle of fire'	190
Fuel	191
Health and Safety Executive (HSE)	201
Health and Safety Executive – what they do	201
Where to get help and advice	202

11 Managing fire safety (fire defence including active and passive fire precautions)

Introduction	206
Means of escape – the principles	208
Means of escape – structural protection	210
Means of escape – measures to facilitate	214
Fire warning	220
Fire alarms	220
Automatic fire detection systems	221
Fire-fighting extinguishers – types of fire	224
Types of fire extinguisher	225
Hose reels	230
Automatic sprinklers	231
Fixed fire suppression systems	234
Foam inlets	235
Dry risers	236
Wet risers	236
Emergency procedures – staff training	236
Evacuation drills	238
Fire action notices	239
Preventative measures – planned maintenance programmes	240

Portable fire-fighting equipment	242
Fixed fire-fighting equipment	243
Fire alarm systems	243
Signs and notices	243
Emergency lighting	244
Security staff and systems	244
Insurance	245

12 Managing fire safety (maintenance, servicing routines and record keeping)

Introduction	246
New systems	247
Fire alarm and automatic fire detection systems	247
Fire alarms and automatic fire detection system: servicing routines and record entries	252
Fire-fighting equipment (portable fire extinguishers)	252
Fire-fighting equipment (portable)	253
Escape or emergency lighting	254
Stand-by generator	257
Automatic sprinkler systems	259
Pressurisation systems (smoke control using pressure differentials)	262
Fire doors, fire shutters and dampers	268
Smoke control systems	270
Conclusion	271

13 Specialist information

Introduction	272
Historic buildings	273
Means of escape	275
Risk management	275
Structural alterations and extensions	277
Legal responsibility	278
Building control	278
Planning control	279
Dealing with contractors	279
Code of practice	280
Advising contractors of your fire safety policy	282
Other potential hazards with contractors	284
Construction (Design and Management) Regulations 1994 (CDM Regulations)	285
To what do the Regulations apply?	286
Health and safety plan	291
Preparation of health and safety plan by the planning supervisor	292
Where specific guidance be found	293
Fire engineering	303
Useful sources of information	305

14	**Fire (before, during and after)**	
	Introduction	307
	Planning	310
	What are 'salvage efforts'?	313
	Review and test your plan	314
15	**The way ahead**	
	Introduction	317
	Possible reform of fire safety legislation	318
	How will these proposals be taken forward, and when will they be implemented?	319
	The current position	321
	Alterations to some legislation	324
	Power to make regulations	334
	The responsible person	335
	Application	337
	Mitigation of the effects of a fire	338
	Requirements of the proposed fire safety regime	340
	Bain Report	360
Appendix	**Sources of further information**	
	Introduction	364
	Means of escape in case of fire	364
	Fire alarms and automatic fire detection systems	365
	Fire extinguishing installations and equipment on premises	367
	Fire tests on building materials and structures	367
	Other British Standards and Codes of Practice relating to fire	368
	Other sources	374
	Websites	375

Table of Cases — 376

Table of Statutes — 377

Table of Statutory Instruments — 381

Table of European Legislation — 386

Index — 387

1 Causes of fire

In this chapter:	
Main causes of fire	1.1
Arson	1.2
Construction and refurbishment	1.7

Main causes of fire

1.1 The main causes of accidental fire in commercial premises, as shown in the current Home Office (now Office of the Deputy Prime Minister) Fire Statistics for England and Wales, 2000, have remained fairly static during recent years. Electrical appliances (about 60,000) are a major ignition source and head the causes of fire, closely followed by cooking appliances (58,000), smoking materials (22,000) and electrical distribution (24,000). The only encouraging aspect is the fact that the percentage of fires attributed to being caused by smoking materials has reduced by more than 40% since 1990. This is presumably due to the fact that many commercial buildings have instituted 'No Smoking' regimes or limited smoking to clearly defined areas.

During this period electrical fires have increased by about 50%, which is almost certainly due to the increase in the number of electrical appliances, eg computers that are now used in commercial premises.

Arson

1.2 Employers, including owners and occupiers, are responsible for the fire safety or security within their company. They should be aware that one of the most serious fire risks faced today is from a fire that is started deliberately.

This worrying increase in the number of deliberate fires resulted in the establishment, in 1991, of the Arson Prevention Bureau. This joint initiative between the Association of British Insurers and the Home Office was set up in order to spearhead and co-ordinate a national campaign to reduce the incidence and cost of arson.

A study was instituted by the Home Office (now Office of the Deputy Prime Minister) in order to provide the Government with a snapshot of the arson

1.3 *Causes of fire*

problems in England and Wales. The study was established in the knowledge that the number of deliberately started fires was rapidly increasing.

The study, published in 1999, showed that the cost of arson fires has now reached over £1.3 billion a year. In the last ten years (to 1999), there have been around 1.7 million arson fires, resulting in 22,000 injuries and 1,100 deaths. Between 1986 and 1996 the number of deliberately started fires increased each year. This now means that in an average week, arson results in 3,500 deliberately started fires, 50 injuries, two deaths and a cost to society of at least £25 million.

The growth in the number of arson fires is also a major problem for the fire service as it accounts for almost 45% of all serious fires.

Arson is a very real and ever-present threat to shops, offices, storage premises, factories, hotels, restaurants, hospitals, schools and churches. It would appear that no building is immune and that arson is often associated with vandalism and burglary.

Who would do it?

1.3 Following several studies in recent years, it is apparent that arson fires are started for a variety of reasons. However, the most common are vandalism, fraud, revenge, and concealment of another crime.

It is important for an employer and the business to have an awareness and an understanding of the type of person who may deliberately start fires.

In order to try to prevent an arson attack on premises and to give an employer an indication of what to look for, set out below are the types of person who may well start fires:

- Vandals with no motive other than an urge to be destructive;
- Thieves and burglars who may start a fire to cover their crime;
- People with mental health problems;
- Employees or ex-employees with a grudge against an employer or business;
- Glory seekers looking for an opportunity to display heroic qualities;
- Excitement seekers (sometimes of a sexual nature) who derive pleasure from watching a fire burning, or seeing the fire service in action;
- Fraudsters hoping to make an insurance claim or;
- Extremists and terrorists.

What to do to prevent it happening

1.4 Preventing an arson attack should be an integral part of the overall fire safety policy.

As there is invariably a link between a potential arson attack and the security of the premises, a high level of security will undoubtedly result in greater protection against arsonists.

An employer should consider the ease with which intruders and potential arsonists could break into the premises and should endeavour to strengthen any identified weaknesses in the defences.

If there have been any small fires on the premises or on neighbouring premises an employer should inform the police immediately as well as calling the fire brigade. A small fire could be a warning of something worse to come.

One of the most common causes of fires on business premises are those started by vandals in rubbish bins and skips located adjacent to buildings, which then spread and involve the fabric of the building. In commercial premises these rubbish containers can be quite large, sometimes made of plastic and contain large volumes of flammable materials. If vandals set them on fire they can generate a lot of radiated heat and flame, which could easily spread to an adjacent building.

The simplest and cheapest method of reducing this risk is to locate your waste bins and skips well away from buildings.

Arson alert! Stop your building becoming an arson statistic

1.5 An employer can do much to reduce the risk to the premises by being aware of the potential threat and adopting the following advice that has been issued by the Arson Prevention Bureau:

- security;
- one entrance is better than two, especially if that one entrance is manned throughout the day;
- outer fences, walls and gates need to be high enough and strong enough to keep out intruders;
- doors and windows must be in good repair and locked when not in use;
- use good quality locks and padlocks;
- gaps under doors should be as small as possible;.
- letterboxes should have metal containers fitted on the inside;

1.5 *Causes of fire*

- know who holds keys to your premise and chase up any that are missing;
- stored material of any kind should not be stacked adjacent to fences or walls where it could be set alight from outside;
- employees;
- warn staff of the threat of arson;
- ask them to challenge anyone who should not be on the premises and report any suspicious activities;
- vet new employees;
- keep an eye on outside contractors;
- visitors;
- the movement of visitors within the building should be controlled;
- fire protection;
- ensure that the fire defence equipment which you have installed, such as extinguishers, hose reels, fire alarms, fire detectors, sprinklers etc, are in good working order and protected against sabotage attempts;
- end of day checks;
- an employer or a named individual must be responsible for securing the building at the end of each working day;
- check that doors and windows are secure;
- check that no combustible materials are left lying around;
- check that no unauthorised people remain on the premises;
- check that security alarms are switched on;
- check that any outside illumination is on; and
- check that any flammable liquids have been locked away in secure store.

Further help and advice on protection of premises is also available from insurers, the fire safety department of the local fire service and the Crime Prevention Department of the local police force.

On 18 October 2002, prior to the fire fighters strike action the Association of British Insurers (ABI) issued a press release in which it is urging all business owners to review their fire prevention and management procedures, and to reduce the risk of suffering an arson attack.

The Association of British Insurers confirmed that every year there are 42,000 malicious fire attacks on businesses. The overall cost commercial fire claims to insurers is £680 million a year – nearly £2 million every day.

Many arson fires can be prevented by businesses paying more attention to basic security steps, such as those outlined below.

Causes of fire 1.5

- Ensuring adequate security on the premises. This includes using good quality locks and padlocks; check that building is secure at the end of each day; make sure that perimeter walls, fences and gates are high and strong enough to deter intruders; flammable material should not be stored next to perimeters, and limit who holds keys to designated staff.
- Make sure that fire detection and reduction equipment works. This includes extinguishers, alarms and sprinklers.
- Have in place a system of visitor identification, so that all visitors have to identify themselves on entry, and do not wander around the premises on their own.
- Ensure that employees are aware of the arson danger, and encourage them to report anything or anyone suspicious, and ensure they know the procedure if there is a fire.

The Association of British Insurers are constantly emphasising, in one way or another the following points:

> 'Arson is a constant threat to all businesses. Yet fire management is too far down the list of priorities of many businesses. Even a small-scale attack can cause extensive and expensive damage and disruption, which can have crippling financial implications. Despite this many arson attacks are preventable, simply by business managers being aware of the danger, and having in place adequate fire reduction policies. It has never been more important for every business to review and if necessary improve their fire management policy'.

This advice complements that issued by the Arson Prevention Bureau. It is most important as it is possible that if the rate at which malicious fires continues to increase there may be practical difficulties in obtaining appropriate insurance cover.

The prevalence of arson in England and Wales fire statistics as shown in the Home Office (now Office of the Deputy prime Minister) Fire Statistics for England and Wales are compiled from the fire report forms completed by the fire brigade for every fire that they attend during the year. However, the fire brigade is not called out to every fire that occurs – many are relatively small, cause little damage and are dealt with by members of the public with recourse to the assistance of the brigade. This under reporting means that the full extent of the arson problem is quite likely to be underestimated.

The Home Office define arson attacks as 'malicious fires' and includes those where malicious or deliberate ignition is merely suspected and recorded by the fire brigade as 'doubtful'.

Fire service data shows that there has been a continued long-term increase in the number of deliberate fires attended by the fire service. It is suggested that

1.5 *Causes of fire*

this may be due in part to better fire investigation techniques by brigades and the concerted effort to move away from attributing the cause of fire as 'unknown' in recent years. Nevertheless, these factors alone probably do not account for the very steep upward movement in the number of malicious or deliberate fires.

In 1989, there were 38,700 fires started deliberately, representing about one-quarter of all primary fires. By 1999, this had risen to a total of 91,000 deliberately set fires – an increase of 135%. This compares to a 25% increase for the overall number of primary fires and a 12% decrease for accidental fires over the same period. A large proportion of the increase is attributable to deliberate car fires as this category has increased by almost 200% since 1989. Over the same period, this compares with increases of 39% in deliberate fires in other buildings; 54% in dwellings and 65% in other outdoor fires.

Analysis of fire service data by fire brigade area shows that deliberate fires are highly concentrated in the metropolitan brigades. Almost half of all deliberate fires within England and Wales were concentrated within the seven metropolitan areas.

The number of deliberate fires as a percentage of total fires showed that all the metropolitan brigades recorded over half of all fires as deliberately started (except London where 40% of all fires were deliberate). West Yorkshire (63%), Merseyside (60%), South Yorkshire (59%) and Greater Manchester (58%) recorded the highest rates. However, the highest rate nationally in 1999 was in Cleveland (64%) – a non-metropolitan brigade. South Wales (61%) recorded the highest rate in Wales. Other high rates among non-metropolitan brigades were recorded by Nottinghamshire (57%), Avon (55%) and Humberside (55%). In contrast, the lowest recorded rates were in Devon (28%) and Cornwall (24%).

These variations in fire service data are mirrored in recorded police data. Arson offences appear to be concentrated in the metropolitan areas with 43% of all arson fires being recorded in these localities.

There are also wide variations in the clear-up rate across the police force areas. The national average in 1997 was 16%. Only twelve police forces cleared up more than one-fifth of all recorded arson offences. The forces with the highest clear-up rates were Dyfed-Powys (38%); Cheshire (36%); Lincolnshire (34%) and Wiltshire (30%). Amongst the lowest were Northumbria (12%); the Metropolitan Police Service (11%); Merseyside (11%); West Yorkshire and Cleveland (9%).

This information, from these official sources, is of vital importance to employers as they must appreciate that not only can their premises suffer a damaging fire, with the real possibility that they may not be able to resume trading, but the perpetrator of the crime may not be apprehended.

How to get further information about arson

1.6 If you would like further information on how to prevent arson in your premises contact:

The Arson Prevention Bureau, http://www.arsonpreventionbureau.org.uk
Telephone: 0207 216 7474, fax: 0207 696 8996

(Source: Arson Scoping Study)

Construction and refurbishment

1.7 This is probably the most dangerous period in the life of a building for a fire to occur. In addition to many of the causes outlined earlier in this Chapter there are at least 1,200 fires, which are attributed to the use of blowlamps, cutting and welding equipment. It is essential for employers to ensure that there is a 'hot work' procedure in their premises.

2 Legislation and Regulations (Fire Precautions Act 1971)

In this chapter:	
Introduction	2.1
What is a fire certificate?	2.6
What does a fire certificate contain?	2.18
Fire authority's power of inspection	2.22
Improvement Notice	2.26
Which premises are exempt from needing a fire certificate?	2.28
Fire authority inspectors' powers of inspection	2.30
Falsification	2.32
Offences by bodies corporate	2.34
Defence of 'due diligence'	2.36
What are 'special premises'?	2.38
Publications providing further guidance on fire safety and Fire Precautions Act 1971	2.40

Introduction

2.1 Like most previous UK fire safety legislation, the *Fire Precautions Act 1971* which came into force at the beginning of the 1970s was as a direct result of a succession of fires which had involved serious loss of life. After several fatal hotel fires, throughout the country, the fire which is generally recognised to have prompted the Government to act, occurred at the Rose and Crown Hotel, Saffron Walden on Boxing Day 1969, when eleven people died and a number of people were injured.

The *Fire Precautions Act 1971* (*FPA 1971*) was a new departure for fire safety legislation, as it was drafted as an 'enabling' Act. In effect, this meant that the legislation enabled the Secretary of State to 'designate', by means of statutory instrument, premises to which *FPA 1971* could be applied. Prior to

Legislation and Regulations (Fire Precautions Act 1971) 2.1

this, Parliament had enacted legislation, in which the specific fire safety detail was prescribed, only to discover that its provisions quickly became out of date.

The Government decided that any future fire safety legislation should be sufficiently flexible to adapt to the changing needs of society.

FPA 1971 therefore enabled the government to target particular types of premises, which it thought were in need of improved fire safety standards, and to quickly 'designate' them under FPA 1971.

To date only two designating orders have been issued by the Secretary of State. The first 'designating' order made under FPA 1971 was the *Fire Precautions (Hotels and Boarding Houses) Order 1972 (SI 1972 No 238)*, which required improved fire safety standards for hotels, boarding houses and similar premises providing sleeping accommodation for guests and/or staff.

The second was the *Fire Precautions (Factories, Offices, Shops and Railway Premises) Order 1989 (SI 1989 No 76)*. This order brought the fire safety provisions, which had previously been contained within *Factories Act 1961*, and *Offices, Shops and Railway Premises Act 1963*, within the scope of FPA 1971.

The first of a number of amendments to FPA 1971 was made by the *Health and Safety at Work etc Act 1974* which added an entirely new category of 'use as a place of work' within the 'designated' section of FPA 1971. This amendment was significant as it allowed the Secretary of State to designate virtually any workplace under FPA 1971.

The main control provision within FPA 1971 is that occupiers of 'designated' premises (or owner(s) in the case of buildings in multiple occupation) are required to apply to the fire authority for a fire certificate. The two designating orders specifically detail the criteria under which the occupier or owner(s) of designated premises must apply for a fire certificate.

Following another fatal fire incident, occurring in 1985 at the Bradford City Valley Parade football stadium in which 56 people died and many more were seriously injured, an enquiry was set up by the Government to examine the operation of the *Safety of Sports Grounds Act 1975*. The enquiry report resulted in the *Fire Safety and Safety of Places of Sport Act 1987*, which also contained sections making a number of amendments to FPA 1971.

The main amendment was to allow fire authorities to relax the requirement for 'designated' premises to be provided with a fire certificate. This relaxation, or 'exemption' can now be made at the discretion of the fire authority, if they think that the risk in case of fire to persons within the premises is not too serious. In deciding whether or not to grant exemption for any premises, the fire authority are required to take into account all the circumstances of the case and, in particular to the degree of seriousness of the risk in case of fire to persons in the premises.

2.2 Legislation and Regulations (Fire Precautions Act 1971)

In addition, two further new amendments were inserted into *FPA 1971*, by *Fire Safety and Safety of Places of Sports Act 1987*. The most significant of these additional amendments was that owners and occupiers of all factory, office, shop and railway premises, which were previously not required to have a fire certificate, now had a duty to provide adequate means of escape in case of fire and also means for fighting a fire.

However, this amendment no longer applies to premises where *Part II* of the *Fire Precautions (Workplace) Regulations 1997 (SI 1997 No 1840)* applies. The mechanics of this are explained more fully in **CHAPTER 3: LEGISLATION AND REGULATION (THE FIRE PRECAUTIONS (WORKPLACE) REGULATIONS)**.

The second amendment to *FPA 1971* gave the fire authority the power to issue Improvement Notices and Prohibition Notices. Details of these notices are provided at **2.24** below.

The *Fire Certificates (Special Premises) Regulations 1976 (SI 1976 No 2003)* deal with very high risk premises and set out the procedures in relation to obtaining fire certificates for these types of premises. In this instance, certificates are issued by the Health and Safety Executive as opposed to the local fire authority. For further guidance and information on whether an employers premises are 'special premises' see **2.38** below.

In this chapter a common sense interpretation and explanation of *FPA 1971* is provided. In addition, where appropriate and for the sake of completeness, the relevant section or regulation from the Act or Statutory Instrument is also included in italics. These extracts include the amendments to the original Act made by subsequent legislation.

Fire Precautions (Workplace) Regulations 1997 and Fire Precautions Act 1971

2.2 These two pieces of legislation are very different in their application to businesses. For example the *Fire Precautions Act 1971* imposes requirements relating to fire precautions upon *occupiers* of premises (or *owners* in the case of premises in multiple occupation), whereas *Fire Precautions (Workplace) Regulations 1997)* place a duty on *employers* to safeguard the safety of employees in case of fire.

In practice, if business premises fall within the 'designated' premises category of the 1971 Act, an employer is required to apply for a fire certificate from the local fire authority, who will prescribe in detail what an employer is required to do to comply. In addition, an 'employer' is required to carry out a fire risk assessment of the workplaces for which they are responsible, in accordance with the *Fire Precautions (Workplace) Regulations 1997* (as amended).

If an employer has a business which employs one or more people and the premises do not fall within the scope of one of the designating orders made under the *Fire Precautions Act 1971*, with few exceptions, the *Fire Precautions (Workplace) Regulations 1997 (SI 1997 No 1840)* will apply to employer and their premises. For more information about these Regulations see **CHAPTER 3: LEGISLATION AND REGULATIONS (THE FIRE PRECAUTIONS (WORKPLACE) REGULATIONS 1997)**.

The criteria for deciding whether the premises are designated or not, are detailed below. If an employer is in any doubt at all on whether *FPA 1971* applies to the premises, an employer is advised to consult a fire safety officer at your local fire service for advice.

Is a fire certificate required?

2.3 Yes, if an employer does not already have a fire certificate and the premises fall within the following criteria, the employer must apply to the local fire authority, for a certificate.

Checklist

2.4

> The criteria for determining whether the premises require a certificate are:
>
> - If an employer is the owner or occupier of a hotel or boarding house, or similar establishment, which is used in the course of a business and there is;
>   - sleeping accommodation for at least six people, whether they are staff or guests; or
>   - there is sleeping accommodation for staff or guests above the first floor or below the ground floor level of the building.
> - If the premises are used as premises in which people are employed to work, and are used either as factory, office, shop or railway premises and in which;
>   - more than 20 persons are at work at any one time;
>   - more than ten persons work at any one time on a floor other than the ground floor; or
>   - explosives or highly flammable materials are stored or used in or under the premises (this refers only to factories).

When counting employees, it is the number of employees within *a building*, which determines if the criteria apply. *FPA 1971* interprets a building as:

2.5 Legislation and Regulations (Fire Precautions Act 1971)

'including a temporary or movable building and also includes any permanent structure and any temporary structure other than a movable one.'

If other employees work entirely in separate buildings within the premises, an employer does not need to take them into account. Each separate building should be treated as a separate entity. A building can only be regarded as being entirely separate if it has space separation between it and another building. If it is joined in any way to another building, in general, it cannot be regarded as being separate.

An employer must take into account any other persons who may occupy other parts of the building. For example there may well be firms or businesses, with their own employees, who occupy other parts of the building. These people must be taken into account in deciding if a building requires a fire certificate or not. In this instance, where the building is multi-occupied, it is the responsibility of the *owner* or *owners* of the building to apply for a fire certificate.

What the law says:

2.5

Fire Precautions (Hotels and Boarding Houses) Order 1972 (SI 1972 No 238)

(3) The following use of premises is hereby designated for the purposes of *section 1* of the Act (which requires fire certificates for premises put to designated uses) that is to say, use for providing, in the course of carrying on the business of a hotel or boarding house keeper, sleeping accommodation for staff or sleeping, dining-room, drawing-room, ball-room or other accommodation for guests:

Provided that the provisions of this Order shall not have effect in relation to any premises unless either–

(a) sleeping accommodation is provided in those premises for more than six persons being staff or guests; or

(b) some sleeping accommodation is provided in those premises for staff or guests on any floor above the first floor of the building which constitutes or comprises the premises; or

(c) some sleeping accommodation is provided in those premises for staff or guests below the ground floor of the building, which constitutes or comprises the premises.

Fire Precautions (Factories, Offices, Shops and Railway Premises) Order 1989 (SI 1989 No 76)

4   Designation of uses of factory, office, shop and railway premises

Subject to article 7 below, the following uses of premises are hereby designated for the purposes of *section 1* of the 1971 Act (which requires fire certificates for premises put to designated uses), that is to say—

(a)   use as factory premises;

(b)   use as office premises;

(c)   use as shop premises; and

(d)   use as railway premises,

being (in each case) a use of premises in which persons are employed to work.

5   Premises exempt from requirement for fire certificate

(1)   Notwithstanding the provisions of article 4 above, a fire certificate shall not by virtue of *section 1* of the 1971 Act be required for any factory premises, office premises, shop premises or railway premises in which—

   (a)   not more than twenty persons are at work at any one time; and

   (b)   not more than ten persons are at work at any one time elsewhere than on the ground floor of the building constituting or comprising the premises,

unless one or more of the conditions specified in paragraph (2) below applies to the premises.

(2)   The conditions referred to in paragraph (1) above are—

   (a)   that the premises are in a building containing two or more sets of premises which are put to any of the uses designated by article 4 above and the aggregate of the persons at work at any one time in both or (as the case may be) all those sets of premises exceeds twenty;

   (b)   that the premises are in a building containing two or more sets of premises which are put to any of such uses and in both or (as the case may be) all those sets of premises the aggregate of the persons at work at any one time elsewhere than on the ground floor of the building exceeds ten;

   (c)   that, in the case of factory premises, explosive or highly flammable materials (other than materials of such a kind and in such a quantity that the fire

2.6 *Legislation and Regulations (Fire Precautions Act 1971)*

> authority have determined that they do not constitute a serious additional risk to persons in the premises in case of fire) are stored or used in or under the premises.
>
> (3) Any reference in this article to persons at work is a reference to any of the following persons:
>
> (a) an individual who works under a contract of employment or apprenticeship;
>
> (b) an individual who works for gain or reward otherwise than under a contract of employment or apprenticeship, whether or not he employs other persons; or
>
> (c) a person receiving training provided pursuant to arrangements made (whether before or after the coming into force of *section 25* of the *Employment Act 1988*) under *section 2* of the *Employment and Training Act 1973*.

What is a fire certificate?

2.6 A fire certificate is a document that places a legal responsibility on the occupier or owner(s) of 'designated' premises, to which it has been issued, to comply with its content.

If the fire authority is satisfied that the premises comply with the *Fire Precautions Act 1971*, they will issue the premises with a fire certificate.

This will normally happen after a detailed inspection of your premises has been carried out by a fire safety officer of the your local fire service and, the satisfactory completion of any requirements in relation to fire safety improvements that the fire authority may have made.

How to apply for a fire certificate

2.7 An employer should apply to the local fire authority on form FP1, which are normally available from the local fire service.

After completion of the form and before an inspection has been carried out by the fire authority (which may be some considerable time after your application) an employer has a legal responsibility to comply with their interim duties.

The employer will probably be reminded of their responsibility by the fire authority, following their receipt of their application.

The 'interim duties' are detailed in **2.9** below.

Who should apply?

2.8 In the case of factories, offices, shops and railway premises, application should normally be made by the occupier. In single occupancy premises that person may also be the owner.

However with the following types of premises the owner or owners of the premises must make the application:

- Premises, which are held under a lease or an agreement for a lease or under a licence and consist of part of a building, all parts of which are owned by the same person. This would apply to multi-occupied premises with a single owner.
- Premises consisting of part of a building, the different parts being owned by different persons. This would apply to multi-occupied premises with more than one owner.

In the case of hotels, boarding houses or similar premises, which are normally in single occupancy, the occupier, who could also possibly be the owner, should apply. It is unusual for these types of premises to be multi-occupied or plurally owned, however, if they are, then the owner or owners should apply.

What to do after having applied for a fire certificate

2.9 After an employer has made an application to the fire authority the *Fire Precautions Act 1971* requires an employer to comply with their 'interim duties'. In effect, after the fire authority has received the application, they will normally notify an employer of what has to be done until they carry out an inspection.

Checklist

2.10

An employer or owners 'interim duties' are to ensure that:

- Any existing means of escape in case of fire with which your premises are provided can be safely and effectively used whenever people are on the premises.
- Any existing fire fighting equipment with which the premises are provided is maintained in efficient working order.

That all the employees in the premises receive instruction or training in what to do in case of fire.

2.11 *Legislation and Regulations (Fire Precautions Act 1971)*

What the Act says:

2.11

> 5   Application for, and issue of, fire certificate
>
>   (2A) Where an application is made for a fire certificate with respect to any premises it is the duty of the occupier to secure that, when the application is made and pending its disposal–
>
>   (a)   the means of escape in case of fire with which the premises are provided can be safely and effectively used at all material times;
>
>   (b)   the means for fighting fire with which the premises are provided are maintained in efficient working order; and
>
>   (c)   any persons employed to work in the premises receive instruction or training in what to do in case of fire.

What happens now?

First stage

2.12

> The process of obtaining a fire certificate can be divided into four distinct stages as follows.
>
> Following receipt of an application, the fire authority has the power to require the applicant to provide them with plans of the premises. If they require an employer to furnish plans of their premises, before going to the expense of instructing an architect to draw up the plans, check to determine if there are already plans of the building in the archives. It is possible that in the past an employer may have carried out alterations or extensions for which plans may have been drawn up for Building Control or planning approval. If an employer owns the building there may be plans attached to the deeds or used as part of the conveyance procedures. It is worth checking with the employer's solicitors to see if there are any plans in their possession. Where an employer leases the building or part of it, the owner or his agent may well have a set of plans of the building. If an employer cannot find any suitable plans it is worth discussing the matter with your local fire safety officer of your local fire authority. Simple single line drawings, which an employer or a member of staff could draw, may be sufficient for their needs.

What the Act says:

2.13

> 5   Application for, and issue of, fire certificate
>
> (2)   On receipt of an application for a fire certificate with respect to any premises the fire authority [shall notify the applicant of his duties under subsection (2A) below and] may require the applicant within such time as they may specify–
>
> (a)   to furnish them with such plans of the premises as they may specify; and
>
> (b)   if the premises consist of part of a building, to furnish them, in so far as it is possible for him to do so, with such plans of such other part or parts of the building as they may specify;
>
> and if the applicant fails to furnish the required plans within that time or such further time as the authority may allow, the application shall be deemed to have been withdrawn at the end of that time or further time, as the case may be.

Second stage

2.14

> Following receipt of the plans of the premises the fire authority have a duty at this stage, taking into account the information supplied on the application form, to consider whether to exempt the premises from the requirement to have a fire certificate. Before deciding whether to issue an exemption, they are required to inspect the premises, if they have not already inspected them during the preceding twelve months. The question of whether premises qualify for exemption, is dealt with below at **2.28** below.

Third stage

2.15

> At this stage, an inspection of the premises will be carried out, which is normally carried out, by appointment, by a fire safety officer from the local fire service. The officer will then carry out a detailed inspection of the premises which will take into account the means of escape provided, the means with which the building is provided for

2.15 Legislation and Regulations (Fire Precautions Act 1971)

> ensuring that the means of escape can be safely and effectively used, any fire-fighting equipment, fire alarm systems and any other fire warning systems that may have been provided.
>
> Following the inspection, a decision will be made on which of the following three courses of action the fire authority will take:
>
> - to issue an exemption from the requirement to have a fire certificate;
>
> - if they are satisfied that the fire precautions are adequate, they will issue a fire certificate; or
>
> - if they are not satisfied that the fire precautions are adequate, they will issue an employer or owner with a notice indicating the 'steps to be taken' to improve the fire precautions, before they will issue a fire certificate.

It is normal for the inspecting officer to give a verbal indication to the applicant, during the visit, on the likely course of action to be taken by the fire authority. The applicant will, in due course, be advised in writing of the decision.

The fire authorities will then decide on the following two options.

- Issue the premises with a fire certificate; or

- Issue an employer or owner with a formal notice advising you that they are not satisfied that the fire precautions provided are adequate and that they will not issue a fire certificate until certain steps are taken to improve matters. The notice will also include a date by which the improvements must be completed. They will also consider a request for an extension of the time allowed to complete the work. This is particularly relevant where structural work is required. There may well be delays in completing the required work in the time allowed, due to Building Control or planning delays, or possible problems with contractors.

If the required improvements are of a structural nature, before issuing a notice, the fire authority is required to consult the local authority (if your premises are located in England and Wales). This is done prior to issuing a formal notice to ensure that the fire authority's requirements do not conflict with building or planning Regulations.

In addition, if the premises are used as a place of work and fall within Part I of the *Health and Safety at Work etc Act 1974*, they are also required to consult the appropriate enforcing authority for those premises.

What the Act says:

2.16

> 5 Application for, and issue of, fire certificate
>
> (3) Where an application for a fire certificate with respect to any premises has been duly made and all such plans (if any) as are required to be furnished under subsection (2) above in connection with it have been duly furnished, it shall be the duty of the fire authority [to consider whether or not, in the case of premises which qualify for exemption under section 5A of this Act, to grant exemption and, if they do not grant it, it shall be their duty] to cause to be carried out an inspection of the relevant building (including any part of it which consists of premises to which any exemption conferred by or under this Act applies), and if the fire authority are satisfied as regards any use of the premises which is specified in the application that–
>
> (a) the means of escape in case of fire with which the premises are provided; and
>
> (b) the means (other than means for fighting fire) with which the relevant building is provided for securing that the means of escape with which the premises are provided can be safely and effectively used at all material times; and
>
> (c) the means for fighting fire (whether in the premises or affecting the means of escape) with which the relevant building is provided; and
>
> (d) the means with which the relevant building is provided for giving to persons in the premises warning in case of fire, are such as may reasonably be required in the circumstances of the case in connection with that use of the premises, the authority shall issue a certificate covering that use.
>
> (4) Where the fire authority, after causing to be carried out under subsection (3) above an inspection of the relevant building, are, as regards any use of the premises specified in the application, not satisfied that the means mentioned in that subsection are such as may reasonably be required in the circumstances of the case in connection with that use, they shall by notice served on the applicant–
>
> (a) inform him of that fact and of the steps which would have to be taken (whether by way of making alterations to any part of the relevant building or of

2.17 *Legislation and Regulations (Fire Precautions Act 1971)*

> otherwise providing that building or, as the case may be, the premises with any of those means) to satisfy them as aforesaid as regards that use; and
>
> (b) notify him that they will not issue a fire certificate covering that use unless those steps are taken (whether by the applicant or otherwise) within a specified time;
>
> and if at the end of that time or such further time as may be allowed by the authority or by any order made by a court on, or in proceedings arising out of, an appeal under section 9 of this Act against the notice, a certificate covering that use has not been issued, it shall be deemed to have been refused.

Fourth stage

2.17

> After an employer or owner has completed any required work detailed in the 'Steps to be taken' notice, they should inform the fire safety officer, who will then carry out a further inspection of the premises to ensure that the work has been carried out satisfactorily. If satisfied, the fire authority will then issue a fire certificate. The fire certificate will detail specifically what an employer or owner is required to do in relation to the maintenance of fire safety in the premises.

What does a fire certificate contain?

2.18 The fire certificate will specify the following information, as it relates to the premises.

- The particular use or uses of the premises that it covers.

- The means of escape in the case of fire, provided for the premises.

- The means of ensuring that the means of escape provided can be safely and effectively used at all relevant times. For example, this could include emergency exit signs, emergency lighting, and fire resisting self-closing fire or smoke-stop doors.

- The type, number and location of the means for fighting fire for use by persons in the building. This section would include any portable fire-fighting equipment or hose reels provided.

- The type, number and location of the means for giving warnings in the case of fire. This section would detail the fire alarm arrangements.

In the case of a factory, the fire certificate will also specify particulars as to any explosives or highly flammable materials stored or used in or under the premises.

In addition, a fire certificate may place requirements on the occupier or owner of the premises in respect of:

- the maintenance of the means of escape provided and to ensure that it is free from obstruction;
- the maintenance of any other fire precautions and fire-fighting equipment set out in the certificate;
- the training of employees on the premises as to what to do in the event of fire and to ensure that suitable records of the training are kept;
- any limitation of the number of persons who at any one time may be on the premises; and
- any other relevant fire precautions.

Fire authority fire certificates are usually accompanied by a plan of the building or premises, which forms an integral part of the certificate. The plan details the fire precautions provided in a diagrammatic or schematic form, with an accompanying legend. The recipient of the certificate is required to maintain the fire precautions detailed, in accordance with the plan.

What to do if premises already have a fire certificate

2.19 If employer or owner has been issued with a fire certificate as a result of a recent application, or the premises have been issued with a certificate sometime during the past as the occupier of named premises within the certificate, an employer or owner have a duty to do the following.

Checklist

2.20

> - keep the certificate in the relevant building;
> - maintain the fire safety provisions in your premises and any other conditions imposed within it, precisely as detailed in your certificate;
> - an employer or owner is also required to give notice to the fire authority *before*:
>   - ○ an employer or owner makes any *material* extensions or structural alteration to the premises;   ➤

2.21 Legislation and Regulations (Fire Precautions Act 1971)

> - an employer or owner makes a *material* alteration to the internal layout of the premises or to the layout of furniture or equipment within your premises; or
> - an employer or owner begins to keep explosive or highly flammable materials in, on or under the premises.
>
> Should an employer or owner be unsure whether or not their proposals constitute 'material' change it is good practice to consult the local fire safety officer for advice on the matter.
>
> If, after informing the fire authority, they (the fire authority) are of the opinion that the changes proposed would affect the fire safety provision within the premises, they will issue the employer or owner with a notice of 'Steps to be taken'. The notice will specify what the employer or owner is required to do, as part of the proposed alterations, to ensure that the fire precautions remain adequate. If the employer or owner does not comply with the requirements contained within the notice, the fire authority may cancel the fire certificate. However, if the employer or owner comply with the notice and there follows a satisfactory inspection of the premises on completion of the alterations, the fire authority will amend the fire certificate or issue a new one.

What the Act says:

2.21

> 8 Change of conditions affecting adequacy of certain matters specified in fire certificate
>
> (2) If, while a fire certificate is in force with respect to any premises—
>
> (a) it is proposed to make a material extension of, or material structural alteration to, the premises; or
>
> (b) it is proposed to make a material alteration in the internal arrangement of the premises or in the furniture or equipment with which the premises are provided; or
>
> (c) the occupier of the premises proposes to begin to keep explosive or highly flammable materials of any prescribed kind anywhere under, in or on the relevant building in a quantity or aggregate quantity greater than the quantity prescribed for the purposes of this paragraph as the maximum in relation to materials of that kind, ➤

Legislation and Regulations (Fire Precautions Act 1971) 2.21

the occupier shall, before the carrying out of the proposals is begun, give notice of the proposals to the fire authority; and if the carrying out of the proposals is begun without such notice having been given, the occupier shall be guilty of an offence.

(3) If, while a fire certificate is in force with respect to any premises not constituting the whole of the relevant building, any person who as occupier of any other part of that building is under section 6(5) of this Act responsible for contraventions of any requirement imposed by the certificate proposes to begin to keep explosive or highly flammable materials of any prescribed kind anywhere under, in or on that building in a quantity or aggregate quantity greater than the quantity prescribed for the purposes of this subsection as the maximum in relation to materials of that kind, that person shall, before the carrying out of the proposals is begun, give notice of the proposals to the fire authority; and if the carrying out of the proposals is begun without such notice having been given, that person shall be guilty of an offence.

(4) If the fire authority are satisfied, as regards any premises with respect to which a notice under subsection (2) above has been given to them, that the carrying out of the proposals notified would result in any of the matters mentioned in section 6(1)(b) to (e) of this Act becoming inadequate in relation to any use of the premises covered by the relevant fire certificate, they may by notice served on the occupier within two months from the receipt of the notice under subsection (2)–

(a) inform the occupier of the steps which would have to be taken in relation to the relevant building (whether by way of making alterations to any part of the relevant building or otherwise) to prevent the matters in question from becoming in their opinion inadequate in relation to that use in the event of the proposals being carried out; and

(b) give him such directions as the fire authority consider appropriate for securing, as regards any of the proposals which may be specified in the directions, that that proposal, or any stage of it which may be so specified, is not carried out until such of those steps as may be so specified in relation to that proposal or stage have been taken (whether by him or otherwise);

and if those steps are duly taken in connection with the carrying out of the proposals, the fire authority shall amend the fire certificate or issue a new one.

2.22 Legislation and Regulations (Fire Precautions Act 1971)

> (7) If any person contravenes a direction given to him in pursuance of subsection (4)(b) above, he shall be guilty of an offence; and the fire authority may cancel the fire certificate issued with respect to any premises if they are satisfied that there has been such a contravention as aforesaid by the occupier, whether or not proceedings are brought in respect of the contravention.

Fire authority's power of inspection

2.22 As long as a fire certificate is in force for the premises an inspector from the fire authority has the power to inspect any part of the premises, at any reasonable time, in order to ensure that the conditions imposed within the fire certificate are being complied with. In particular, they will wish to establish if any change of conditions has taken place and resulted in the fire safety measures provided becoming inadequate. This includes the means of escape, the means with which the building is provided for ensuring that the escape routes can be safely and effectively used, any first aid fire-fighting equipment, fire alarm systems and any other fire warning systems.

When the fire authority consider that following an inspection and, as a consequence of a change of conditions, the fire safety provision has become inadequate in relation to any use of the premises covered by the certificate, they may serve a notice on the occupier, requiring them to carry out improvements to the relevant building. The notice will also state that if the required improvements are not carried out within a specified time period, the fire authority may cancel the fire certificate.

What the Act says:

2.23

> 8 Change of conditions affecting adequacy of certain matters specified in fire certificate
>
> (1) So long as a fire certificate is in force with respect to any premises, the fire authority may cause any part of the relevant building to be inspected at any reasonable time for the purpose of ascertaining whether there has been a change of conditions by reason of which any of the matters mentioned in section 6(1)(b) to (e) of this Act have become inadequate in relation to any use of the premises covered by the certificate; but where a building or part of a building is used as a dwelling or consists of premises of any other description prescribed for the purposes of this subsection, an inspection of the building or, as the case may be, of such a ➤

Legislation and Regulations (Fire Precautions Act 1971) 2.23

part shall not be made under this subsection as of right unless twenty-four hours' notice has been given to the occupier of the building or, as the case may be, of the part in question.

For the purposes of this subsection a description of premises may be framed in any of the ways mentioned in section 1(4) of this Act.

(5) If the fire authority are satisfied (whether as a result of an inspection made under subsection (1) above or otherwise) that, as regards any premises with respect to which a fire certificate is in force, any of the matters mentioned in section 6(1)(b) to (e) of this Act has, in consequence of a change of conditions, become inadequate in relation to any use of the premises covered by the certificate, they may by notice served on the occupier–

(a) inform him of that fact and of the steps which would have to be taken in relation to the relevant building (whether by way of making alterations to any part of the relevant building or otherwise) to make the matter in question adequate in their opinion in relation to that use; and

(b) notify him that if those steps are not taken (whether by him or otherwise) within such period as may be specified in the notice, the fire certificate may be cancelled;

and if those steps are duly taken, the fire authority shall, if necessary, amend the fire certificate or issue a new one.

(6) If the fire authority consider (whether as a result of an inspection made under subsection (1) above or otherwise) that, as regards any premises with respect to which a fire certificate is in force, it would, in consequence of a change of conditions or of the coming into force of any Regulations made under section 12 of this Act, be appropriate to amend the certificate for any of the following purposes, that is to say–

(a) to vary or revoke any requirement which the certificate imposes by virtue of section 6(2) of this Act; or

(b) to add to the requirements which the certificate so imposes; or

(c) to alter the effect of the certificate as to the person or persons responsible under or by virtue of section 6(5) of this Act for contraventions of any requirement imposed (whether by virtue of section 6(2) or otherwise) by the certificate, the authority may, subject to section 6(6) of this Act, make such amendments in the

> certificate as they think appropriate for that purpose or issue a new certificate embodying those amendments.
>
> (9) Where a notice has been served under subsection (5) above in connection with any premises and the steps mentioned in it in accordance with paragraph (a) of that subsection are not taken within the period specified in the notice in accordance with paragraph (b) of that subsection or such longer period as may be allowed by the fire authority or by any order made by a court on, or in proceedings arising out of, an appeal under section 9 of this Act against the notice, the fire authority may cancel the fire certificate in force with respect to the premises or, if it covers two or more uses of the premises, may either cancel it or amend it so as to remove from those uses one or more of them (and in that case may make in it all such amendments as they think appropriate in connection with the removal of the use or uses in question).

What are Prohibition and Improvement Notices?

2.24 A Prohibition Notice is a formal notice issued by the fire authority to prohibit or restrict the use of virtually any premises, when they consider that a dangerous situation exists. The issue of a Prohibition Notice is quite a rare event. However, when a notice is issued it is because the fire authority are of the opinion, that there is a serious risk to persons on the premises in the case of fire and is such that the premises ought to be prohibited or restricted in some way, until the matter is resolved. As the prohibition or restriction takes effect from the moment it is issued, it is in effect, an instant solution to a dangerous situation, where people's lives are at risk.

The notice when issued will state:

- that the fire authority are of the opinion that use of the premises involves or will involve a risk to persons on the premises in case of fire so serious that use of the premises ought to be prohibited or restricted;

- specify the matters which in their opinion give or, as the case may be, will give rise to that risk; and

- direct that the use to which the prohibition notice relates is prohibited or restricted to such extent as may be specified in the notice until the specified matters have been remedied.

A Prohibition Notice may include directions as to the steps, which must be taken to remedy the matters specified in the Notice. The recipient of a notice has the right of appeal to a court within twenty-one days, from the date on which the Prohibition Notice is served. However, bringing an appeal will not automatically suspend the operation of the notice. Only a court, on the application of the appellant can suspend the operation of the notice.

It is an offence for any person to contravene any prohibition or restriction imposed by a Prohibition Notice.

What the Act says:

2.25

> 10   Special procedure in case of serious risk: prohibition notices
>
>   (1)   This section applies to–
>
>   (a)   any premises which are being or are proposed to be put to a use (whether designated or not) which falls within at least one of the classes of use mentioned in section 1(2) of this Act, other than premises of the description given in section 2 of this Act; and
>
>   (b)   any premises to which section 3 of this Act for the time being applies.
>
>   (2)   If as regards any premises to which this section applies the fire authority are of the opinion that use of the premises involves or will involve a risk to persons on the premises in case of fire so serious that use of the premises ought to be prohibited or restricted, the authority may serve on the occupier of the premises a notice (in this Act referred to as 'a prohibition notice').
>
>   (3)   The matters relevant to the assessment by the fire authority, for the purposes of subsection (2) above, of the risk to persons in case of fire include anything affecting their escape from the premises in that event.
>
>   (4)   A prohibition notice shall–
>
>   (a)   state that the fire authority are of the opinion referred to in subsection (2) above;
>
>   (b)   specify the matters which in their opinion give or, as the case may be, will give rise to that risk; and
>
>   (c)   direct that the use to which the prohibition notice relates is prohibited or restricted to such extent as may be specified in the notice until the specified matters have been remedied.
>
>   (5)   A prohibition notice may include directions as to the steps which will have to be taken to remedy the matters specified in the notice.
>
>   (6)   A prohibition or restriction contained in a prohibition notice in pursuance of subsection (4)(c) above shall take effect immediately it is served if the authority are of the

opinion, and so state in the notice, that the risk of serious personal injury is or, as the case may be, will be imminent, and in any other case shall take effect at the end of a period specified in the prohibition notice.

(7) Where a prohibition notice has been served under subsection (2) above the fire authority may withdraw the notice at any time.

10A Rights of appeal against prohibition notices

(1) A person on whom a prohibition notice is served may, within twenty-one days from the date on which the prohibition notice is served, appeal to the court.

(2) On an appeal under this section, the court may either cancel or affirm the notice, and, if it affirms it, may do so either in its original form or with such modifications as the court may in the circumstances think fit.

(3) Where an appeal is brought under this section against a prohibition notice, the bringing of the appeal shall not have the effect of suspending the operation of the notice, unless, on the application of the appellant, the court so directs (and then only from the giving of the direction).

10B Provision as to offences

(1) It shall be an offence for any person to contravene any prohibition or restriction imposed by a prohibition notice.

(2) In any proceedings for an offence under subsection (1) above where the person charged is a person other than the person on whom the prohibition notice was served, it shall be a defence for that person to prove that he did not know and had no reason to believe the notice had been served.

(3) Any person guilty of an offence under subsection (1) above shall be liable—

(a) on summary conviction, to a fine not exceeding the statutory maximum,

(b) on conviction on indictment, to a fine, or imprisonment for a term not exceeding two years, or both.

Improvement Notice

2.26   An Improvement Notice is a formal notice issued by the fire authority when it believes that the duty imposed by *section 9A* of the *Fire Precautions Act 1971* on the occupier or owner(s) of premises, to provide adequate means of escape and means for fighting fire, has been contravened.

Legislation and Regulations (Fire Precautions Act 1971) **2.27**

This notice is used more frequently than Prohibition Notices as it is used to improve fire safety in all of the other premises covered by *FPA 1971*, that are not required to have a fire certificate.

The notice when issued will state:

- that the fire authority consider that *FPA 1971* has been contravened;
- specify, by reference to a Code of Practice if necessary, what is required to be done to remedy the contravention; and
- require the occupier to remedy the contravention within a stated time period.

The recipient of an Improvement Notice has the right of appeal to a court within twenty-one days from the date on which the Improvement Notice is served. The bringing of the appeal shall have the effect of suspending the operation of the notice until the appeal is finally disposed of or until the appeal is withdrawn. It should be noted that this is distinctly different from an appeal against the receipt of a Prohibition Notice where only a court can suspend the operation of a notice on the application of the appellant.

It is an offence to contravene any requirement imposed by an Improvement Notice.

What the Act says:

2.27

> 9D Improvement notices
>
> (1) Where a fire authority are of the opinion that the duty imposed by section 9A of this Act has been contravened in respect of any premises to which that section applies, they may serve on the occupier of those premises a notice (in this Act referred to as 'an improvement notice') which–
>
> (a) states they are of that opinion;
>
> (b) specifies, by reference to a code of practice under section 9B of this Act if they think fit, what steps they consider are necessary to remedy that contravention; and
>
> (c) requires the occupier to take steps to remedy that contravention within such period (ending not earlier than the period within which an appeal against the improvement notice can be brought under section 9E of this Act) as may be specified in the notice.
>
> 9E Rights of appeal against improvement notices

> (1) A person on whom an improvement notice is served may, within twenty-one days from the date on which the improvement notice is served, appeal to the court.
>
> (2) On an appeal under this section, the court may either cancel or affirm the notice, and, if it affirms it, may do so either in its original form or with such modifications as the court may in the circumstances think fit.
>
> (3) Where an appeal is brought under this section against an improvement notice, the bringing of the appeal shall have the effect of suspending the operation of the notice until the appeal is finally disposed of or, if the appeal is withdrawn, until the withdrawal of the appeal.
>
> 9F Provision as to offences
>
> (1) It is an offence for a person to contravene any requirement imposed by an improvement notice.
>
> (2) Any person guilty of an offence under subsection (1) above shall be liable–
>
> > (a) on summary conviction, to a fine not exceeding the statutory maximum;
> >
> > (b) on conviction on indictment, to a fine, or imprisonment for a term not exceeding two years, or both.

Which premises are exempt from needing a fire certificate?

2.28 An employer or owner will only be aware if the premises are exempt from the requirement to have a fire certificate if they have been advised in writing, by the fire authority, that the premises have been granted an exemption.

The Act allows the fire authorities to relax the requirement for 'designated' premises to be provided with a fire certificate. This relaxation, or 'exemption' can be made at the discretion of the fire authority, if it is thought by them, having regard to all the circumstances, that the seriousness of the risk to persons in the premises, in case of fire, is low. If an Exemption Notice has been issued it will state:

- that an exemption from the requirement to have a fire certificate has been granted for a particular use or uses of the premises; and

- may specify the greatest number of persons who can safely be in the premises at any one time.

An exemption is likely to be issued as part of the application process for a fire certificate, detailed at **2.11** above.

The decision will probably be made during the course of an inspection by a fire safety officer from the local fire service, who may decide that despite the premises meeting the criteria for requiring a certificate, that they are low risk and can therefore be exempted.

What the Act says:

2.29

> 5A Powers for fire authority to grant exemption in particular cases
>
> (1) A fire authority may, if they think fit as regards any premises which appear to them to be premises qualifying for exemption under this section as respects any particular use, grant exemption from the requirement to have a fire certificate covering that use.
>
> (2) Exemption under this section for any premises as respects any use of them may be granted by the fire authority, with or without the making of an application for the purpose–
>
> (a) on the making of an application for a fire certificate with respect to the premises covering that use; or
>
> (b) at any time during the currency of a fire certificate with respect to the premises which covers that use.
>
> (3) In deciding whether or not to grant exemption under this section for any premises the fire authority shall have regard to all the circumstances of the case and in particular to the degree of seriousness of the risk in case of fire to persons in the premises.
>
> (4) For the purpose of making that decision the fire authority may–
>
> (a) require the applicant or, as the case may be, the occupier of the premises to give such information as they require about the premises and any matter connected with them; and
>
> (b) cause to be carried out an inspection of the relevant building.
>
> (5) The fire authority shall not grant exemption under this section for any premises without causing an inspection to be carried out under subsection (4) above unless they have caused the premises to be inspected (under that or any other power) within the preceding twelve months.

2.29 *Legislation and Regulations (Fire Precautions Act 1971)*

(6) The effect of the grant of exemption under this section as respects any particular use of premises is that, during the currency of the exemption, no fire certificate in respect of the premises is required to cover that use and accordingly–

    (a) where the grant is made on an application for a fire certificate, the grant disposes of the application or of so much of it as relates to that use; and

    (b) where the grant is made during the currency of a fire certificate, the certificate shall wholly or as respects that use cease to have effect.

(7) On granting an exemption under this section, the fire authority shall, by notice to the applicant for the fire certificate or the occupier of the premises, as the case may be, inform him that they have granted exemption as respects the particular use or uses of the premises specified in the notice and of the effect of the grant.

(8) A notice of the grant of exemption for any premises as respects a particular use of them may include a statement specifying the greatest number of persons of a description specified in the statement for the purposes of that use who, in the opinion of the fire authority, can safely be in the premises at any one time.

(9) Where a notice of the grant of exemption for any premises includes a statement under subsection (8) above, the fire authority may, by notice served on the occupier of the premises, direct that, as from a date specified in the notice, the statement–

    (a) is cancelled; or

    (b) is to have effect as varied by the notice;

and, on such a variation the statement shall be treated, so long as the variation remains in force, as if the variation were specified in it.

5B   Withdrawal of exemptions under section 5A

(1) A fire authority who have granted an exemption under section 5A of this Act from the requirement to have a fire certificate covering any particular use of premises may, if they think fit, at any time, withdraw the exemption in accordance with subsections (2) to (4) below.

(2) In deciding whether or not to withdraw an exemption they have granted the fire authority shall have regard to all the circumstances of the case and in particular to the degree of seriousness of the risk in case of fire to persons in the premises.

(3) The fire authority may withdraw an exemption they have granted as respects any particular use of premises without exercising any of the powers of inspection or inquiry conferred by section 19 of this Act but they shall not withdraw the exemption without first giving notice to the occupier of the premises that they propose to withdraw it and the reasons for the proposal and giving him an opportunity of making representations on the matter.

(4) An exemption shall be withdrawn by serving a notice on the occupier of the premises to which the exemption relates stating that the exemption will cease to have effect as respects the particular use or uses of the premises specified in the notice on such date as is so specified, being a date not earlier than the end of the period of fourteen days beginning with the date on which service of the notice is effected.

(5) If premises cease to qualify for exemption under section 5A of this Act a fire authority who have granted an exemption under that section shall notify the occupier of the premises of the fact and date of the cessation of the exemption.

[8A Change of conditions affecting premises for which exemption has been granted]

[(1) If, during the currency of an exemption granted under section 5A of this Act for any premises, it is intended to carry out in relation to those premises any proposals to which this section applies, the occupier shall, before the carrying out of the proposals is begun, give notice of the proposals to the fire authority; and if the carrying out of the proposals is begun without such notice having been given, the occupier shall be guilty of an offence.

(2) This section applies to the following proposals, namely, any proposal–

(a) to make–

(i) an extension of, or structural alteration to, the premises which would affect the means of escape from the premises; or

(ii) an alteration in the internal arrangement of the premises, or in the furniture or equipment with which the premises are provided, which would affect the means of escape from the premises; or

(b) on the part of the occupier, to begin to keep explosive or highly flammable materials of any prescribed kind anywhere under, in or on the building which constitutes or comprises the premises in a quantity or aggregate quantity greater than the quantity prescribed for the purposes of this paragraph as the maximum in relation to materials of that kind; or

(c) in a case where the notice of exemption under section 5A of this Act includes a statement under subsection (8) of that section, to make such a use of the premises as will involve there being in the premises at any one time a greater number of persons in relation to whom the statement applies than is specified or treated as specified in the statement.

(3) A person guilty of an offence under subsection (1) above shall be liable—

(a) on summary conviction, to a fine not exceeding the statutory maximum;

(b) on conviction on indictment, to a fine or to imprisonment for a term not exceeding two years, or both.]

9 Right of appeal as regards matters arising out of sections 5 to 8

(1) A person who is aggrieved—

(a) by anything mentioned in a notice served under section 5(4) of this Act as a step which would have to be taken as a condition of the issue of a fire certificate with respect to any premises, or by the period allowed by such a notice for the taking of any steps mentioned in it; or

(b) by the refusal of the fire authority to issue a fire certificate with respect to any premises; or

(c) by the inclusion of anything in, or the omission of anything from, a fire certificate issued with respect to any premises by the fire authority; or

(d) by the refusal of the fire authority to cancel or to amend a fire certificate issued with respect to any premises; or

(e) by any direction given in pursuance of section 8(4)(b) of this Act; or

(f) by anything mentioned in a notice served under section 8(5) of this Act with respect to any premises as a step which must be taken if the fire authority are not to become entitled to cancel the fire certificate relating to the premises, or by the period allowed by such a notice for the taking of any steps mentioned in it; or

(g) by the amendment or cancellation in pursuance of section 8(6),(7) or (9) of this Act of a fire certificate issued with respect to any premises, may, within twenty-one days from the relevant date, appeal to the court; and on any such appeal the court may make such order as it thinks fit.

(2) In this section 'the relevant date' means—

(a) in relation to a person aggrieved by any such refusal, direction, cancellation or amendment as is mentioned in subsection (1) above or by any matter mentioned in paragraph (a) or (f) of that subsection, the date on which he was first served by the fire authority with notice of the refusal, direction, cancellation, amendment or matter in question;

(b) in relation to a person aggrieved by the inclusion of anything in, or the omission of anything from, a fire certificate issued with respect to any premises, the date on which the inclusion or omission was first made known to him;

and for the purposes of paragraph (b) above a person who is served with a fire certificate or a copy of, or of any part of, a fire certificate shall be taken to have had what the certificate or that part of it does and does not contain made known to him at the time of the service on him of the certificate or copy.

(3) Where an appeal is brought under this section against the refusal of the fire authority to issue a fire certificate with respect to any premises or the cancellation or amendment in pursuance of section 8(7) or (9) of this Act of a fire certificate issued with respect to any premises, a person shall not be guilty of an offence under section 7(1) or (2) of this Act by reason of the premises in question being put to a designated use or used as a dwelling at a time between the relevant date and the final determination of the appeal.

(4) Where an appeal is brought under this section against the inclusion in a fire certificate of anything which has the effect of making the certificate impose a requirement, a person shall not be guilty of an offence under section 7(4) of this Act by reason of a contravention of that requirement which occurs at a time between the relevant date and the final determination of the appeal.

(5) Where an appeal is brought under this section against—

(a) the inclusion in a fire certificate, in pursuance of sub-section (5) of section 6 of this Act, of a provision making any person responsible for contraventions of any requirement imposed by the certificate; or

(b) the omission from a fire certificate of a provision which, if included in pursuance of that subsection,

2.30 Legislation and Regulations (Fire Precautions Act 1971)

> would prevent any person from being, as the occupier of any premises, responsible under that subsection for contraventions of any requirement imposed by the certificate,
>
> that person shall not be guilty of an offence under section 7(4) of this Act by reason of a contravention of that requirement which occurs at a time between the relevant date and the final determination of the appeal.

Fire authority inspectors' powers of inspection

2.30 Inspectors appointed by the fire authority under the *Fire Precautions Act 1971*, have wide ranging powers of inspection, which are clearly defined in *FPA 1971, s 19* below. In addition, it should be noted that it is an offence to intentionally obstruct an inspector in the exercise or performance of their powers or duties as it is to fail to comply with any requirement imposed by an inspector, without reasonable excuse.

What the Act says:

2.31

> 19   Powers of inspectors
>
> (1) Subject to the provisions of this section, any of the following persons (in this section referred to as 'inspectors') namely an inspector appointed under section 18 of this Act and a fire inspector, may do anything necessary for the purpose of carrying this Act and Regulations thereunder into effect and, in particular, shall, so far as may be necessary for that purpose, have power to do at any reasonable time any of the following things, namely–
>
> (a) to enter any such premises as are mentioned in subsection (2) below, and to inspect the whole or any part thereof and anything therein;
>
> (b) to make such inquiry as may be necessary for any of the purposes mentioned in subsection (3) below;
>
> (c) to require the production of, and to inspect, any fire certificate in force with respect to any premises or any copy of any such certificate;
>
> (d) to require any person having responsibilities in relation to any such premises as are referred to in paragraph (a) above (whether or not the owner or occupier of the premises or a person employed to work therein) to

Legislation and Regulations (Fire Precautions Act 1971) **2.31**

give him such facilities and assistance with respect to any matters or things to which the responsibilities of that person extend as are necessary for the purpose of enabling the inspector to exercise any of the powers conferred on him by this subsection.

(2) The premises referred to in subsection (1)(a) above are the following, namely–

(a) any premises requiring a fire certificate or to which any Regulations made under section 12 of this Act apply;

(aa) any premises in respect of which there is in force an exemption under section 5A of this Act from the requirement for a fire certificate with respect to them;

(b) any premises such as are mentioned in section 10(1)(a) of this Act;

(c) any premises to which section 3 of this Act for the time being applies;

(d) any premises not falling within any of the foregoing paragraphs which form part of a building comprising any premises so falling; and

(e) any premises which the inspector has reasonable cause to believe to be premises falling within any of the foregoing paragraphs.

(3) The purposes referred to in subsection (1)(b) above are the following, namely–

(a) to ascertain, as regards any premises, whether they fall within any of paragraphs (a) to (d) of subsection (2) above;

(b) to identify the owner or occupier of any premises falling within any of those paragraphs;

(c) to ascertain whether, in the case of any premises to which section 3 of this Act for the time being applies, any person has the overall management of the building constituting or comprising the premises and, if so, to identify that person;

(d) to ascertain, as regards any premises falling within any of the said paragraphs (a) to (d), whether the provisions of this Act and Regulations made under section 12 thereof are complied with, and, where a fire certificate is in force in respect of any such premises, whether the requirements imposed by the certificate are complied with.

2.32 *Legislation and Regulations (Fire Precautions Act 1971)*

> (4) An inspector shall, if so required when visiting any premises in the exercise of powers conferred by this section, produce to the occupier of the premises some duly authenticated document showing his authority.
>
> (5) In the case of premises used as a dwelling or premises of any other description prescribed for the purposes of this subsection, no power of entry conferred by subsection (1) above shall be exercised as of right unless twenty-four hours' notice has been given to the occupier; and for the purposes of this subsection a description of premises may be framed in any of the ways mentioned in section 1(4) of this Act.
>
> (6) A person who–
>
> (a) intentionally obstructs an inspector in the exercise or performance of his powers or duties under this Act; or
>
> (b) without reasonable excuse fails to comply with any requirement imposed by an inspector under subsection (1)(d) above,
>
> shall be guilty of an offence and liable on summary conviction to a fine not exceeding [level 3 on the standard scale].

Falsification

2.32 Section 22 of the *Fire Precautions Act 1971* below, sets out the offences in relation to the falsification of documents, the making false statements and the giving of false information.

What the Act says:

2.33

> 22 Falsification of documents, false statements etc
>
> (1) If a person–
>
> (a) with intent to deceive . . . makes or has in his possession a document so closely resembling a fire certificate as to be calculated to deceive; or
>
> (b) for the purpose of procuring the issue of a fire certificate, makes any statement or gives any information which he knows to be false in a material particular or recklessly makes any statement or gives any information which is so false; or

> (c) in purported compliance with any obligation to give information to which he is subject under or by virtue of this Act, or in response to any inquiry made by virtue of section 19(1)(b) of this Act, gives any information which he knows to be false in a material particular or recklessly gives any information which is so false; or
>
> (d) makes in any register, book, notice or other document required by or by virtue of Regulations made under this Act to be kept, served or given, an entry which he knows to be false in a material particular,
>
> he shall be guilty of an offence and liable on summary conviction to a fine not exceeding [level 5 on the standard scale].
>
> (2) If a person with intent to deceive pretends to be–
>
> (a) an inspector within the meaning of section 19 of this Act, or
>
> (b) a person by whom the powers conferred by that section on a fire inspector are exercisable by virtue of section 20 of this Act,
>
> he shall be guilty of an offence and liable on summary conviction to a fine not exceeding level 3 on the standard scale.

Offences by bodies corporate

2.34 Section 23 of the *Fire Precautions Act 1971* below sets out the liability of the officers of a body corporate to be prosecuted, where an offence under FPA 1971, has been committed by a body corporate.

What the Act says:

2.35

> 23 Offences by bodies corporate
>
> (1) Where an offence under this Act committed by a body corporate is proved to have been committed with the consent or connivance of, or to be attributable to any neglect on the part of, any director, manager, secretary or other similar officer of the body corporate, or any person purporting to act in any such capacity, he as well as the body

> corporate shall be guilty of that offence, and shall be liable to be proceeded against and punished accordingly.
>
> (2) Where the affairs of a body corporate are managed by its members, this section shall apply in relation to the acts and defaults of a member in connection with his functions of management as if he were a director of the body corporate.

Defence of 'due diligence'

2.36 During any prosecution under the *Fire Precautions Act 1971*, it is a defence for a person charged, to prove that they took all reasonable precautions and exercised all due diligence to prevent an offence being committed. The term 'due diligence' is explained more fully in **CHAPTER 6: LEGISLATION AND REGULATIONS (HEALTH AND SAFETY)**.

What the Act says:

2.37

> Defence available to persons charged with offences
>
> In any proceedings for an offence under this Act or under Regulations made thereunder, it shall be a defence for the person charged to prove that he took all reasonable precautions and exercised all due diligence to avoid the commission of such an offence.

What are 'special premises'?

2.38 'Special premises' are premises, which either operate hazardous processes or manufacture, store or even have facilities provided for the storage of certain specified hazardous materials in quantities, which present a significant hazard to employees and the general public. These types of premises are required to comply with the *Fire certificates (Special Premises) Regulations 1976 (SI 1976 No 2003)*.

In the case of premises containing hazardous materials or processes, covered by the Regulations, a fire certificate must be obtained from the Health and Safety Executive ('HSE'), as opposed to the local fire authority. This applies even if only a small number of persons are employed on the premises. Exemption from this requirement may be granted by the HSE, where the Regulations are inappropriate, or not reasonably practicable to implement.

The duty to apply for a fire certificate under these Regulations falls upon 'the responsible person', as defined in *SI 1976 No 2003, Reg 2(1)(a)*. The procedure to be followed insofar as the information to be supplied, the inspection and issue of a certificate is very similar to that under *FPA 1971*.

After a fire certificate has been issued by the HSE the occupier of the premises covered by the certificate must post a notice on the premises that is easily seen and read by any person who might be affected by any of the provisions of the certificate.

Checklist

2.39

> The notice must state:
>
> - that the certificate has been issued;
> - where the certificate (or a copy) can be inspected; and
> - the date of the posting of the notice.
>
> The 'special premises' for which a fire certificate is required from the HSE are set out below:
>
> - any premises at which are carried on any manufacturing processes in which the total quantity of any highly flammable liquid under pressure greater than atmospheric pressure and above its boiling point at atmospheric pressure may exceed 50 tonnes;
> - any premises at which there is stored, or there are facilities provided for the storage of, liquefied petroleum gas in quantities of (or in excess of) 100 tonnes except where the liquefied petroleum gas is kept for use at the premises either as a fuel, or for the production of an atmosphere for the heat-treatment of metals;
> - any premises at which there is stored, or there are facilities provided for the storage of, liquefied natural gas in quantities of, or in excess of, 100 tonnes except where the liquefied natural gas is kept solely for use at the premises as a fuel;
> - any premises at which there is stored, or there are facilities provided for the storage of, any liquefied flammable gas consisting predominantly of methyl acetylene in quantities of, or in excess of, 100 tonnes except where the liquefied flammable gas is kept solely for use at the premises as a fuel;
> - any premises at which oxygen is manufactured and at which there are stored, or there are facilities provided for the storage of, quantities of liquid oxygen of (or in excess of) 135 tonnes;

2.39 *Legislation and Regulations (Fire Precautions Act 1971)*

- any premises at which there are stored, or there are facilities provided for the storage of, quantities of chlorine of (or in excess of) 50 tonnes except when the chlorine is kept solely for the purpose of water purification;

- any premises at which artificial fertilisers are manufactured and at which there are stored, or there are facilities provided for the storage of, quantities of ammonia of (or in excess of) 250 tonnes;

- any premises at which there are in process, manufacture, use or storage at any one time, or there are facilities provided for such processing, manufacture, use or storage of, quantities of any of the materials listed below in (or in excess of) the quantities specified–

- Ethylene Oxide   20 tonnes
- Acrylonitrile   50 tonnes
- Ethylene   100 tonnes
- Any highly flammable liquid not otherwise specified   4,000 tonnes
- Phosgene   5 tonnes
- Carbon Disulphide   50 tonnes
- Hydrogen Cyanide   50 tonnes
- Propylene   100 tonnes

- explosives, factories or magazines that are required to be licensed under the *Explosives Act 1875*;

- any building on the surface of any mine within the meaning of the *Mines and Quarries Act 1952*;

- any premises in which there is comprised–

  (a) any undertaking on a site for which a licence is required in accordance with *section 1* of the *Nuclear Installations Act 1965* or for which a permit is required in accordance with *section 2* of that Act;

  (b) any undertaking which would, except for the fact that it is carried on by the United Kingdom Atomic Energy Authority, or by, or on behalf of, the Crown, be required to have a licence or permit in accordance with the provisions mentioned in sub-paragraph;

  (c) (a) above;

- any premises containing any machine or apparatus in which charged particles can be accelerated by the equivalent of a voltage of not less than 50 mega volts except where the premises are used as a hospital;

Legislation and Regulations (Fire Precautions Act 1971) 2.39

> - premises to which *Regulation 26* of the *Ionising Radiation Regulations 1985 (SI 1985 No 1333)* applies;
> - any building, or part of a building, which either–
>   (a) is constructed for temporary occupation for the purposes of building operations or works of engineering construction; or
>   (b) is in existence at the first commencement there of any further such operations or works and which is used for any process of work ancillary to any such operations or works.
>
> It should be noted that providing each condition, set out below, in Parts I and II of Schedule 1 of the *Fire certificate (Special Premises) Regulations 1976 (SI 1976 No 2003)* is satisfied, then a building or part of a building is exempt from needing to obtain a fire certificate.

Conditions to be satisfied if a fire certificate is not to be required for premises specified in paragraph 15:

> (Paragraph 15 states)–
>
> 15 Any building, or part of a building, which either–
>   (a) is constructed for temporary occupation for the purposes of building operations or works of engineering construction; or
>   (b) is in existence at the first commencement there of any further such operations or works;
>   and which is used for any process or work ancillary to any such operations or works.
>
> 16 Not more than twenty persons are employed at any one time in the building or part of the building.
>
> 17 Not more than ten persons are employed at any one time elsewhere than on the ground floor of the building or part of the building.
>
> 18 No explosive or highly flammable material is stored or used in or under the building or part of the building.
>
> 19 The building or part of the building is provided with such means of escape in case of fire for the persons employed there as may reasonably be required in the circumstances of the case.
>
> 20 In the building or part of the building there is provided and maintained appropriate means for fighting fire which are so placed as to be readily available for use.

21. While any person is in the building or part of the building for the purpose of employment or meals, the doors of the building or part and of any room therein in which he is, and any doors which afford a means of exit for persons employed in the building or part are not locked or fastened in such a manner that they cannot easily and immediately be opened from the inside.

22. Any doors opening on to any staircase or corridor from any room in the building or part of the building in which more than ten persons are employed, except in the case of sliding doors, are constructed to open outwards.

23. Every window, door or other exit affording a means of escape from the building or part of the building in case of fire or giving access thereto, other than the means of exit in ordinary use, is distinctively and conspicuously marked by a notice of adequate size.

24. The contents of any room in the building or part of the building in which persons are employed are so arranged or disposed that there is a free passage way for all persons employed in the room to a means of escape in case of fire.

Publications providing further guidance on fire safety and Fire Precautions Act 1971

2.40

Fire Safety – An employers guide.	The Stationary Office ISBN 0 11 341229 0
Fire Precautions Act 1971. Guide to Fire Precautions in Existing Places of Work that require a Fire certificate; Factories, Offices, Shops and Railway Premises.	The Stationery Office ISBN 0 11 341079 4
Fire Precautions Act 1971. Guide to Fire Precautions in Premises used as Hotels and Boarding Houses which require a Fire certificate.	The Stationery Office ISBN 0 11 341005 0
Fire Precautions Act 1971. Fire Safety Management in Hotels and Boarding Houses.	The Stationery Office ISBN 0 11 340980
Guide to Fire Precautions in Existing Places of Entertainment and Like Premises.	The Stationery Office ISBN 0 11 340907 9

The Building Regulations 1991 Approved Document B: Fire Safety.	DETR Publications ISBN 1 85112 3512
Technical Standards for compliance with the Building Standards (Scotland) Regulations 1990 as amended.	The Stationery Office ISBN 0 11 49 5866 1
Fire and the Design of Educational Buildings: Building Bulletin 7.	The Stationery Office ISBN 0 11 270585 5
FIRECODE (and FIRECODE Scotland). A suite of documents aimed at healthcare premises.	The Stationery Office ISBN various – contact NHS Estates (0113 254 7000) for details.
Heritage under Fire. A guide to the protection of historic buildings.	Fire Protection Association ISBN 0 90 216790 1

3 Legislation and Regulations (The Fire Precautions (Workplace) Regulations 1997 and 1999)

In this chapter:	
Introduction	3.1
European approach	3.2
What do the Regulations cover?	3.5
Co-operation and co-ordination	3.10
Enforcement	3.12
Serious cases: offence	3.14
Serious cases: Prohibition Notices	3.16
Enforcement Notices	3.19
Enforcement Notices: rights of appeal	3.21
Enforcement Notice: offence	3.23
Where the Regulations do not apply	3.25
Other major legislation relating to fire safety in the workplace	3.27

Introduction

3.1 The *Fire Precautions (Workplace) Regulations 1997 (SI 1997 No 1840)* came into force on 1 December 1997, as a result of an EC initiative, which led to the drafting of various EC Directives, designed to introduce a common approach to the health and safety of workers across all member states within the Union.

The 1997 Regulations were subsequently amended by the *Fire Precautions (Workplace) (Amendment) Regulations 1999 (SI 1999 No 1877)*, which came in to effect on 1 December 1999. These two sets of regulations are known collectively as the Fire Regulations.

Implementation of these regulations in the UK introduced the concept of self-compliance with fire safety legislation. Prior to these Regulations, owners and occupiers of premises had been use to a typical British approach whereby the enforcing authority inspected the premises then produced a report detailing what an employer or occupier was required to change or improve in order to comply with the law.

Any necessary work had to be completed within a specified time scale, at the end of which a further inspection would be made to verify that the required work had been carried out satisfactorily. Once the enforcing authority (which was usually the local fire service) was satisfied that the premises complied with the requirements, the occupier or owner had a responsibility to maintain the fire defence systems and to ensure that staff were trained in what to do in the event of a fire. In addition, if owners and occupiers wanted to carry out structural alterations or extensions to their premises, they were required to seek the approval of enforcing authority of their proposals before carrying out the work.

European approach

3.2 These Regulations are now modelled on a typical European approach insofar as premises are no longer policed in the same way by the enforcing authority. Employers are now required to find out how to comply with the legislation. It is entirely their responsibility. Employers should also be aware that, following an investigation by the enforcing authority, if they were found to have broken the law they would be prosecuted.

The Government introduced the Regulations in an attempt to rationalise the existing fire safety law. Such existing law had evolved in a very piece-meal fashion, often as a result of tragic incidents in which many lives were lost. The consequence of much of this 'stable door' legislation was that fire safety requirements were included in more than 60 separate pieces of legislation. This inevitably resulted in inconsistencies, which made an understanding of the application of the appropriate legislation extremely difficult, even for fire safety professionals. For a lay person attempting to comply with the legislation, the law was bewildering and often incomprehensible.

Under these new Regulations, an assessment of the risk of fire within premises has to be carried out by those who exert 'control' over the particular premises that constitute the workplace. Depending upon the particular circumstances, this will be the owner, the occupier or the employer. The assessment is carried out in order to assess whether the existing fire precautions are adequate and proportionate to the perceived risk. This risk assessment must be carried out under the *Management of Health and Safety at Work Regulations 1999 (SI 1999 No 3242)*.

The *Management of Health and Safety at Work Regulations 1999 (SI 1999 No 3242)* place a requirement on every employer to assess the risks to the health

3.3 *Legislation and Regulations (Fire Prec. (Workplace) Regs. 1997 & 1999)*

and safety of his employees whilst at work. The employer must also assess the risks to the health and safety of persons not in his employment arising from the way he conducts his 'undertaking'. The employer is required to assess the workplace and implement any measures necessary to eliminate or reduce the risks that have been identified. The Regulations also require employers to establish and implement procedures to be followed in the event of serious and imminent danger, which of course includes fire. Basically, these Regulations require employers to demonstrate that they have adopted a systematic and controlled approach in dealing with health and safety and risk assessment.

Within this chapter, a common sense interpretation of the Regulations is provided. In addition, where appropriate and for the sake of completeness, the relevant extract from the Regulations will be included in italics. The full names of the Regulations in question are:

- *Fire Precautions (Workplace) Regulations 1997 (SI 1997 No 1840)* (as amended by the *Fire Precautions (Workplace) (Amendment) Regulations 1999 (SI 1999 No 1877)*; and

- *Management of Health and Safety at Work Regulations 1999 (SI 1999 No 3242)*.

The specific references to the fire precautions requirements contained within *Part II of SI 1997 No 1840* Regulations are discussed within **CHAPTER 7: HOW TO CARRY OUT A FIRE RISK ASSESSMENT** which should be read in conjunction with this chapter.

To whom do the Regulations apply?

3.3   Virtually everyone. The Regulations apply to those who are, or are treated as, controlling workplaces. This includes:

- every employer; and

- every person, other than the employer, who has control of a workplace.

Further, persons who by virtue of any contract or tenancy have an obligation in relation to the maintenance, repair or safety of a workplace are treated as having control of the workplace to the extent of their obligations.

A list of 'excepted' workplaces, and places to which the Regulations do not apply are detailed at **3.26** below.

What the Fire Precautions (Workplace) Regulations 1997 say:

3.4

> **What is a 'workplace'?**
>
> 2 Interpretation
>
> 'Workplace' means any premises or part of premises, not being domestic premises, used for the purposes of an employer's undertaking and which are made available to an employee of the employer as a place of work and includes–
>
> (a) any place within the premises to which such employee has access while at work; and
>
> (b) any room, lobby, corridor, staircase, road or other place–
>
> > (i) used as a means of access to or egress from that place of work; or
> >
> > (ii) where facilities are provided for use in connection with that place of work, other than a public road.
>
> **What is the responsibility of those who 'control' the workplace?**
>
> 3 Application of Part II
>
> (1) Every employer shall ensure that the requirements of this Part of these Regulations are complied with in respect of every workplace, other than an excepted workplace, which is to any extent under his control.
>
> (2) Every person, other than the employer referred to in paragraph (1), who has, to any extent, control of a workplace, other than an excepted workplace, shall ensure that, so far as relates to matters within his control, the workplace complies with any applicable requirement of this Part of these Regulations.
>
> (3) Where a person has, by virtue of any contract or tenancy, an obligation of any extent in relation to–
>
> > (a) the maintenance or repair of any workplace; or
> >
> > (b) the safety of any workplace,
> >
> > that person shall be treated, for the purposes of paragraph (2), as being a person who has control of the workplace to the extent that his obligation so extends.
>
> (4) Any reference in this regulation to a person having control of any workplace is a reference to a person having control of

3.5 *Legislation and Regulations (Fire Prec. (Workplace) Regs. 1997 & 1999)*

> the workplace in connection with the carrying on by him of a trade, business or other undertaking (whether for profit or not).

What do the Regulations cover?

3.5   They are designed to ensure that a minimum standard of fire safety is provided in every place *where people work*. This includes shared or common areas and facilities as well as any access areas that staff need to use in order to reach their place of work. For example, if a firm occupies part of a building with other companies, those areas such as the access corridors, staircases, canteens and other shared facilities have to comply. However, if an employer and other occupiers of the building do not have control, or a legal obligation for these areas, there is a responsibility on the owner or landlord to make sure those parts comply with the Regulations.

What has to be done?

3.6   The law requires an employer to carry out a fire risk assessment of the workplace to establish if existing arrangements for fire safety for employees, and all other people who may be affected by a fire in the workplace, are adequate or not.

An employer will need to look at the type of activity that they carry out, the risks, hazards and circumstances that apply to the *workplace.*

CHAPTER 7: HOW TO CARRY OUT A FIRE RISK ASSESSMENT provides detailed guidance on carrying out this task.

What the Management of Health and Safety at Work Regulations 1999 say:

3.7

> 3   Risk assessment
>
> (1)   Every employer shall make a suitable and sufficient assessment of—
>
> (a)   the risks to the health and safety of his employees to which they are exposed whilst they are at work; and
>
> (b)   the risks to the health and safety of persons not in his employment arising out of or in connection with the conduct by him of his undertaking,

Legislation and Regulations (Fire Prec. (Workplace) Regs. 1997 & 1999) **3.8**

> for the purpose of identifying the measures he needs to take to comply with the requirements and prohibitions imposed upon him by or under the relevant statutory provisions and by Part II of the Fire Precautions (Workplace) Regulations 1997.
>
> 5 Health and safety arrangements
>
> (1) Every employer shall make and give effect to such arrangements as are appropriate, having regard to the nature of his activities and the size of his undertaking, for the effective planning, organisation, control, monitoring and review of the preventive and protective measures.
>
> (2) Where the employer employs five or more employees, he shall record the arrangements referred to in paragraph (1).
>
> 8 Procedures for serious and imminent danger and for danger areas
>
> (1) Every employer shall–
>
> (a) establish and where necessary give effect to appropriate procedures to be followed in the event of serious and imminent danger to persons at work in his undertaking.

What to do after completing a fire risk assessment

3.8 The first priority after having carried out an assessment is to implement the necessary measures in order to eliminate any risks identified or to reduce them to acceptable levels.

After completing a risk assessment, the employer should write it down and keep a record of the findings. If an employer employs five or more employees they must record the significant findings of the assessment and any group of employees identified as being especially at risk.

It is good practice to tell staff or their representatives about the risk assessment findings. In addition to a formal risk assessment report, the employer may wish to make it available to employees, if they request it.

However, an employer should be aware that under *Regulation 10* of the *Management of Health and Safety at Work Regulations 1999 (SI 1999 No 3242)* they have a legal requirement to provide employees with 'comprehensive and relevant information' on:

- risks to their health and safety identified in the risk assessment;
- preventative and protective measures;
- fire fighting measures;

3.9 *Legislation and Regulations (Fire Prec. (Workplace) Regs. 1997 & 1999)*

- persons nominated to implement the fire fighting measures; and
- other risks notified to the employer by other employers sharing the workplace.

What the Management of Health and Safety at Work Regulations 1999 say:

3.9

---

10   Information for employees

(1)   Every employer shall provide his employees with comprehensible and relevant information on–

(a)   the risks to their health and safety identified by the assessment;

(b)   the preventive and protective measures;

(c)   the procedures referred to in regulation 8(1)(a) and the measures referred to in regulation 4(2)(a) of the Fire Precautions (Workplace) Regulations 1997;

(d)   the identity of those persons nominated by him in accordance with regulation 8(1)(b) and regulation 4(2)(b) of the Fire Precautions (Workplace) Regulations 1997; and

(e)   the risks notified to him in accordance with regulation 11(1)(c).

Co-operation and co-ordination

3.10   If the workplace premises are shared by other businesses the employer has a responsibility to make the other employers aware of any significant risks that may have been identified as part of the assessment and of any action that has been taken to reduce that risk. Employers are also required to co-operate with other employers who may share the workplace to enable them to discharge their responsibilities under the *Fire Precautions (Workplace) Regulations 1997 (SI 1997 No 1840)*. In addition, an employer should take all reasonable steps to co-ordinate their fire safety measures with those of any other employers who may share the workplace.

There are obvious financial advantages in co-operating and mutually sharing any fire safety measures provided within the individual occupancies and by other firms, within employers' premises.

A simple example of this is that a fire risk assessment may reveal that an employer requires an additional fire exit, which could be solved by accessing a fire exit within an adjoining business. By making that exit available to this employers' business, the other business may also be solving their own fire exit problems, by gaining access to the first employer's premises and exits.

What the Management of Health and Safety at Work Regulations 1999 say:

3.11

> 11 Co-operation and co-ordination
>
> (1) Where two or more employers share a workplace (whether on a temporary or a permanent basis) each such employer shall–
>
> (a) co-operate with the other employers concerned so far as is necessary to enable them to comply with the requirements and prohibitions imposed upon them by or under the relevant statutory provisions and by Part II of the Fire Precautions (Workplace) Regulations 1997;
>
> (b) (taking into account the nature of his activities) take all reasonable steps to co-ordinate the measures he takes to comply with the requirements and prohibitions imposed upon him by or under the relevant statutory provisions and by Part II of the Fire Precautions (Workplace) Regulations 1997 with the measures the other employers concerned are taking to comply with the requirements and prohibitions imposed upon them by that legislation; and
>
> (c) take all reasonable steps to inform the other employers concerned of the risks to their employees' health and safety arising out of or in connection with the conduct by him of his undertaking.
>
> (2) Paragraph (1)(except in so far as it refers to Part II of the Fire Precautions (Workplace) Regulations 1997) shall apply to employers sharing a workplace with self-employed persons and to self-employed persons sharing a workplace with other self-employed persons as it applies to employers sharing a workplace with other employers; and the references in that paragraph to employers and the reference in the said paragraph to their employees shall be construed accordingly.

3.12 *Legislation and Regulations (Fire Prec. (Workplace) Regs. 1997 & 1999)*

Enforcement

3.12 The local fire authority has the responsibility for supervision and for ensuring that workplaces within their area comply with the Regulations.

What the Management of Health and Safety at Work Regulations 1999 say:

3.13

---

10 Enforcement

(1) It shall be the duty of every fire authority to enforce within their area the workplace fire precautions legislation.

(2) A fire authority may perform their functions under these Regulations through inspectors appointed by them pursuant to section 18(1) of the 1971 Act.

(6) In sections 24 and 33(1) of the Fire Services Act 1947 (which authorise the appointment of inspectors and the holding of inquiries for obtaining information as to the performance by fire authorities of their functions under that Act) the references to that Act shall be read as including references to these Regulations.

(7) Nothing in this regulation shall be taken to authorise a fire authority in Scotland to institute proceedings for any offence.

Serious cases: offence

3.14 It is a criminal offence, for any person who is under a requirement to comply with the Regulations, to fail to comply with any provision of the workplace fire precautions legislation and that failure places one or more employees at risk of death or serious injury in the event of a fire.

An example of an offence, might be the locking of an emergency fire exit, which places employees at risk should a fire occur.

In any proceedings for such an offence, it is a defence for the person charged to prove that they took all reasonable precautions and exercised all 'due diligence', to avoid the commission of the offence.

It should also be noted that if a person is the subject of an Enforcement Notice (under *Regulation 13* of the *Fire Precautions (Workplace) Regulations 1997*

(*SI 1997 No 1840)*) for failure to comply with workplace fire precautions legislation, they cannot be guilty of an offence under this Regulation (*SI 1997 No 1840, Reg 11*).

What the Management of Health and Safety at Work Regulations 1999 say:

3.15

> 11 Serious cases: offence
>
> (1) A person shall be guilty of an offence if–
>
> (a) being under a requirement to do so, he fails to comply with any provision of the workplace fire precautions legislation; and
>
> (b) that failure places one or more employees at risk of death or serious injury in case of fire.
>
> (2) Any person guilty of an offence under this regulation shall be liable–
>
> (a) on summary conviction to a fine not exceeding the statutory maximum; or
>
> (b) on conviction on indictment, to a fine, or to imprisonment for a term not exceeding two years, or both.
>
> (2A) In any proceedings for an offence under this regulation it shall be a defence for the person charged to prove that he took all reasonable precautions and exercised all due diligence to avoid the commission of the offence.
>
> (3) A person is not guilty of an offence under this regulation in respect of any failure to comply with the workplace fire precautions legislation which is the subject of an Enforcement Notice.

Serious cases: Prohibition Notices

3.16 In very serious cases, where a serious threat to life exists, the fire authority can immediately serve you with a notice under *Fire Precautions Act 1971, section 10*. This notice can prohibit or restrict the use of a workplace until the risk to employees or other people has been reduced. Failure to comply with the notice is a criminal offence. An employer can appeal to a Magistrates' Court (in Scotland, the Sheriff Court) against this notice but it will remain in force until such time as the court says otherwise or the workplace has been made safe.

3.17 *Legislation and Regulations (Fire Prec. (Workplace) Regs. 1997 & 1999)*

For more detailed information see **2.23** above.

What the Fire Precautions Act 1971 says:

3.17

> 10 Special procedure in case of serious risk: prohibition notices
>
> (1) This section applies to—
>
>   (a) any premises which are being or are proposed to be put to a use (whether designated or not) which falls within at least one of the classes of use mentioned in section 1(2) of this Act, other than premises of the description given in section 2 of this Act; and
>
>   (b) any premises to which section 3 of this Act for the time being applies.
>
> (2) If as regards any premises to which this section applies the fire authority are of the opinion that use of the premises involves or will involve a risk to persons on the premises in case of fire so serious that use of the premises ought to be prohibited or restricted, the authority may serve on the occupier of the premises a notice (in this Act referred to as 'a prohibition notice').
>
> (3) The matters relevant to the assessment by the fire authority, for the purposes of subsection (2) above, of the risk to persons in case of fire include anything affecting their escape from the premises in that event.
>
> (4) A prohibition notice shall—
>
>   (a) state that the fire authority are of the opinion referred to in subsection (2) above;
>
>   (b) specify the matters which in their opinion give or, as the case may be, will give rise to that risk; and
>
>   (c) direct that the use to which the prohibition notice relates is prohibited or restricted to such extent as may be specified in the notice until the specified matters have been remedied.
>
> (5) A prohibition notice may include directions as to the steps which will have to be taken to remedy the matters specified in the notice.
>
> (6) A prohibition or restriction contained in a prohibition notice in pursuance of subsection (4)(c) above shall take effect immediately it is served if the authority are of the opinion, and so state in the notice, that the risk of serious

> personal injury is or, as the case may be, will be imminent, and in any other case shall take effect at the end of a period specified in the prohibition notice.
>
> (7) Where a prohibition notice has been served under subsection (2) above the fire authority may withdraw the notice at any time.

What the Fire Precautions (Workplace) Regulations 1997 say:

3.18

> 12 Serious cases: prohibition notices
>
> (1) Sections 10 to 10B of the 1971 Act (special procedure in case of serious risk: prohibition notices) shall apply to–
>
> (a) tents and other movable structures (other than vessels);
>
> (b) places of work in the open air; and
>
> (c) vessels remaining moored or remaining on dry land, which are relevant workplaces.
>
> (2) For the purposes of paragraph (1), a relevant workplace is a workplace other than an excepted workplace.

Enforcement Notices

3.19 Providing a workplace is subject to the Regulations, the fire authority can issue a notice requiring an employer to improve the fire precautions within their workplace. They may issue a notice if they are of the opinion that an employer has not complied with any one of the provisions of the workplace fire precautions legislation, in respect of the workplace, or the employees who work there. Such a notice is known as an Enforcement Notice and failure to comply with it is a criminal offence.

The Enforcement Notice must notify the person who is served with it that the fire authority is of the opinion that he is in breach of the Regulations and the reasons why they consider it so.

The notice must also:

- specify the steps required to remedy the breach;
- require those steps to be taken within a given time; and

3.20 *Legislation and Regulations (Fire Prec. (Workplace) Regs. 1997 & 1999)*

- provide details of the appeals procedure relating to Enforcement Notices.

A person served with an Enforcement Notice must be given a reasonable amount of time to comply and can appeal against the notice to a Magistrates' Court (in Scotland, the Sheriff Court) within 21 days from the day on which the notice is served.

If the steps to be taken to remedy the breach require an alteration to be made to a building or structure the fire authority must consult the relevant person or authority whose consent to the alteration is required before the Enforcement Notice can be served. This is detailed in *Regulation 13(5)* of the *Fire Precautions (Workplace) Regulations 1997 (SI 1997 No 1840)* below.

Where a fire authority has issued an Enforcement Notice, they may withdraw the notice at any time before the end of the given time for completion. In addition, if an appeal against the notice is not pending, they may extend and further extend the original time given for completion of the improvements.

What the Fire Precautions (Workplace) Regulations 1997 say:

3.20

---

13   Enforcement Notices

(1)   Where a fire authority are of the opinion that—

(a)   a person, being under an obligation to do so, has failed to comply with any provision of the workplace fire precautions legislation in respect of a workplace, or employees who work in a workplace, situated in the area for which they perform the functions of fire authority;

the authority may serve on that person a notice (in these Regulations referred to as 'an Enforcement Notice') which—

(i)   states that they are of that opinion and why;

(ii)   specifies what steps they consider are necessary to remedy that failure;

(iii)   requires that person to take steps to remedy the failure within such period from the date of service of the notice (not being less than 21 days) as may be specified in the notice; and

(iv) explains how, where, within what period and on what grounds an appeal may be brought against the Enforcement Notice.

(2) Where a fire authority are of the opinion that a person's failure to comply with the workplace fire precautions legislation also extends to a workplace, or employees who work in a workplace, situated outside the area for which they perform the functions of fire authority, the notice served by them under paragraph (1) may include requirements concerning that workplace or those employees; but before including any such requirements the authority shall consult the fire authority for the area in which the workplace is situated.

(5) Before serving an Enforcement Notice which would oblige a person to make an alteration to a building or structure, the fire authority shall consult–

(a) such persons as they would have been required to consult under section 17 of the 1971 Act (duty of fire authorities to consult other authorities before requiring alterations to buildings) if the proposed Enforcement Notice had been an improvement notice proposed to be issued under section 9D of that Act; and

(aa) in the case of a building or structure in England or Wales in relation to all or any part of which an initial notice given under section 47 of the Building Act 1984 is in force, the approved inspector who gave that initial notice;

(ab) in the case of a workplace which is, includes or forms part of–

(i) a designated sports ground, or

(ii) a sports ground at which there is a regulated stand, the local authority, and in this sub-paragraph 'sports ground', 'designated sports ground' and 'local authority' have the same meaning as in the Safety of Sports Grounds Act 1975 and 'regulated stand' has the same meaning as in the Fire Safety and Safety of Places of Sport Act 1987'; and

(b) any other person whose consent to the alteration would be required by or under any enactment.

(6) Where an Enforcement Notice has been served–

3.21 Legislation and Regulations (Fire Prec. (Workplace) Regs. 1997 & 1999)

> (a) the fire authority may withdraw the notice at any time before the end of the period specified in the notice; and
>
> (b) if an appeal against the notice is not pending, the fire authority may extend or further extend the period specified in the notice.
>
> (8) Without prejudice to the power of the court to cancel or modify an Enforcement Notice under regulation 14, no failure on the part of a fire authority to [consult under paragraph (2) or (5)] shall make an Enforcement Notice void.

Enforcement Notices: rights of appeal

3.21 A person who has had an Enforcement Notice served upon them may appeal to a magistrate's court (in Scotland, the Sheriff Court) within 21 days from the day on which the Notice was served.

Following an appeal, the court may either cancel or affirm the Enforcement Notice. If it affirms the Notice, it may do so either in its original form or with such modifications as the court may decide.

If an appeal is made against an Enforcement Notice, the appeal has the effect of suspending the operation of the Notice, until the appeal is finally disposed of, or the appeal is withdrawn.

What the Fire Precautions (Workplace) Regulations 1997 say:

3.22

> 14 Enforcement Notices: rights of appeal
>
> (1) A person on whom an Enforcement Notice is served may, within 21 days from the day on which the Enforcement Notice is served, appeal to the court.
>
> (2) On an appeal under this regulation the court may either cancel or affirm the Enforcement Notice, and, if it affirms it, may do so either in its original form or with such modifications as the court may in the circumstances think fit.
>
> (4) Where an appeal is brought under this regulation against an Enforcement Notice, the bringing of the appeal shall have the effect of suspending the operation of the notice until the

> appeal is finally disposed of or, if the appeal is withdrawn, until the withdrawal of the appeal.
>
> (5) In this regulation 'the court' means—
>
> > (a) in England and Wales, a magistrates' court acting for the petty sessions area in which any relevant workplace is situated; and
> >
> > (b) in Scotland, the sheriff within whose jurisdiction any relevant workplace is situated,
> >
> > and for this purpose a 'relevant workplace' is a workplace in respect of which the Enforcement Notice was served, other than a workplace covered by the notice by virtue of regulation 13(2).
>
> (6) An appeal to the sheriff under this regulation shall be by summary application.

Enforcement Notice: offence

3.23 It is an offence for a person to contravene any requirement imposed by an Enforcement Notice. However, in any proceedings for such an offence, it is a defence for the person charged to prove that they took all reasonable precautions and exercised all 'due diligence', to avoid the commission of the offence.

What the Fire Precautions (Workplace) Regulations 1997 say:

3.24

> 15 Enforcement Notices: offence
>
> (1) It is an offence for a person to contravene any requirement imposed by an Enforcement Notice.
>
> (2) Any person guilty of an offence under paragraph (1) shall be liable—
>
> > (a) on summary conviction to a fine not exceeding the statutory maximum; or
> >
> > (b) on conviction on indictment, to a fine, or to imprisonment for a term not exceeding two years, or both.
>
> (3) In any proceedings for an offence under this regulation it shall be a defence for the person charged to prove that he

3.25 *Legislation and Regulations (Fire Prec. (Workplace) Regs. 1997 & 1999)*

> took all reasonable precautions and exercised all due diligence to avoid the commission of the offence.

Where the Regulations do not apply

3.25 There are a number of specific types of workplaces to which the Regulations do not apply or which are 'excepted' from the requirements of the Regulations. These include:

- workplaces used only by the self-employed;
- private dwellings;
- construction sites – any workplace to which the *Construction (Health, Safety and Welfare) Regulations 1996 (SI 1996 No 1592)* apply;
- ships within the meaning of the *Merchant Shipping Act 1995*, other than a ship in the course of construction or repair;
- mines, with the exception of any building on the surface of a mine;
- offshore installations – workplaces to which the *Offshore Installations and Pipelines Work (Management and Administration) Regulations 1995 (SI 1995 No 738)* apply; and
- any workplace on an aircraft, locomotive or rolling stock, trailer or semi-trailer, used as a means of transport.

Agricultural or forestry land workplaces situated away from the undertaking's main buildings.

If you are unsure if your workplace is covered by the Regulations you are advised to consult a Fire Safety Officer at your local Fire Service.

What the Fire Precautions (Workplace) Regulations 1997 say:

3.26

> 3(5) For the purposes of these Regulations, an 'excepted workplace' is—
>
> (d) any workplace which is or is on a construction site within the meaning of regulation 2(1) of the Construction (Health, Safety and Welfare) Regulations 1996 and to which those Regulations apply;
>
> (e) any workplace which is or is in or on a ship within the meaning of section 313(1) of the Merchant Shipping Act 1995 other than a ship which is—

(i) in the course of construction; or

(ii) in the course of repair by persons who include persons other than the master and crew of the ship;

(g) any workplace which forms part of a mine, other than any building on the surface at a mine;

(h) any workplace which is or is in or on an offshore installation within the meaning of regulation 3 of the Offshore Installations and Pipelines Works (Management and Administration) Regulations 1995;

(i) any workplace which is or is in or on an aircraft, locomotive or rolling stock, trailer or semi-trailer used as a means of transport or a vehicle for which a licence is in force under the Vehicle Excise and Registration Act 1994 or a vehicle exempted from duty under that Act; and

(j) any workplace which is in fields, woods or other land forming part of an agricultural or forestry undertaking but which is not inside a building and is situated away from the undertaking's main buildings.

(6) The requirements of this Part of these Regulations shall not have effect to the extent that they would prevent–

(a) any member of the armed forces of the Crown or of any visiting force (within the meaning of regulation 19);

(b) any constable or any member of a police force not being a constable; or

(c) any member of any emergency service, from carrying out their duties.

(7) Without prejudice to paragraph (6), regulation 5(2)(f) shall not apply to any premises falling within the scope of section 40(2)(a) or (b) of the 1971 Act (prisons) or any part of any other premises used for keeping persons in lawful custody or detention.

(8) Where paragraph (6) or (7) applies, the safety of employees in case of fire shall nevertheless be ensured so far as is possible.

Other major legislation relating to fire safety in the workplace

3.27 Although there are a number of other pieces of legislation and specific Regulations relating to fire safety in the workplace, the major legislation which an employer should be aware of, is briefly described below:

Fire Precautions Act 1971

3.28 The *Fire Precautions Act 1971* imposes requirements relating to fire precautions upon occupiers of *premises*, as distinct from *Fire Precautions (Workplace) Regulations 1997*, which imposes requirements upon the *employer* or *persons who 'control' the workplace*. The main control provision under the Act is that certain 'designated' premises require a Fire Certificate.

For more detailed information concerning the Act, refer to **CHAPTER 2: LEGISLATION AND REGULATIONS (FIRE PRECAUTIONS ACT 1971)**.

Fire Certificates (Special Premises) Regulations 1976

3.29 'Special premises' are premises, which:

- operate hazardous processes;
- manufacture or store certain specified hazardous materials in quantities which present a significant hazard to employees and the general public; or
- have facilities provided for the storage of certain specified hazardous materials in quantities which present a significant hazard to employees and the general public.

These types of premises are required to comply with the *Fire Certificates (Special Premises) Regulations 1976 (SI 1976 No 2003)*.

For more detailed information concerning special premises, refer to **2.37** above.

Health and Safety at Work etc Act 1974

3.30 These following sections relate to an employer's responsibilities, which include protecting employees and other persons who might be on the premises, from the risk of fire.

HSWA 1974, s 2(1) identifies and places a *'general duty'* on every employer to ensure, so far as is reasonably practicable, the health, safety and welfare at work of their employees.

HSWA 1974, s 3 places a *'general duty'* on every employer to conduct their undertaking in such a way as to ensure, so far as is reasonably practicable, that persons not in their employment who may be affected thereby are not thereby exposed to risks to their health or safety.

For more detailed information concerning the Act, refer to **CHAPTER 6: LEGISLATION AND REGULATIONS (HEALTH AND SAFETY)**.

Health and Safety (Safety Signs and Signals) Regulations 1996.

3.31 These Regulations require the employer, if risks to health and safety cannot be avoided or controlled adequately by other means, to provide and maintain safety signs.

Any signs that are provided cannot be solely composed of text, but should also be provided with a pictogram. Examples of signs used to help control fire risk are 'NO SMOKING' or 'FIRE POINTS'.

Where any of the existing signs currently provided in the workplace do not include a pictogram, they should have been replaced by a new sign, or alternatively have had a pictogram added. This should have been implemented by 24 December 1998.

It is also necessary for an employer to ensure that all of the safety signs provided in the workplace, which includes those provided for fire safety, should be maintained. In addition, employees should be advised on the meaning of any signs provided, in order that they know what action to take in relation to them.

Building Regulations

3.32 The following Building Regulations apply in:

- England and Wales – *Building Regulations 1991 (SI 1991 No 2768)*;
- Scotland – *Building Standards (Scotland) Regulations 1991 (SI 1991 No 158) (as amended)*; and
- Northern Ireland – *Building Regulations (Northern Ireland) 1994 (SI 1994 No 243)*.

These Regulations relate to employers and their premises when adding a new building or making extensions or any significant alterations to existing buildings. They also apply where there is a material change of use. Recent amendments to the 1991 rules have been made by the *Building Regulations (Amendment) (No 2) Regulations 1999 (SI 1999 No 3410)*.

Updated guidance on how to meet the revised Building Regulations on fire safety in England and Wales has been issued by the DETR. The guidance, entitled *Approved Document to Support Part B: Fire Safety*, came into force on 1 July 2000 and contains the requirements relating to fire safety. For the first time, guidance is included on fire alarm and fire detection systems within a wide range of buildings.

3.32 *Legislation and Regulations (Fire Prec. (Workplace) Regs. 1997 & 1999)*

An Approved Document is similar to an Approved Code of Practice. This means that it is not an offence in itself to fail to comply with the Code but it can be used as evidence of a failure to comply with the Regulations unless you have achieved compliance with them in some other way.

Enforcement of the *Building Regulations* is by the Building Control Department of the local authority.

The main changes in the 2000 edition of the *Approved Document to Support Part B: Fire Safety*, which replaces the 1992 edition were:

- General introduction
  - (a) Hospitals: HTM 81 can be used instead of the Approved Document.
- B1
  - (b) Fire alarms:
    - (i) the Requirement has been expanded to include fire alarm and fire detection systems; and
    - (ii) the guidance forms a new Section 1 and has been extended to loft conversions and buildings other than dwellings.
  - (c) Alternative approaches: this guidance has been moved from Section 3 to the 'Introduction' and expanded.
  - (d) Door width: the definition has been modified to align with that given in Approved Document M and corresponding reductions made to Table 5 'Widths of escape routes and exits'.
  - (e) Means of escape:
    - (i) Dwellings – storeys not more than 4.5m above ground level need to be provided with emergency egress windows;
    - (ii) Single escape routes and exits – the limit of 50 persons has been increased to 60;
    - (iii) Alternative escape routes – the 45 degree rule has been changed;
    - (iv) Minimum number of escape routes – Table 4 has been simplified;
    - (v) Mixed use buildings – the guidance has been modified;
    - (vi) Door fastenings – more guidance is given;
    - (vii) Escape lighting – changes have been made in Table 9 regarding toilet accommodation;
    - (viii) Storeys divided into different uses – guidance has been added to deal with storeys which are also used for the consumption of food and/or drink by customers;

Legislation and Regulations (Fire Prec. (Workplace) Regs. 1997 & 1999) **3.32**

        (ix) Shop store rooms – guidance is given on when these need to be enclosed in fire-resisting construction.

- B2

    (f) Special applications: guidance is given on the use of air supported structures, structures covered with flexible membranes and PTFE based materials.

- B3

    (g) Places of special fire hazard: these need to be enclosed in fire-resisting construction.

    (h) Compartments: maximum compartment dimensions have been extended to single storey Schools and to the Shop/Commercial purpose group.

- B4

    (i) Rooflights: separate Tables are given for Class 3 and TP(a)/(b) plastics rooflights and the provisions relating to Class 3 rooflights on industrial buildings has been modified.

- B5

    (j) Vehicle access:

        (i) specific guidance is now included for single family houses and for blocks of flats and maisonettes;

        (ii) the 9m height in Table 20 has been increased to 11m.

    (k) Personnel access:

        (i) modifications have been made to the heights at which fire fighting shafts are needed, with corresponding reductions to the 20m height in B1 (access lobbies & corridors), B3 (Table 12), B4 (Diagram 40) and Appendix A (Tables A2 and A3);

        (ii) guidance is given regarding fire fighting shafts in blocks of flats and maisonettes.

Appendix A

    (l) Uninsulated glazed elements: table A4 has been modified and extended.

    (m) Notional designations of roof coverings:

- bitumen felt pitched roof coverings have been deleted from table A5.

Appendix B

    (n) Compartment walls: limits are now specified on the use of uninsulated doors.

3.33 *Legislation and Regulations (Fire Prec. (Workplace) Regs. 1997 & 1999)*

Appendix E

(o) Fire separating element: this new definition has been added to support Sections 9 to 11 in B3.

Appendix F

(p) This new Appendix gives guidance on insulating core panels.

Publications providing further guidance on fire precautions in the workplace

3.33

Fire Safety: An employer's guide	The Stationery Office	ISBN 0 11 341229 0
Fire Precautions Act 1971. Guide to Fire Precautions in Existing Places of Work that require a Fire Certificate; Factories, Offices, Shops and Railway Premises	The Stationery Office	ISBN 0 11 341079 4
Fire Precautions Act 1971. Guide to Fire Precautions in Premises used as Hotels and Boarding Houses which require a Fire Certificate	The Stationery Office	ISBN 0 11 341005 0
Fire Precautions Act 1971. Fire Safety Management in Hotels and Boarding Houses	The Stationery Office	ISBN 0 11 340980
Guide to Health, Safety and Welfare at Pop Concerts and Similar Events	The Stationery Office	ISBN 0 11 341072 7
Guide to Fire Precautions in Existing Places of Entertainment and Like Premises	The Stationery Office	ISBN 0 11 340907 9
The Building Regulations 1991: Approved Document B: Fire Safety	DETR Publications	ISBN 1 85112 3512
Technical Standards for compliance with the Building Standards (Scotland) Regulations 1990 as amended	The Stationery Office	ISBN 0 11 49 5866 1
Fire and the Design of Educational Buildings: Building Bulletin 7	The Stationery Office	ISBN 0 11 270585 5

FIRECODE (and FIRECODE Scotland). A suite of documents aimed at healthcare premises	The Stationery Office	ISBN various – contact NHS Estates (0113 254 7000) for details.
Heritage Under Fire: A guide to the protection of historic buildings	Fire Protection Association	ISBN 0 90 216790 1

For further guidance on particular risks within workplaces see **CHAPTER 13: SPECIALIST INFORMATION.**

4 Legislation and Regulations (Disability Discrimination Act 1995)

In this chapter:	
Disability Discrimination Act 1995	4.1
How does 'Part III – Access to Goods and Services' apply?	4.2
What are auxiliary aids and services?	4.18

Disability Discrimination Act 1995

4.1 The *Disability Discrimination Act 1995* (*DDA 1995*), which came into force in December 1996, introduced new laws aimed at ending the discrimination that many disabled people suffer.

The intention of *DDA 1995* was that it should protect disabled people in the areas of employment, access to goods, facilities and services and the management, buying or renting of land or property.

Some of the measures contained within *DDA 1995* came into force immediately, whereas other sections are being phased in over a period of time, as follows:

- since 2 December 1996, it has been unlawful for businesses and service providers to treat disabled people less favourably than other people for a reason, which is related to their disability;

- since 1 October 1999, businesses and service providers are required to make reasonable adjustments for disabled people, such as providing extra help or making changes to the way they provide their services; and

- from 2004 service providers will also have to consider making reasonable adjustments to the physical features of their premises to overcome physical barriers to access.

The meaning of 'disability' and 'disabled person' under *DDA 1995* is as follows:

'Subject to the provisions of Schedule 1, a person has a disability for the purposes of this Act if he has a physical or mental impairment, which has a substantial and long-term adverse effect on his ability to carry out normal day-to-day activities.'

DDA 1995 prohibits a disabled person from being treated less favourably (because of their disability) than a person who is not disabled. In effect *DDA 1995* gives disabled people new rights of access to goods, facilities and services, as well as in employment and buying or renting property.

DDA 1995 makes it unlawful for employers with 20 or more staff to discriminate against current or prospective employees with disabilities because of a reason relating to their disability, this applies to all employment matters, including recruitment and retention of employees, training and development, promotion and transfers, and the dismissal process.

Where, or if, their employment arrangements or premises substantially disadvantage a disabled employee/applicant, employers will have to look at what changes they could make to the workplace or the way work is done and make any changes that are reasonable. Employers will be able to take into account how much the changes would cost, and how much they would help. Employers will not be expected to make any changes, which would break health and safety laws.

As this handbook is primarily concerned with fire safety management, this chapter covers only the implications for businesses contained within *DDA 1995, Part III*, which deals with access to goods and services for the disabled.

The employment provisions, detailed in *DDA 1995, Part II* are not dealt with in this chapter although the definition of a disabled person is exactly the same.

How does 'Part III – Access to Goods and Services' apply?

Goods, facilities, services and property

4.2   The *Disability Discrimination Act 1995* applies to anyone who provides goods, facilities or services to members of the public whether paid for or free. It applies both to commercial and public services. These could range from buying bread in a supermarket, using the facilities in a bank or launderette, borrowing a book from a public library or using a hospital or railway station. Private clubs and associations are not included.

Providing the same standard of service to everyone

- It is against the law to refuse to serve someone who is disabled.

4.2 Legislation and Regulations (Disability Discrimination Act 1995)

For example, it is against the law for a supermarket owner to refuse to serve someone whose disability means that they shop slowly.

- It is against the law to offer a disabled person a service, which is not as good as the service, being offered to other people.

For example, it is against the law for a restaurant owner to insist that a person with a facial disfigurement or other disability sits out of sight of the other customers.

- It is against the law to provide a service to a disabled person on terms that are different from the terms given to other people.

For example, it is against the law to ask a disabled person for a bigger deposit when they are booking a holiday.

Exceptions

- If the health or safety of the disabled person or other people would be in danger, it would not be against the law to refuse to provide the service to a disabled person or to provide it on different terms.

For example, a fairground operator could refuse to let a person using a wheel chair onto a ride, if the person could not stand unaided and it was necessary for people on the ride to stand.

Other exceptions would arise if:

- the customer was not capable of understanding the terms of a contract or giving informed consent; or

- providing the service or the same standard of service would deny service to other customers.

For example, a cinema owner who has set aside an adequate area for wheel chair users would not have to make every seat in the auditorium accessible for disabled people. That is because the cost could lead to the cinema closing down.

Making changes to the way goods, facilities and services are provided

- It is against the law for someone to run a service, or provide goods or facilities, in a way that makes it impossible or unreasonably difficult for a disabled person to use the service or goods.

For example, a restaurant that does not allow animals is not able to refuse admission to a disabled person with a guide dog.

- It is not against the law, however, to run a service in a way that is not favourable to a disabled person if that way of operating the service is fundamental to the business.

For example, dim lighting could be considered essential to a nightclub even though it causes difficulties for someone with poor eyesight.

- People must provide equipment or other helpful items that will make it easier for disabled people to use their service, if it is reasonable to do so.

  For example, an induction loop may make it easier for people who use hearing aids or a handrail for people who find walking up stairs difficult.

- Service providers will have to remove or alter features of their premises that make it unreasonably difficult, or impossible, for a disabled person to use the service (for example, widening entrance doors) or they must provide other ways of enabling the disabled to use their services, if it is reasonable to do so.

  For example, if a library's reference section is on the first floor and there is no lift, library staff could offer to bring the reference books to the disabled person.

- Service providers are not able to charge a disabled person more to meet the cost of making it easier for them to use their service.

Who has rights under the Act?

4.3   DDA 1995 is designed to protect the rights of a wide range of people with physical, sensory, or mental disabilities. This includes people who use wheelchairs, those who are blind and partially sighted, deaf people, people with arthritis, people with long-term illnesses and people with learning disabilities. The Act also covers people with severe disfigurements and, in certain circumstances, people who have had a disability in the past – for example, someone who had severe depression, but has since recovered.

Are there any exemptions to service providers under the Act?

4.4   Yes – DDA 1995, Part III excludes education and the use of any means of transport from its provisions. It also excludes services not available to the public, such as those provided by private clubs to their members. Manufacturers and designers of goods are also excluded, unless they provide a service direct to the public.

An example of a 'direct to the public service' might be where the public is allowed into the workplace to view how a product is made. Glass blowing and pottery making are common examples of this. In these types of instances, the exemption would not apply.

What are the legal responsibilities as a service provider?

4.5   As a service provider, an employer must not discriminate against a disabled person:

4.6 *Legislation and Regulations (Disability Discrimination Act 1995)*

- by treating him or her 'less favourably' than other customers who are not disabled because of their disability; or
- by not making reasonable adjustments to the way you deliver your services, so that disabled people can use them.

What does 'less favourably' mean?

4.6   There are three ways in which you might, as a service provider, treat a disabled person 'less favourably' than you treat other customers:

- by refusing to serve him or her;
- by providing him or her with a lower standard of service; or
- by providing him or her with a service on less favourable terms than other customers.

What is an employer required to do?

4.7   Where it is impossible or unreasonably difficult for disabled people to use the services' an employer may be required to undertake the following.

- Take reasonable steps to change current practices, policies or procedures.
- Provide a reasonable alternative method of making services available to disabled people.
- An employer may also have to take reasonable steps to provide an auxiliary aid or service to assist or enable disabled people to use services.
- An employer must consider taking reasonable steps to change their *practices* (what they do), *policies* (what they intend to do) or *procedures* (how they plan to go about it), if they make it impossible or unreasonably difficult for disabled people to use their services.

   For example, a guest house or hotel owner should offer wheelchair users those bedrooms with level access to the lift, or bedrooms on lower floor levels to facilitate means of escape in case of fire. Without that change of policy, the disabled person would otherwise be unable to use the guest-house or hotel and this change is necessary in order to provide the service to the disabled guest.

- An employer must take reasonable steps to provide a reasonable alternative method for making services available to disabled people, where a physical feature makes it impossible or unreasonably difficult for them to use these services.

   For example, an independent cinema has a steep flight of stairs at its front entrance, making it impossible or unreasonably difficult for visitors with a mobility impairment to access the cinema. A side entrance for staff use only is fully accessible and always open. The cinema decides to allow disabled people to use this side entrance.

- An employer must take reasonable steps to provide an auxiliary aid or service if it enables disabled people to use the services or helps them to do so.

What are auxiliary aids and services?

4.8 This will depend on the situation, but it might be a piece of equipment or just extra assistance to disabled people from trained staff. Auxiliary aids and services should make it easier for disabled people to use the employer's service.

For example, a guesthouse or hotel may install an audio-visual type of fire alarm in one of its guest bedrooms to accommodate visitors with sensory impairments.

Another example, a cinema or theatre manager makes a member of staff available to assist a wheelchair user in the event of a fire evacuation of the building.

How can employers know what special aids to provide?

4.9 It is good practice to consult disabled customers about their needs. An employer could also talk to local or national disability groups, or trade associations and find out what other firms have done in similar circumstances.

Does an employer have to make physical alterations to their premises?

4.10 At present, an employer does not have to remove or change physical features of the premises, or provide a reasonable means of avoiding a particular feature. These duties will be introduced in 2004. However, there is nothing in *DDA 1995* to prevent an employer making physical changes to their premises before then. For example, where employers are planning building work or refurbishment they could include these physical changes at the same time.

However, where a physical feature of their premises prevents a disabled person from using the services, then an employer is obliged to consider whether they can take steps to provide an alternative way of making the service available to disabled people.

Means of escape for the disabled

4.11 Ensuring the means of escape of disabled people in case of fire can present problems to the owners of business premises.

4.12 *Legislation and Regulations (Disability Discrimination Act 1995)*

As a consequence of this the owners or occupiers of business premises should consider the needs of the disabled when drawing up their evacuation plans.

Listed below are some suggestions on how the safe evacuation of people with specific types of disability could be achieved by means of forward planning and the provision of special equipment.

- Mental illness – trained employees should be available to lead and generally assist these people to evacuate the building to a safe location.

- Impaired vision – consideration should be given to providing brightly illuminated means of escape routes within the building. Again trained employees should be available to lead and generally assist these people to evacuate the building.

- Impaired hearing – consideration should be given to providing both an audio and visual type of fire alarm system to ensure that people with hearing difficulties are aware of a fire alarm warning.

- Users of wheelchairs – they should be prevented from using lifts in the event of a fire. The exception to this is if the lift has been designed and fire protected in accordance with British Standard 5588: '*Fire precautions in the design, construction and use of buildings Part 8: 1999 – Code of practice for means of escape for disabled people*'.

- Again, if staircases have to be negotiated, employees who have been suitably trained to carry the disabled should be made available specifically for this purpose.

If there are specific problems in relation to means of escape for the disabled that an employer cannot resolve, it is good practice to consult the fire safety officer of your local fire service for advice.

Checklist of good practice

4.12 The following amended checklist, issued by the Department for Education and Employment, should be considered as a best practice guide.

Checklist

4.13

> Listed below are some steps an employer can take, to ensure that they comply with the spirit of *DDA 1995*, as well as meet the legal obligations.
>
> - Think and plan ahead to meet the requirements of disabled customers.

Legislation and Regulations (Disability Discrimination Act 1995) **4.14**

- Do not make assumptions about disabled people based on speculation or stereotypes. Think about the wide range of disabilities that there are when planning adjustments.

- If in doubt, ask disabled people themselves how they can best be served. Listen carefully and respond to what they really want. An employer could also consult with disabled staff and disability organisations.

- Think about the way to treat disabled customers. Let them know how to request assistance, and have a customer complaints procedure that is easy for them to use.

- Ensure that the dignity of a disabled person is respected when providing them with services.

- Establish a positive policy on providing services to ensure it includes disabled people. Communicate this policy to staff and monitor its effectiveness.

- Consider putting in place positive practices, which will encourage disabled people and others to use services.

- Make sure staff training includes company policy towards disabled people and their legal rights, disability awareness and disability etiquette training.

- Regularly review whether services are accessible to disabled people.

Do not wait until 2004 to remove any physical features in buildings that create a barrier to access for disabled people. Consider doing this at the same time as any building or refurbishment work that is being planned, which could help reduce costs and disruption.

Some smaller businesses may find some of these steps more important than others in making their smaller-scale services accessible to disabled people. However, adopting best practice can only help to ensure that an employer complies with *DDA 1995*.

Publications providing further guidance on the Disability Discrimination Act 1995

4.14 *Disability Discrimination Codes of Practice (Education) (Appointed Day) Order 2002 – Statutory Instrument 2002 (SI 2002 No 2216)*

Department for Education and Employment booklet entitled *'An Introduction for Small and Medium-sized Businesses – Rights of Access to Goods, Facilities, Services and Premises – Disability Discrimination Act 1995'*: ISBN 1 84185 086 1

4.15 *Legislation and Regulations (Disability Discrimination Act 1995)*

Department of the Environment (now Office of the Deputy Prime Minister): *Approved Document M – Access and facilities for the disabled*: (1999 Edition) ISBN 0 11 753469 2

This document took effect from 25 October 1999, it does not cover means of escape but contains sound guidance regarding spectator seating, ramped approaches, corridor and door widths etc.

Approved Document B-Fire Safety (2000 Edition) ISBN 1 85112 351 2

This document gives valuable guidance on fire safety in general but does not cover means of escape for the disabled in any great detail. It refers the reader mainly to British Standard 5588 *'Fire precautions in the design, construction and use of buildings'* Part 8: 1999 – Code of Practice for means of escape for disabled people

Disability Discrimination Act 1995: Code of Practice; Rights of Access to Goods, Facilities, Services and Premises

The Code gives practical guidance on how to prevent discrimination against disabled people in accessing services. It describes the duties of service providers and others, helps disabled people understand the law, and seeks to encourage good practice: ISBN 0 11 271055 7.

Published by The Stationery Office and available from:

The Publications Centre, PO Box 276, London SW8 5DT (also from The Stationery Office Bookshops). Telephone orders/general enquiries: 0870 600 5522, fax orders: 0870 600 5533

Publications providing further guidance on means of escape for the disabled

British Standard 5588: *'Fire precautions in the design, construction and use of buildings'* Part 8: 1999 – Code of Practice for means of escape for disabled people

This Code of Practice provides guidance on the provision of adequate means of escape for the disabled and applies to new buildings and existing buildings, which are being substantially altered. It also permits the use of lifts for means of escape provided they are suitably constructed as evacuation lifts.

Other useful sources of information

4.15   Department of Education and Skills website: www.disability.gov.uk

Disability Rights Commission website: www.drc-gb.org

5 Legislation and Regulations (British and European Standards)

In this chapter:	
Introduction	5.1
What are they?	5.2
Relationship with International and European Standards	5.3
Contacting the British Standards Institution	5.4

Introduction

5.1 The British Standards Institution ('BSI') is now considered to be the world's leading standards and quality services organisation. Formed in 1901 and incorporated under Royal Charter in 1929, the British Standards Institution is also the oldest national standards-making body in the world.

Independent of government, industry and trade associations it is a non-profit distributing organisation. The British Standards Institution is recognised, globally, as an independent and impartial body serving both the private and public sectors.

The British Standards Institution works with manufacturing and service industries, businesses and governments to facilitate the production of British, European and International Standards. As well as facilitating the writing of British Standards, the British Standards Institution represents UK interests across the full range of European and International Standards committees.

There are now over 17,000 British Standard publications covering all industry sectors, with some 2,000 or so, new or revised standards being issued or re-issued each year.

The British Standards Institution 'Kitemark' logo is probably the UK's most recognised product certification mark. It is used to indicate to the purchaser that a product conforms to a published specification.

5.2 *Legislation and Regulations (British and European Standards)*

What are they?

5.2 The development of common standards evolved as business organisations needed to trade with each other both in the UK and abroad, using commonly agreed, mutually respected, operating standards in which all participants were aware of the 'ground rules'.

Standards have not only been produced to specify how a given product should be manufactured, but they also set out 'best practice' methods by means of various Codes of Practice. In relation to fire safety, standards have been produced which specify best practice for fire precautions in the design and construction of buildings, the provision and siting of fire fighting equipment, emergency lighting, fire alarm systems and fire doors etc.

In recent years, British Standards have been increasingly used in conjunction with legislation. For example, the *Building Regulations* in their Approved Documents publications specify certain codes or standards as being a suitable to satisfy the Regulations. In effect, this means, that providing an employer, occupier or owner ensures that they provide or install equipment, or carry out building work in accordance with a particular Standard, the work will be considered to comply with the Regulations.

A relatively recent development has been the emphasis on 'fire engineering' solutions as an alternative to the use of the prescriptive codes in dealing with fire safety in new buildings. A British Standard Draft for Development *(DD) 240 Fire safety engineering in buildings* was introduced in 1997. This document is in the course of being replaced by British Standard *7974: 2001 Code of Practice on the Application of Fire Safety Engineering Principles to the Design of Buildings*.

This code is to be supported by a number of Published Documents and/or other Codes of Practice. The extent to which the use of this document will be relevant to individual buildings has still to be assessed.

There is further comment regarding this approach in **CHAPTER 13: SPECIALIST INFORMATION** (see **13.69** below) as this approach can only apply at the time that a building is erected. Should a fire risk assessment reveal shortcomings in the design, eg. changes in manufacturing processes, fire load, number of persons within a building etc additional structural alterations or installation of fire safety may be required to ensure the safety of the occupants of a building.

Another example, relating specifically to fire, is where the Standards are used to indicate how to comply with the *Fire Precautions Act 1971*. There are a number of Guides to this Act (which are detailed in **CHAPTER 2: FIRE PRECAUTIONS ACT 1971**), which advise that by employing the various British Standards, an employer, occupier or owner can 'virtually' ensure compliance with *FPA 1971*.

Committees consisting of representatives who have a particular interest in the subject, including manufacturers, users, research organisations, government departments, local authorities and consumers, generally prepare or 'oversee' the drafting all new Standards. This is invariably the method that is adopted with regards to those standards that have a particular bearing upon fire safety.

New or revised standards are issued annually to ensure that the technical content is 'up to date' and to encompass new materials, processes and technologies. Before any standard is published, the draft document is made available for public comment.

Relationship with International and European Standards

5.3   Internationally, there are now more than 100 similar organisations to the BSI, which are members of the International Organisation for Standardisation ('ISO') and the International Electrotechnical Commission ('IEC'). These bodies produce harmonised world standards. The BSI ensures that the views of British industry are represented on these bodies.

The BSI also represents the British viewpoint to the three European standards organisations, which are the European Committee for Standardisation ('CEN'), the Committee for Electrotechnical Standardisation ('CENELEC') and the European Telecommunications Standards Institute ('ETSI'). Together they prepare European Standards in specific sectors of activity and the three make up the 'European Standardisation System'. These bodies develop the harmonised European standards as part of the process of achieving the single market for Europe. The concept of harmonisation is intended to diminish trade barriers, promote safety, allow interchange of products, systems and services, and promote common technical understanding.

Most standards are prepared at the request of industry. In addition the European Commission can also request the standards bodies to prepare standards in order to implement European legislation.

One of the most obvious manifestations to the consumer of the harmonisation of standards is the 'CE' logo, which is increasingly seen on consumer products. The CE mark is a European proof of conformity and can also be regarded as a 'passport' that allows manufacturers and exporters to circulate products freely within the EU. The letters, 'CE' is French, for 'Conformité Européenne,' which indicates that the manufacturer has satisfied all assessment procedures specified by law for its product.

Unlike the British Standards Institution 'Kitemark', it should be emphasised that the CE mark is not a quality mark; it only indicates that the manufacturer of the product is declaring that it complies with the relevant EU directive.

5.4 *Legislation and Regulations (British and European Standards)*

EU product directives are limited to essential safety, health or other performance requirements in the general public interest. The technical details of how to meet these requirements are left to manufacturers who self-certify products.

The European Commission has published the book, *Guide to the implementation of Community harmonisation directives based on the new approach and the global approach*. This document can be obtained via the EC-DGIII Information Desk.

For more information, see the 'EC-DGIII Documents and Publications' web page at: www.europa.eu.int/comm/dg03/public.htm

In addition a list of 'harmonised standards' cited in the *Official Journal of the European Communities* is found on the EC-DGIII website at:

www.europa.eu.int/comm/dg03/directs/dg3b/newapproa/eurstd/harmstds/index.html

Contacting the British Standards Institution

5.4 See **APPENDIX: FURTHER SOURCES OF INFORMATION** for a list of the most common British Standards relevant to fire related matters, currently in use in the UK.

To request further information on any of BSI's products and services or to purchase standards, publications or merchandise, contact: BSI, 389 Chiswick High Road, London W4 4AL. Tel: 0208 996 9000; fax: 0208 996 7400. Website: www.bsi.org.uk.

6 Legislation and Regulations (Health and Safety)

In this chapter:	
Introduction	6.1
Recent trends in United Kingdom health and safety law	6.2
High profile cases and future trends	6.8
Risk assessments	6.9
Recent changes in legislation	6.11
Amendments to existing Regulations	6.28
Criminal law	6.30
Levels of duty	6.32
'Six pack'	6.33
Other important Regulations	6.35
Management of health and safety at work	6.43
Application of the 'management cycle'	6.46

Introduction

6.1 It is impossible to look at the legal implications for any business, with regard to fire safety legislation in isolation, as fire safety forms just one facet of a very much wider picture of health and safety legislative requirements to which businesses in the United Kingdom are subject. There would seem to be a number of changes that are to be made in the *relatively* near future particularly with regard to fire Regulations. The general aims of these expected changes are outlined in **CHAPTER 15: THE WAY AHEAD**.

More than 28 years have passed since the major legislative enactment, the *Health and Safety at Work etc Act 1974* (*HSWA 1974*) came into force. Since then, health and safety legislation has been constantly evolving and growing until it now relates to every aspect of a business and its workforce.

Nor can it be expected that this growth will to come to a standstill, as with advances in technology, new unforeseen hazards will arise and further legislation will be enacted to counteract the perceived hazards.

An appropriate analogy for this situation is, 'safety is a journey and not a destination'. This describes how, in any given situation, you never actually arrive at *the totally* safest point. What an employer can hope to achieve by taking health and safety precautions is that in any given situation the workplace will be as safe as an employer can possibly make it.

This chapter sets out some of the developments and the background to United Kingdom health and safety legislation, together with the legal implications for businesses of various health and safety initiatives. Such initiatives include the highly publicised 'corporate manslaughter' prosecutions and the possible new offence of 'corporate killing'.

These developments may or could have major implications for company directors and owners of businesses in which fatalities occur as a result of a fire and in which subsequent investigations reveal a failure to comply with fire safety legislation.

The chapter also lists and explains some recent changes in legislation, most of which are not directly related to fire safety. However they have been included as a useful source of reference, as it is appreciated that many fire safety managers will also have a wider remit for all health and safety matters.

This information can also be found in *Tolley's Health and Safety at Work* Looseleaf and CD Rom.

Recent trends in United Kingdom health and safety law

Role of the European Union

6.2 For the past decade the European Union ('EU') has provided the principal motive for changes in United Kingdom health and safety legislation. The majority of recent health and safety legislation has been enacted due to the United Kingdom's obligations to implement European Union directives. A huge increase in the number of directives resulted from the *Single European Act 1987 (SEA 1987)*.

Prior to 1987 only six directives exclusively on health and safety at work had emerged from the European Commission. *SEA 1987* inserted a new Article, *SEA 1986, Art 138* (which was previously numbered *118A*) into the *Treaty of Rome*. *Article 138* enabled the European Union to adopt health and safety directives by means of qualified majority voting amongst member states in the Council of Ministers. Prior to *Article 138* unanimity amongst member states was required.

Following the introduction of what is now *Article 138*, the EC Council issued a 'Framework' Directive (*Directive 89/391/EEC*) that sets out key principles applying to all areas of work and imposing general duties on employers.

The details in relation to discrete areas of work activity are provided by a series of 'daughter' directives.

The United Kingdom has largely chosen to implement European Union health and safety directives by means of Regulations made under the *Health and Safety at Work etc Act 1974, section 15*.

Implementation of European Union directives

6.3 The United Kingdom, along with other Member States, must implement European Union directives by designated dates. This is normally overseen by the Health and Safety Commission ('HSC'), which first circulates consultative documents incorporating the proposed Regulations, together with a related Approved Code of Practice and/or guidance to interested parties (trades unions, employers' organisations, professional bodies, local authorities etc).

Following the consultation process, the Health and Safety Commission submits a final version to the Secretary of State who then lays the Regulations before Parliament. It is at this stage that a date for the implementation of the Regulation(s) coming into force is determined which is usually that stipulated by the original European Union directive.

'Six pack' and beyond

6.4 This process peaked on 1 January 1993 when six sets of Regulations came into operation on the same day. These were:

- *Management of Health and Safety at Work Regulations 1992 (SI 1992 No 2051)* (these have now been replaced by the *Management of Health and Safety at Work Regulations 1999 (SI 1999 No 3242)*);
- *Provision and Use of Work Equipment Regulations 1992 ('PUWER')(SI 1992 No 2932)*;
- *Workplace (Health, Safety and Welfare) Regulations 1992 (SI 1992 No 3004)*;
- *Personal Protective Equipment at Work Regulations 1992 (SI 1992 No 2966)*;
- *Manual Handling Operations Regulations 1992 (SI 1992 No 2793)*; and

6.5 Legislation and Regulations (Health and Safety)

- Health and Safety (Display Screen Equipment) Regulations 1992 (SI 1992 No 2792).

Although the pace has now slackened, new Regulations and changed requirements continue to appear on a regular basis. These have included a new version of *SI 1998 No 2306* and *SI 1999 No 3242*, replacing the 1992 Regulations.

A summary of the more significant changes during the last couple of years is provided below at **6.11** below. Application of the requirements of some of the more specific European Union directives has resulted in the production of some quite detailed Regulations or Approved Code of Practices. Concerns have been expressed (not least within the United Kingdom) in relation to the variable standards of implementation and enforcement within European Union Member States. There have been moves recently to address these concerns and also to evaluate the impact of European Union directives.

Deregulation

6.5 The latter stages of the Conservative Government of 1979–1997 saw a general drive towards deregulation in Britain because of the perceived burden of legislation on business, particularly on small employers. As part of this process the Health and Safety Commission was asked to review health and safety Regulations and its report was published in May 1994.

However, the scope of its recommendations was limited by the need to comply with European Union directives and by the requirement contained in the *Health and Safety at Work etc Act 1974, section 1(2)* that any new Regulations or Orders must maintain or improve health and safety standards.

Consequently the report recommended no major changes. It has resulted merely in the repeal of some outdated legislation (which had limited application because alternative legislation already existed) and also in a call for the Health and Safety Executive to make their Approved Code(s) of Practices and Guidance publications more 'user friendly'.

Many people, both inside and outside the Health and Safety Commission and Health and Safety Executive, were concerned that political pressure was being brought to bear on the enforcement of health and safety legislation. Criticisms particularly centred on the so-called 'minded to' procedure, which required inspectors to advise employers in writing of their intention to issue an improvement notice.

Labour Government

6.6 One of the early initiatives of the present Labour Government was to remove the 'minded to' procedure. Responsibility for the Health and Safety

Executive passed to the newly created Department of Environment, Transport and the Regions ('DETR') early in 1998, which announced an increase in Health and Safety Executive funding of £4.5 million.

At the same time the Department of Environment, Transport and the Regions stated that more efficient ways of enforcing the law would be discussed and particular attention would be paid to employers who evade their health and safety responsibilities by ostensibly employing home workers and self-employed people.

The Government also announced that the Health and Safety Executive would be developing a long-term strategy for occupational health. Asbestos, work-related stress, hand-arm vibration and solvents have all been mentioned as specific areas of interest.

In 1999 the Minister responsible for health and safety announced that the Government would be marking the 25th anniversary of the *Health and Safety at Work etc Act 1974* with a full review of the law relating to health and safety. Many observers expect this review to extend to the penalties available for breaches of the law, particularly in the light of several high profile prosecutions.

Manslaughter and the proposed offence of 'corporate killing'

6.7 There have been well-publicised failures in recent years to obtain convictions for manslaughter against companies such as P&O Ferries (following the Zeebrugge disaster in 1987) and Great Western Trains (following the Southall rail crash in 1997).

The difficulty is that, under current law, a company can be found guilty of manslaughter only if the prosecution is able to prove gross negligence on the part of a director or senior executive who can be identified with the control or management of the company. Such a 'controlling officer' must himself be guilty of manslaughter for the company to be liable.

As a result very few prosecutions for manslaughter have succeeded and generally such prosecutions involved smaller companies where the 'controlling officer' has been more readily identified. Thus, the managing director of a one-man company was clearly the 'controlling officer' and therefore prosecuted after four teenagers drowned in a canoeing accident in Lyme Bay in 1993.

In 1996 the Law Commission recommended changes to the law of manslaughter and suggested an offence of 'corporate killing'. A corporation would be guilty of the offence if:

87

- a management failure by the corporation was the cause or one of the causes of a person's death; and
- that failure constituted conduct falling far below what could reasonably be expected of the corporation in the circumstances.

There would be a 'management failure' if the way in which the company's activities were managed or organised failed to ensure the health and safety of persons employed in or affected by those activities.

In terms of securing more prosecutions, the advantage of the offence lies in the fact that there would be no need to identify someone within the organisation, whose conduct can be attributed to the company. Instead, the offence emphasises the company's 'management failure'.

Such reform would make successful prosecutions of large companies easier and convictions for corporate manslaughter more likely. As a result of the Law Commission's recommendation, the Government announced in May 2000 their intention to introduce the offence of 'corporate killing'.

High profile cases and future trends

6.8 Recently the courts have shown an increasing willingness to impose significant fines. This trend is now equally applicable to the more typical health and safety prosecutions as a result of the decision by the Court of Appeal in *R v F Howe & Son (Engineers) Limited [1999] 2 All ER 249* in 1998. The Court of Appeal said that magistrates should think carefully before accepting jurisdiction in health and safety cases where the fine may exceed the limit of their jurisdiction or where death or serious injury has resulted.

As a consequence, it is likely that in future more health and safety cases will be committed to the Crown Court for trial rather than being dealt with by magistrates. This will inevitably result in an increase in the levels of fines imposed.

A record fine on an individual company for a health and safety offence of £1.5 million was imposed on the Great Western Trains Company Limited for the railway accident that occurred at Southall in 1997. Seven passengers were killed and 147 were taken to hospital when a high-speed train operated by the company went through a red signal and collided with an empty freight train. The Great Western Trains Company pleaded guilty to contravening *Health and Safety at Work etc Act 1974, section 3(1)* on the basis that they failed to ensure that the public was not exposed to risks to their health and safety.

In September 1994 six people were killed and seven seriously injured when a pedestrian walkway at the port of Ramsgate collapsed. After a twenty-five day trial early in 1997, two Swedish companies that had constructed and designed the walkway were fined £750,000 and £250,000 respectively. Lloyds Register of Shipping (who inspected the walkway) was fined £500,000 and Port Ramsgate, the client, was fined £200,000. With costs added to the total fines of £1.7 million, the bill to the defendants for the criminal proceedings alone came to a total of just under £2.5 million.

The collapse in October 1994 of a rail tunnel that was under construction at Heathrow did not result in any injuries but was described in court as one of the 'biggest near misses' in years. When the prosecution came to court in 1999, Balfour Beatty Civil Engineering were fined £700,000 plus £100,000 costs, with the tunnelling sub-contractor, Geoconsult GES MBH, fined £500,000 plus £100,000 costs.

Both of these cases show how responsibilities for health and safety are often shared between companies, with the Ramsgate prosecution illustrating, in particular, the importance of exerting control throughout the 'supply chain' see **6.10** below.

A recent case (*City Logistics Ltd v the Northamptonshire County Fire Officer [2001] EWCA Civ 1216*) will certainly influence the way fire safety legislation is interpreted. There is further comment on this aspect in **CHAPTER 15: THE WAY AHEAD**.

The basic facts of the case revolved around a warehouse that was completed early in 1998. This warehouse is about 175 metres long and about 80 metres wide and 16 metres high. Apparently, high racking has been installed in the building and the warehouse is used for storage, ie a variety of goods on pallets. The pallets were lifted onto the racking by forklift trucks.

On 12 April 1998 the developers received a Buildings Regulations Certificate from the Building Authority, ie the Northampton Borough Council. On the 12 June 1998 the appellants, having installed the racking sought further Buildings Regulations approval. A fire officer visited the building prior to the issue of a fire certificate.

The fire officer proposed that City Logistics first install a sprinkler system. The cost of this installation would have been in the order of half a million pounds. City Logistics balked at this proposal as there were only 30 staff working in the building, and there were 46 portable fire extinguishers in place. The fire officer indicated that he was unwilling to issue a certificate unless the sprinklers were installed.

Using the provisions of the acts, City Logistics appealed to the local magistrates court. The court supported the fire officer. At the subsequent appeal to the Crown Court, the fire officer acknowledged that if a fire broke out in this building, the worst-case scenario would still result in the 30 people having time to evacuate. The fire officer's legal representatives said that it was one thing to accept that people could escape, but that the *Fire Precautions Act 1971* was also about fighting fire. The Crown Court judge disagreed and held that the act is about escape only.

The County fire officer appealed to the High Court and won a decision in his favour. City Logistics Ltd. took the issue to the Court of Appeal, where the three Lord Justices concluded that the duty of the fire authority when inspecting such buildings is to be satisfied that the means of escape in a fire is reasonable. They stated that this 'means' is provided and actually works; and that the building has proper warning notices in case of fire.

The Lord Justices unanimously rejected the idea that *FPA 1971* required that the building must also be safe for those outside. The court rejected the notion that a building must not endanger the fire crews who attend a fire. It also rejected the suggestion that there was a statutory duty to install fire-fighting equipment to protect the stock or building. Furthermore, they concluded that *FPA 1971* did not oblige the warehouse owners to install equipment to protect neighbouring people and property. It was not right that City Logistics had to put in sprinklers to avoid polluting the atmosphere in the event of a fire, nor to minimise the economic affects of a fire on the property owner's insurers.

This means that this important Act, which has been in use by fire officers for several years had been wrongly applied. In effect they, the Appeal Court Judges, concluded that there is only one test for the issue of a fire certificate, ie can the occupants escape?

In their judgement, the Lord Justices indicated that it was not for the court to decide whether the law was right or wrong – that was the responsibility of Parliament.

Risk assessments

6.9   Much recent United Kingdom legislation has included a requirement for some type of risk assessment. The concept was introduced in the early 1980s in Regulations applying to asbestos and lead but it came to wider prominence as a core requirement of the *Control of Substances Hazardous to Health Regulations 1988* ('COSHH'). The 'six pack' of Regulations, which came into force on 1 January 1993, continued this trend with four sets of Regulations requiring an assessment of one kind or another.

The most important of these was the general requirement for risk assessment contained in *Management of Health and Safety at Work Regulations 1992* ('Management Regulations'), *Regulation 3*. Other important Regulations requiring more specific types of risk assessment relate to noise, manual handling operations, personal protective equipment, display screen equipment and fire precautions. The subject of fire risk assessments is dealt with more fully in **CHAPTER 7: HOW TO CARRY OUT A FIRE RISK ASSESSMENT**.

In practice a less formal type of risk assessment was already required by health and safety legislation, an assessment of the level of risk, the adequacy of existing precautions and the costs of additional precautions were necessary in order to determine what was 'reasonably practicable'.

Risk assessment takes the concept of self-regulation somewhat further. Employers must be able to demonstrate that they have identified relevant risks together with appropriate precautions and in most cases must have records available to prove this.

Supply chain Regulation

6.10 Following the introduction of the *Health and Safety at Work etc Act 1974*, many court decisions have emphasised that employers have responsibilities for the activities of employees of other organisations such as independent contractors. These responsibilities arise from *HSWA 1974, s 3(1)*. The employers' responsibility in relation to independent contractors was clearly defined by the definitive interpretation of 'conduct his undertaking' provided by the House of Lords in *R v Associated Octel Co Ltd [1996] 4 All ER 846*. The Lords' interpretation states that where an employer engages an independent contractor to do work which forms part of the conduct of the employer's undertaking, (and such is a question of fact in each case), the employer must stipulate those conditions which are needed to avoid risks to the health or safety of persons not in his employment and are reasonably practicable.

The structure of *HSWA 1974* and much subsidiary legislation is such that responsibilities overlap between employers rather than being neatly apportioned between them. However, employers now have a clearly defined responsibility for health and safety failings of the contractors that they employ. They are required to take a pro-active interest in the health and safety standards of those contractors.

This responsibility was illustrated by the Ramsgate prosecution (at **6.8** above). Despite having engaged specialists to carry out the design, construction and checking of the passenger walkway, Port Ramsgate was still convicted of an offence. Mr Justice Clark stated that:

> 'the jury has found, in my judgement correctly, that an owner and operator of a port cannot simply sit back and do nothing and rely on others, however expert'.

In recent years a growing number of larger companies, local authorities and other public bodies have had increasingly formalised procedures for checking the health and safety standards of contractors wishing to work for them. This process has been accelerated by the demands of the *Construction (Design and Management) Regulations 1994* ('CDM') that require clients to satisfy themselves (via their planning supervisors) that potential principal contractors are capable of dealing with the health and safety issues associated with projects. The Regulations also place responsibilities on principal contractors in respect of subcontractors.

6.11 *Legislation and Regulations (Health and Safety)*

Consequently contractors are frequently required to provide details of their health and safety policies and generic risk assessments together with risk assessments and/or method statements for specific projects or activities. Many clients also take an extremely hands-on approach in policing the work of contractors on their premises. This regulation of health and safety through the supply chain seems likely to increase.

Recent changes in legislation

6.11 Summarised below in alphabetical order are the more significant changes in health and safety legislation in recent years.

Confined Spaces Regulations 1997 (SI 1997 No 1713)

6.12 These provide a definition of 'confined space' in which various defined foreseeable 'specified risks' may arise. Entry into confined spaces must be avoided unless this is not reasonably practicable. Where entry is made, it must be in accordance with a safe system of work (arrived at through a process of risk assessment) and with suitable emergency arrangements in place. Health and Safety Executive booklet L101 '*Safe Work in Confined Spaces*' contains an Approved Code of Practice and much practical guidance.

Construction (Health, Safety and Welfare) Regulations 1996 (SI 1996 No 1592)

6.13 These consolidate, modernise and simplify most of the previous Regulations relating to construction as well as introducing some new requirements resulting from a European Commission Directive.

Although detailed in some areas, the Regulations adopt an objective-setting approach in others. They are supported by a Health and Safety Executive booklet, Health and Safety (G) 150 '*Health and Safety in Construction*' and several other Health and Safety Executive publications.

Lifting operations in construction are now covered by the *Lifting Operations and Lifting Equipment Regulations 1998* (see **6.20** below).

Control of Lead at Work Regulations 1998 (SI 1998 No 543)

6.14 These set stricter levels of lead in blood at which employers must remove employees from work. The Health and Safety Executive has published a revised edition of its Approved Code of Practice – COP 2 '*Control of lead at work*'.

Control of Major Accident Hazards Regulations 1999 ('COMAH 1999') (SI 1999 No 743)

6.15   Control of Major Accident Hazards Regulations 1999 ('COMAH 1999') replaces the Control of Industrial Major Accident Hazard Regulations 1984 ('CIMAH'). The Regulations apply where specified quantities of dangerous substances are present, requiring the preparation of a Major Accident Prevention Policy ('MAPP'). Additional requirements are placed on 'top tier sites' where greater quantities of dangerous substances are kept. Enforcement will be co-ordinated between the Health and Safety Executive and the Environment Agency (in England and Wales) or the Scottish Environment Protection Agency, with the Health and Safety Executive for the first time having to charge companies for inspections. 'A Guide to the Control of Major Accident Hazards Regulations' was published by the Health and Safety Executive in 1999.

Control of Substances Hazardous to Health Regulations 1999 ('COSHH 1999') (SI 1999 No 437)

6.16   These Regulations require an assessment to be made of all substances hazardous to health in order to identify means of preventing or controlling exposure. There are also requirements for the proper use and maintenance of control measures and for workplace monitoring and health surveillance in certain circumstances.

SI 1999 No 437 Regulations replace SI 1994 No 3246 Regulations (and subsequent amendment Regulations) and make the following main changes:

- Schedule listing substances assigned maximum exposure limits ('MELs') has been removed; and

- Definition of 'substances hazardous to health' is expanded to include trigger limits for 'total inhalable dust' or 'respirable dust'.

What constitutes suitable personal protective equipment is redefined by reference to the Personal Protective Equipment Regulations 1992 (SI 1992 No 3139).

A revised General Approved Code of Practice on the Control of Substances Hazardous to Health Regulations 1999 is published in a single volume together with the existing Approved Code of Practices on the Control of Carcinogenic Substances and on the Control of Biological Agents.

Diving at Work Regulations 1997 (SI 1997 No 2776)

6.17   These replace and modernise previous requirements for commercial diving and also apply to the United Kingdom Continental Shelf. Some previously exempt groups (the media, researchers and recreational instructors) are now included. Practical information on compliance is provided in five Approved Code of Practices:

- L103: Commercial diving projects offshore;
- L104: Commercial diving projects inland/inshore;
- L105: Recreational diving projects;
- L106: Media diving projects; and
- L107: Scientific and archaeological diving projects.

Fire Precautions (Workplace) Regulations 1997 (SI 1997 No 1840)

6.18 *SI 1997 No 1840* is dealt with in considerable detail in CHAPTER 3: LEGISLATION AND REGULATIONS (THE FIRE PRECAUTIONS (WORKPLACE) REGULATIONS 1997 AND 1999).

Ionising Radiations Regulations 1999 ('IRR 1999') (SI 1999 No 3232)

6.19 *SI 1999 No 3232* replaces the *Ionising Radiations Regulations 1985 (SI 1985 No 1333)* ('IRR 1985') and makes the following main changes:

- introduces the Health and Safety Executive ('HSE') criteria of competence for individuals or organisations wishing to act as radiation protection advisors ('RPAs');
- contains a new requirement for employers to carry out a suitable and sufficient risk assessment before they first start work activities with ionising radiation;
- contains a new requirement for employers to be authorised before they use accelerators or X-ray sets for certain specified purposes;
- contain enhanced requirements for restricting exposure as far as reasonably practicable;
- set out revised dose limits; and
- allows more flexibility in the designation of controlled and supervised areas.

SI 1999 No 3232 contains a number of transitional provisions, the most notable being that employers who have notified the Health and Safety Executive under *SI 1985 No 1333* concerning working with ionising radiation, will not need to make a new notification under *SI 1999 No 3232*.

Lifting Operations and Lifting Equipment Regulations 1998 ('LOLER 1998') (SI 1998 No 2307)

6.20 These replace most of the previous sector-based law relating to lifting equipment (including hoists and lifts). Employers now have a choice between ensuring thorough examinations of lifting equipment either at fixed intervals or in accordance with an examination scheme. A key requirement in *SI 1998 No 2307, Reg 8* is for lifting operations to be 'properly planned by a competent person; appropriately supervised; and carried out in a safe manner'.

Health and Safety Executive booklet L113 'Safe Use of Lifting Equipment' includes an Approved Code of Practice and guidance.

Management of Health and Safety at Work Regulations 1999 (SI 1999 No 3242)

6.21 *SI 1999 No 3242* came into force on 29 December 1999. These Regulations revoke *Management of Health and Safety at Work Regulations 1992 (SI 1992 No 2051), Management of Health and Safety at Work (Amendment) Regulations 1994 (SI 1994 No 2865), Health and Safety (Young Persons) Regulations 1997 (SI 1997 No 135)* and *Fire Precautions (Workplace) Regulations 1997 (SI 1997 No 1840), Part III*. They re-enact *SI 1992 No 2051*, with the following modifications.

- A new *SI 1999 No 3242, Reg 4* requires an employer to implement preventive and protective measures on the basis of the principles specified in *Schedule 1*.

    These are:
    - avoiding risks;
    - evaluating the risks that cannot be avoided;
    - combating the risks at source;
    - adapting the work to the individual, especially as regards the design of workplaces, the choice of work equipment and the choice of working and production methods, with a view, in particular, to alleviating monotonous work and work at a predetermined work-rate and to reducing their effect on health;
    - adapting to technical progress;
    - replacing the dangerous by the non-dangerous or the less dangerous;
    - developing a coherent overall prevention policy that covers technology, organisation of work, working conditions, social relationships and the influence of factors relating to the working environment;

○ giving collective protective measures priority over individual protective measures; and

○ giving appropriate instructions to employees.

- *SI 1999 No 3242, Reg 7(1)* provides that every employer must appoint one or more competent persons to assist him in undertaking the measures he needs to take for the purposes of complying with the statutory health and safety requirements and prohibitions imposed upon him. A new *SI 1999 No 3242, Reg 7(8)* provides that, where there is a competent person in the employer's employment, that person must be appointed for the purposes of *SI 1999 No 3242, Reg 7(1)* in preference to a competent person not in his employment. A person is to be regarded as competent where he has sufficient training and experience or knowledge and other qualities to enable him properly to assist in undertaking the measures referred to in *SI 1999 No 3242, Reg 7(1)*.

- A new *SI 1999 No 3242, Reg 9* provides that employers must ensure that any necessary contacts with external services are arranged, particularly as regards first aid, emergency medical care and rescue work.

- Under a new *SI 1999 No 3242, Reg 21*, an employer is not to be afforded a defence for contravention of the relevant statutory provisions as defined in *Health and Safety at Work etc Act 1974, section 53* by reason of any act or default caused by his employee or by a person appointed by him under *SI 1999 No 3242, Reg 7*.

- *SI 1999 No 3242, Reg 24* revokes *Health and Safety (First-Aid) Regulations 1981, Regulation 6*, which confers power on the HSE to grant exemptions from those *Regulations*.

- *SI 1999 No 3242, Reg 25* amends *Offshore Installations and Pipeline Works (First-Aid) Regulations 1989* to limit the scope of the exemptions that may be granted by the HSE to those specified in *SI 1999 No 3242, Reg 5(1)(b)*, *SI 1999 No 3242, Reg 5(1)(c)* and *SI 1999 No 3242, Reg 5(2)(a)* of those Regulations.

- *SI 1999 No 3242, Reg 27* amends *Construction (Health, Safety and Welfare) Regulations 1996 (SI 2996 No 1592), Regulation 20(2)* so that arrangements made for dealing with foreseeable emergencies on construction sites include the designation of persons to implement the arrangements and the inclusion of necessary contacts with external emergency services, particularly in respect of rescue work and fire-fighting.

Offshore Electricity and Noise Regulations 1997 (SI 1997 No 1993)

6.22 *Electricity at Work Regulations 1989 (SI 1989 No 635)* and *Noise at Work Regulations 1989 (SI 1989 No 1790)* were amended by *SI 1997 No 1993* so that they applied offshore from 21 February 1998.

Police (Health and Safety) Regulations 1999 (SI 1999 No 860)

6.23 *SI 1999 No 860* ensures that all health and safety Regulations made under the *Health and Safety at Work etc Act 1974*, apply to police officers and complete the work started by the *Police (Health and Safety) Act 1997*.

Provision and Use of Work Equipment Regulations 1998 ('PUWER 1998') (SI 1998 No 2306)

6.24 These make some changes to the general requirements of *Provision and Use of Work Equipment Regulations 1998 (SI 1992 No 2932)* (which they replace) including requiring inspections of work equipment where relevant (*SI 1998 No 2306, Reg 6*).

Specific sections concerning mobile work equipment (*SI 1998 No 2306, Reg 25–30*) and power presses (*SI 1998 No 2306, Reg 31–35*) have been added.

The remaining provisions of several old Regulations have been revoked including those relating to unfenced machinery, power presses, abrasive wheels and woodworking machines. The Health and Safety Executive has published several booklets on the Regulations:

- L22: Safe use of work equipment (Approved Code of Practice and guidance);
- L112: Safe use of power presses (Approved Code of Practice and guidance); and
- L114: Safe use of woodworking machinery (Approved Code of Practice and guidance).

Railway Safety (Miscellaneous Provisions) Regulations 1997 (SI 1997 No 553) and Level Crossings Regulations 1997 (SI 1997 No 487)

6.25 These two sets of Regulations modernised and strengthened previous legislation, some of which was 150 years old. The *Health and Safety Executive* has published two related booklets:

- L97: A Guide to the Level Crossings Regulations; and
- L98: Railway safety miscellaneous provisions: Guidance on the Regulations.

Transport of Dangerous Goods (Safety Advisers) Regulations 1999 (SI 1999 No 257)

6.26 Employers who are involved in the transport of dangerous goods by road or rail must have appointed qualified safety advisers by 31 December 1999. The advisers must hold a relevant vocational training certificate, acquired by passing an examination based on a specified syllabus.

The *Health and Safety Executive* has produced a series of booklets *(Health and Safety (G) 160–164)* on various aspects of transporting different types of dangerous goods by road or rail.

Working Time Regulations 1998 (SI 1998 No 1883)

6.27 These implement the European Commission *Working Time Directive (93/104/EC)* and the working time provisions of the *Young Workers Directive (94/33/EC)*. They came into operation on 1 October 1998 and are enforced by the *Health and Safety Executive* in respect of weekly working time and night work, and by employment tribunals in relation to rest periods, breaks and paid annual leave. The Department of Trade and Industry has published a free booklet, *A Guide to Working Time Regulations*.

Amendments to existing Regulations

6.28 Various minor amendments have also been made to existing Regulations. These have involved:

- *Asbestos (Licensing) Regulations 1983 (SI 1983 No 1649)*

  Work with asbestos insulation board has been brought into the scope of the Regulations from February 1999. Contractors will now need a licence from the Health and Safety Executive before working with this material.

- *Asbestos (Prohibitions) (Amendment) Regulations 1999 (SI 1999 No 2373)*

  These Regulations amend the *Asbestos (Prohibitions) Regulations 1992* and ban the importation, supply and use in Great Britain of chrysotile (white asbestos) with effect from 24 November 1999.

- *Carriage of Dangerous Goods (Amendment) Regulations 1999 (SI 1999 No 303)*

  These implemented various amendments resulting from changes to EC Directives. The Health and Safety Executive leaflet IND (G) 234 summarises the present legislation governing carriage of dangerous goods.

- *Control of Asbestos at Work Regulations 1987 (SI 1987 No 2115)*

    The Regulations now apply to all types of work, which may lead to exposure. The exposure level has been lowered for white asbestos fibres, and asbestos-analysing laboratories must be accredited to EN 45001.

- *Chemicals (Hazard Information and Packaging for Supply) (Amendment) Regulations 1999 (CHIP 1999) (SI 1999 No 3165)*

    CHIP 99(2) came into force on 1 January 2000. This introduced a 5th edition of the *Approved Supply List*, (replacing the 4th edition and its supplement) and a 4th edition of the *Approved Classification and Labelling Guide*. Both are available from Health and Safety Executive Books.

Structure of United Kingdom health and safety law

6.29 Health and safety in the workplace involves two different branches of the law, criminal law and civil law.

Criminal law

6.30 Criminal law is the process by which society, through the courts, punishes organisations or individuals for breaches of its rules. These rules, known as 'statutory duties', are comprised in Acts passed by Parliament (eg *Health and Safety at Work etc Act 1974*) or Regulations which are made by Government Ministers using powers given to them by virtue of Acts (eg the *Manual Handling Operations Regulations 1992 (SI 1992 No 2793)*).

Cases involving breaches of criminal law may be brought before the courts by the enforcement authorities, which, in the case of health and safety law, are the Health and Safety Executive and local authorities via their environmental health departments. Magistrates' Courts hear the vast majority of health and safety prosecutions although more serious cases can be heard by the Crown Courts.

The maximum fine, which can be imposed by the Magistrates' Courts is £20,000 for breaches of certain sections of *Health and Safety at Work etc Act 1974* and, £5,000 for most other offences. (Current information as at January 2002 is listed at the end of this clause).

Where cases are heard by the Crown Court there is no limit on the fines that can be imposed.

There are also a limited number of health and safety offences that can result in prison sentences of up to two years. These include:

6.31 *Legislation and Regulations (Health and Safety)*

- contravention of licensing requirements (eg for asbestos removal);
- explosives-related offences;
- contravention of an improvement or prohibition notice; and
- contravention of a court remedy order.

As in all criminal prosecutions the case must be proved 'beyond reasonable doubt'. There is a right of appeal to the Court of Appeal (Criminal Division) and eventually to the House of Lords or even the European Union Courts, although in practice very few health and safety cases go to appeal. A death involving work activities might result in manslaughter charges, which could lead to more severe penalties, as discussed above at **6.7** above.

As at January 2002. These penalties can change from time to time.

Penalties for Health and Safety Offences

6.31   *HSWA 1974, s 33* (as amended) sets out the offences and maximum penalties under health and safety legislation.

Failing to comply with an improvement or prohibition notice, or a court remedy order (issued under *HSWA 1974, ss 21, 22 and 42* respectively):

- lower court maximum £20,000 and/or six months' imprisonment; or
- higher court maximum unlimited fine and/or two years' imprisonment.

Breach of *HSWA 1974, ss 2–6*, which set out the general duties of employers, self-employed persons, manufacturers and suppliers to safeguard the health and safety of workers and members of the public who may be affected by work activities:

- lower court maximum £20,000; or
- higher court maximum unlimited fine.

Other breaches of *HSWA 1974*, and breaches of 'relevant statutory provisions' under the Act, which include all health and safety regulations. These impose both general and more specific requirements, such as requirements to carry out a suitable and sufficient risk assessment or to provide suitable personal protective equipment:

- lower court maximum £5,000;
- higher court maximum unlimited fine; or
- contravening licence requirements or provisions relating to explosives.

Licensing requirements apply to nuclear installations, asbestos removal, and storage and manufacture of explosives. All entail serious hazards which must be rigorously controlled:

- lower court maximum £5,000; or
- higher court maximum Unlimited fine and/or two years' imprisonment.

The Health and Safety at Work etc Act 1974 ('HSWA 1974')

6.32   *HSWA 1974* is the most important Act of Parliament relating to health and safety. It applies to everyone 'at work' – employers, self-employed and employees (with the exception of domestic servants in private households). In addition, it also protects the general public who may be affected by work activities. The Health and Safety Executive has published a booklet, *'A Guide to the Health and Safety at Work etc Act 1974'*.

Some of the key sections of the Act are listed below.

Section 2 – Duties of employers

Section 2(1), HSWA 1974 is the catch-all provision:

'It shall be the duty of every employer to ensure, so far as is reasonably practicable, the health, safety and welfare at work of all his employees.'

See **6.33** below for further discussion of the term 'reasonably practicable'.

Section 2(2), HSWA 1974 goes on to detail more specific requirements relating to:

- the provision and maintenance of plant and systems of work;
- the use, handling, storage and transport of articles and substances;
- the provision of information, instruction, training and supervision;
- places of work and means of access and egress; and
- the working environment, facilities and welfare arrangements.

These are also qualified by the term *'reasonably practicable'*.

Section 2(3), HSWA 1974 provides that an employer with five or more employees must prepare a written health and safety policy statement, together with the organisation and arrangements for carrying it out, and bring this to the notice of employees.

Section 3 – Duties to others

Section 3(1), HSWA 1974 provides:

> 'It shall be the duty of every employer to conduct his undertaking in such a way as to ensure, so far as is reasonably practicable, that persons not in his employment who may be affected thereby are not thereby exposed to risks to their health or safety.'

Employers thus have duties to contractors (and their employees), visitors, customers, members of the emergency services, neighbours, passers-by and the public at large. This may extend to include trespassers, particularly if it is 'reasonably foreseeable' that they could be endangered, for example, where high-risk workplaces are left unfenced.

Individuals who are self-employed are placed under a similar duty and must also take care of themselves. (If they have employees, they must comply with *HSWA 1974, s 2*).

Section 4 – Duties relating to premises

Under this section persons in total or partial control of work premises (and plant or substances within them) must take 'reasonable' measures to ensure the health and safety of those who are not their employees but use the premises as a place of work or as a place where they may use plant or substances provided for their use there. These responsibilities might be held by landlords or managing agent's etc, even if they have no presence on the premises.

Section 6 – Duties of manufacturers, suppliers etc

Those who design, manufacture, import, supply, erect or install any article, plant, machinery, equipment or appliances for use at work, or who manufacture, import or supply any substance for use at work, have duties under this section.

Section 7 – Duties of employees

'It shall be the duty of every employee while at work:

(a) to take reasonable care for the health and safety of himself and of other persons who may be affected by his acts or omissions at work; and

(b) as regards any duty or requirement imposed on his employer or any other person by or under any of the relevant statutory provisions, to co-operate with him so far as is necessary to enable that duty or requirement to be complied with.'

Consequently employees must not do anything, which could endanger themselves or others. Equally they must not fail to do something, which could endanger themselves or others. It should be noted that managers and supervisors also hold these duties as employees.

Section 8 – Interference and misuse

'No person shall intentionally or recklessly interfere with or misuse anything provided in the interests of health, safety or welfare in pursuance of any of the relevant statutory provisions.'

Section 9 – Duty not to charge

'No employer shall levy or permit to be levied on any employee of his, any charge in respect of anything done or provided in pursuance of any specific requirement of the relevant statutory provisions.'

Levels of duty

6.33 Health and safety law contains different levels of duty.

Absolute

Absolute requirements must be complied with whatever the practicalities of the situation or the economic burden.

Practicable

The term 'practicable' means that measures must be possible in the light of current knowledge and invention.

Reasonably practicable

This term is contained in the main sections of the *Health and Safety at Work etc Act 1974* and many important Regulations. It requires the risk to be weighed against the costs necessary to avert it (including time and trouble as well as financial cost). If, compared with the costs involved, the risk is small then the precautions need not be taken. It should be noted that such a comparison should be made before any incident has occurred. The burden of proof, however, rests on the person with the duty (usually the employer), they must prove why something was not reasonably practicable at a particular point in time. The duty holder's ability to meet the cost is not a factor to be taken into account.

6.34 *Legislation and Regulations (Health and Safety)*

In effect, considering what is 'reasonably practicable' requires that a risk assessment be carried out. The existence of a well-documented and carefully considered risk assessment would go a long way towards determining what was or was not reasonably practicable. Neither risks nor costs remain the same forever and what is practicable or reasonably practicable will change with time, hence the need to keep risk assessments up to date.

'Six pack'

6.34 The term 'six pack' is often used to describe the six Regulations which all came into operation on 1 January 1993 to implement EC Directives. They are:

Management of Health and Safety at Work Regulations 1992 (SI 1992 No 2051)

These Regulations, now superseded by *Management of Health and Safety at Work Regulations 1999 (SI 1999 No 3242)*, required employers and the self-employed to manage the health and safety aspects of their activities in a systematic and responsible way. The Regulations include requirements for risk assessment, the availability of competent health and safety advice and emergency procedures. The management aspects of these Regulations are explored below at **6.44** below.

PUWER 1992 (SI 1992 No 2932) (See also **6.24** above).

These Regulations, which have now been superseded by *PUWER 1998, (SI 1998 No 2306)* covered equipment safety, including the guarding of machinery. The definition of 'work equipment' also includes hand tools, vehicles, laboratory apparatus, lifting equipment, access equipment etc.

Workplace (Health, Safety and Welfare) Regulations 1992 (SI 1992 No 3004)

Physical working conditions, safe access for pedestrians and vehicles, and welfare provisions are covered by these Regulations.

Personal Protective Equipment at Work Regulations 1992 (SI 1992 No 2966)

Employers must assess the personal protective equipment ('PPE') needs created by their work activities, provide the necessary personal protective equipment, and take reasonable steps to ensure its use.

Manual Handling Operations Regulations 1992 (SI 1992 No 2793)

Manual handling operations involving risk of injury must either be avoided or be assessed by the employer with steps taken to reduce the risk, so far as is reasonably practicable.

Health and Safety (Display Screen Equipment) Regulations 1992 (SI 1992 No 2792)

Where there is significant use of display screen equipment ('DSE'), employers must assess display screen equipment workstations and offer 'users' eye and eyesight tests (which may necessitate provision of spectacles for display screen equipment work).

Dangerous Substances and Explosive Atmospheres Regulations 2002 (DSEAR) (SI 2002 No 2776)

6.35 These regulations have been introduced progressively from 9 December 2002 and all of the regulations will be in force from 30 June 2003. Existing premises will have to be altered and fully conform to these regulations from 2006. All new premises or where premises are altered will, after 30 June 2003 need to conform to these regulations.

Enforcement of these regulations is the responsibility of the following agencies:

The Health and Safety Executive or local authorities depending on the allocation of premises under the *Health and Safety (Enforcing Authority) Regulations 1998 (SI 1998 No 494)*. In the main, the Health and Safety Executive will enforce the regulations at industrial premises and local authorities (environmental health officers) elsewhere eg in retail premises.

Fire brigades at most premises subject to *Dangerous Substances and Explosive Atmospheres Regulations 2002* (DSEAR) in relation to general fire precautions such as means of escape.

At retail petrol filling stations in relation to storage and dispensing of petrol, Liquefied Petroleum Gas (LPG) and any other fuel subject to DSEAR (SI 2002 No 2776), the Petroleum Licensing Authorities.

Other important Regulations

6.36 *Construction (Design and Management) Regulations 1994 (SI 1994 No 3140) as amended by SI 1996 No 1592*

The broad definition of 'construction work' used in the Regulations means that they apply to many medium-sized engineering and maintenance projects as well as to traditional construction activities and all demolition work. The Regulations provide for specific duties to be carried out by the 'client' (including the appointment of a 'planning supervisor') and by a 'principal contractor' upon whom specific duties are also placed. Key requirements of

6.36 *Legislation and Regulations (Health and Safety)*

the Regulations are for the development and implementation of a formal 'health and safety plan' and the creation of a 'health and safety file' for the project.

Control of Substances Hazardous to Health Regulations 1999 (COSHH) (SI 1999 No 437)

These require an assessment to be made of all substances hazardous to health in order to identify means of preventing or controlling exposure.

There are also requirements for the proper use and maintenance of control measures and for workplace monitoring and health surveillance in certain circumstances.

Electricity at Work Regulations 1989 (SI 1989 No 635) as amended by *SI 1996 No 192* and *SI 1997 No 1993*

These Regulations contain requirements relating to the construction and maintenance of all electrical systems and work activities on or near such systems. They apply to all electrical equipment, from a battery-operated torch to a high-voltage transmission line.

Health and Safety (Consultation with Employees) Regulations 1996 (SI 1996 No 1513)

These Regulations extended the previous requirements (contained in the *Safety Representatives and Safety Committees Regulations 1977 (SI 1977 No 500)*) so that employers must now also consult workers not covered by trade union safety representatives.

Health and Safety (First-Aid) Regulations 1981 (SI 1981 No 917) as amended

Basic first-aid equipment controlled by an 'appointed person' must be provided for all workplaces. Higher risk activities or larger numbers of employees may require additional equipment and fully trained first-aiders.

Health and Safety (Safety Signs and Signals) Regulations 1996 (SI 1996 No 341)

These Regulations require safety signs to be provided, where appropriate, for risks which cannot adequately be controlled by other means. Signs must be of the prescribed design and colours.

Health and Safety (Training for Employment) Regulations 1990 (SI 1990 No 1380)

Those receiving 'relevant training' (through training for employment schemes or work experience programmes) are treated as being 'at work' for the purposes of health and safety law. The provider of the 'relevant training' is deemed to be their employer. Youth trainees and students on work experience placements therefore have the status of employees and must be protected accordingly.

Noise at Work Regulations 1989 (SI 1989 No 1790) as amended

Employers must carry out an assessment to determine their employees' level of exposure to noise. The precautions required include noise reduction measures, provision of hearing protection and the establishment of hearing protection zones.

Reporting of Injuries, Diseases and Dangerous Occurrences Regulations 1995 ('RIDDOR') (SI 1995 No 3163)

Fatal accidents, major injuries (as defined) and dangerous occurrences (as defined) must be reported immediately to the enforcing authority 'by the quickest practicable means', and reported in writing within ten days.

Accidents involving four or more consecutive days' absence must be reported in writing as soon as practicable and in any event within ten days.

Safety Representatives and Safety Committees Regulations 1977 (SI 1977 No 500)

Recognised trade unions may appoint safety representatives to formally represent employees in consultations with their employer in respect of health and safety issues. The functions and rights of safety representatives are detailed in the Regulations. The employer must establish a safety committee if at least two representatives request this in writing.

Civil law

6.37 A civil action can be initiated by an employee who has suffered injury or damage to health caused by their work. This may be based upon the law of negligence, ie where the employer has been in breach of the duty of care, which they owe to the employee. Being part of the common law, the law of negligence has evolved, and continues to evolve, by virtue of decisions in the courts. Parliament has had virtually no role to play in its development.

Civil actions may also be brought on the grounds of breach of statutory duty. It should be noted however, that the *Health and Safety at Work etc Act 1974* does not confer a right of civil action, (although the statutory duties owed by employers to employees under the *Health and Safety at Work etc Act 1974* have their equivalent obligations at common law). A breach of a duty imposed by Health and Safety Regulations passed under the *Health and Safety at Work etc Act 1974* is, so far as it causes damage, actionable except in so far as the Regulations provide otherwise. However, a civil cause of action for breach of

6.38 *Legislation and Regulations (Health and Safety)*

Management of Health and Safety at Work Regulations 1999 (SI 1999 No 3242) is specifically excluded by *SI 1999 No 3242, Reg 22*, save for specified exceptions.

Duty of care

6.38 Every member of society is under a 'duty of care', ie they must take reasonable care to avoid acts or omissions that they can reasonably foresee are likely to injure their neighbour (anyone who ought reasonably to have been kept in mind). What is 'reasonable' will depend upon the circumstances.

Employers owe a duty of care not only to employees but also to such people as contractors, visitors, customers, and people on neighbouring property. In the case of the duty of care owed by employers to employees, it includes the duty to provide:

- safe premises;
- a safe system of work;
- safe plant, equipment and tools; and
- safe fellow workers.

Vicarious liability

6.39 Employers are liable to persons injured by the wrongful acts of their employees, if such acts are committed in the course of their employment. Thus if an employee's careless driving of a forklift truck injures another employee (or a contractor or customer), the employer is likely to be liable. There is no vicarious liability if the act is not committed in the course of employment, thus the employer is not likely to be held liable if one employee assaults another.

Civil procedure

6.40 Civil actions must commence within three years from the time of knowledge of the cause of action. In an action for negligence, this will be the date on which the claimant knew or should have known that there was a significant injury and that it was caused by the employer's negligence. The claimant must be prepared to prove his case in the courts, but in practice most cases are settled out of court following negotiations between the claimant's legal representatives and the employer's insurers or their representatives. The *Employers' Liability (Compulsory Insurance) Act 1969* requires employers to be insured against such actions, although some public bodies, for example local authorities, are exempt from the provisions of this Act.

As the result of recommendations made by Lord Woolf in his '*Access to Justice*' report of 1996, the *Civil Procedure Rules 1998* introduced widespread changes to civil procedure on 26 April 1999, affecting the progress of civil claims from their commencement to their conclusion.

The rules involve a 'pre-action protocol' and govern the conduct of litigation in a way that is intended to limit delay. In most cases a single expert, medical or non-medical, will be instructed rather than each party using separate experts. Even if the case goes to court, the expert's report will usually be in writing, with both parties able to ask written questions of the expert and to see the replies. The new arrangements include a fast track system for personal injury claims up to a value of £15,000.

Damages

6.41 Damages are assessed under a number of headings including:

- loss of earnings (prior to trial);
- damage to clothing, property etc;
- pain and suffering (before and after trial);
- future loss of earnings;
- disfigurement;
- medical or nursing expenses; and
- inability to pursue personal or social interests or activities.

Defences

6.42 The claimant must prove breach of a statutory duty or of the duty of care on a balance of probabilities. However, a number of defences are available to the employer, including:

> **Contributory negligence**
>
> The employer may claim that the injured person was careless or reckless and that his own negligence contributed to the damage. For example, that he ignored clear safety rules or disobeyed instructions. Accidental errors are distinguished from a failure to take reasonable care. Damages will be reduced by the percentage of contributory negligence established, which will vary with the facts of each case.
>
> **Injuries not reasonably foreseeable**
>
> The employer may claim that the claimant suffered harm of a type or kind that was not reasonably foreseeable. In cases of noise-induced

hearing damage, mesothelioma (an asbestos-related cancer) or vibration-induced white finger, the courts have established dates after which a reasonable employer should have been aware of the relevant risks and taken precautions.

Voluntary assumption of risk

If an employee consents to take risks as part of the job, the employer may escape liability. However, this defence (*volenti non fit injuria*) is rarely successful and, in any event, cannot be used for cases involving breach of statutory duty. No one can contract out of their statutory obligations or be deprived of statutory protection.

Other civil actions

6.43   Other health and safety-related situations may result in civil actions by employees. Employment protection legislation has recently been strengthened in relation to dismissals or redundancies resulting from health and safety activities (including refusal to work in situations of serious and imminent danger). Suspension or dismissal on maternity or medical grounds may also give a right of action.

Management of health and safety at work

6.44   The Robens doctrine of self-regulation has been developed considerably in recent years by the Health and Safety Executive as it has paid increasing attention to the way health and safety is managed. Its Accident Prevention Advisory Unit ('APAU') worked with a number of large organisations from the late 1970s onwards, steadily building up its expertise. This culminated in the publication by the Health and Safety Executive 1991 of '*Successful Health and Safety Management*'. A revised edition of the booklet (reference HS(G) 65) was published in 1997.

'*Successful Health and Safety Management*' drew together much of what was already known about management techniques and the principles on which they are based. The introduction to the booklet included the paragraph:

> 'Many of the features of effective health and safety management are indistinguishable from the sound management practices advocated by proponents of quality and business excellence. Indeed, commercially successful companies often also excel at health and safety management, precisely because they bring efficient business expertise to bear on health and safety as on all other aspects of their operations. The general principles of good management are therefore a sound basis for deciding how to bring about improved health and safety performance.'

However, the then Chief Inspector of Factories, Tony Linehan, stated in his Foreword:

'The path described is neither easy nor short. There are no short cuts to successful health and safety management. It cannot be sidelined. It must not be delegated out of sight. The clearest lesson from practical experience is that the starting point is the genuine and thoughtful commitment of top management. I believe firmly that such commitment is beneficial and worthwhile.'

Since the publication of '*Successful Health and Safety Management*', several of its principles have found their way into statutory obligations, particularly through the requirements for risk assessments included in a number of Regulations, and in the *Management of Health and Safety at Work Regulations 1992, Regulation 4*, which requires the application of the 'management cycle' to health and safety precautions.

The costs of accidents

6.45 Their own humanitarian attitudes together with social pressures and the possibility of legal sanctions have for many years acted as powerful motivational factors for employers to avoid accidents at work. '*Successful Health and Safety Management*' drew attention to the high costs to employers of failing to manage health and safety effectively. This theme was developed further by the publication by the *Health and Safety Executive* in 1993 of '*The Costs of Accidents at Work*'. A revised edition of the booklet (reference HS (G) 96) was published in 1997.

The Accident Prevention Advisory Unit (APAU) carried out five detailed case studies in different industrial sectors in order to determine the full cost of accidents. Employers often believe that most accident costs are covered by insurance but the studies demonstrated the opposite. The ratio between insured and uninsured costs varied from 1:8 to 1:36.

Whilst insurance is likely to cover employer's liability and public liability claims and major damage costs, together with major losses due to business interruption, there are a host of uninsured costs associated with accidents, many of which go unrecorded.

Employee sick pay can usually be quantified but the costs of minor accident repairs are often hidden within much larger maintenance figures, and loss or damage of product due to accidents is seldom separated out from other wastage statistics. Indirect losses due to the unavailability of staff or equipment while both are being 'repaired' are difficult to quantify, as are other indirect costs such as the administrative time involved and the possible damage to the employer's image (to staff, customers or the wider public, including investors).

Statistical analysis has shown that for all businesses the number of serious accidents is small in relation to the number of minor accidents and damage incidents, although these latter types may not be fully recorded or investigated. Whilst minor accidents and damage incidents seldom result in insurance claims, their cumulative effects result in significant uninsured costs to the employer. Accidents were found to be costing one of the companies in the Accident Prevention Advisory Unit (APAU) study 37 per cent of its annual profits.

Health and safety policies

6.46

> '*Successful Health and Safety Management*' stated that 'accidents are caused by the absence of adequate management control' and stressed the importance of effective health and safety policies in establishing such control. The *Health and Safety at Work etc Act 1974*, section 2(3) requires employers to prepare in writing:
>
> - a statement of their general policy with respect to the health and safety at work of their employees; and
> - the organisation and arrangements for carrying out the policy.
>
> It also requires the statement and any revision to be brought to the notice of all employees. Employers with fewer than five employees are exempt from this requirement.
>
> Policies are normally divided into three sections, to meet the three separate demands of *HSWA 1974*.
>
> **(1) The statement of intent**
>
> This involves a general statement of good intent, usually linked to a commitment to comply with relevant legislation. Many employers extend their policies so as to relate also to the health and safety of others affected by their activities. In order to demonstrate clearly that there is commitment at a high level, the statement should preferably be signed by the chairman, chief executive or someone in a similar position of seniority.
>
> **(2) Organisational responsibilities**
>
> It is vitally important that the responsibilities for putting the good intentions into practice are clearly identified. In a small organisation this may be relatively simple but larger employers should identify the responsibilities held by those at different levels in the management structure. Whenever possible, key individuals or their appointments

should be named and their responsibilities defined. Whilst reference to employees' responsibilities may be included, it should be emphasised that the law requires the employer's organisation to be detailed in writing.

Types of responsibilities to be covered in the policy might include:

- making adequate resources available to implement the policy;
- setting health and safety objectives;
- developing suitable procedures and safe systems;
- delegating specific responsibilities to others;
- monitoring the effectiveness of others in carrying out their responsibilities;
- monitoring standards within the workplace; and
- feeding concerns up through the organisation.

(3) Arrangements

The policy need not contain all of the organisation's arrangements relating to health and safety but should contain information as to where they might be found, for example in a separate health and safety manual or within various procedural documents. Topics that may require detailed arrangements to be specified are:

- operational procedures relating to health and safety;
- training;
- personal protective equipment;
- health and safety inspection programmes;
- accident and incident investigation arrangements;
- fire and other emergency procedures;
- first aid;
- occupational health;
- control of contractors and visitors;
- consultation with employees; and
- audits of health and safety arrangements.

Employees must be aware of the policy and, in particular, must understand the arrangements, which affect them and what their own responsibilities might be. They may be given their own copy (for example, within an employee handbook) or the policy might be

> displayed around the workplace. With regard to some arrangements detailed briefings may be necessary, for example as part of induction training.
>
> Employers must revise their policies as often 'as may be appropriate'.
>
> Larger employers are likely to need to arrange for formal review and, where necessary, for revision to take place on a regular basis (eg by way of an ISO 9000 procedure). Dating of the policy document is an important part of this process.

Application of the 'management cycle'

6.47 The requirement for employers to carry out a risk assessment has been included in many recent health and safety Regulations. Even where there is no explicit requirement, it is often implicit within the wording of Regulations that compliance will involve a risk assessment process. This is frequently stated in the accompanying Approved Code of Practice or guidance – the *'Provision and Use of Work Equipment Regulations 1998 (SI 1998 No 2306)'*, the *'Lifting Operations and Lifting Equipment Regulations 1998 (SI 1998 No 2307)'* and the *'Confined Spaces Regulations 1997 (SI 1997 No 1713)'* all provide examples of this.

Whilst risk assessment involves the identification of risks and an assessment of their significance, its purpose is to identify the measures necessary to eliminate the risks or to control them to a satisfactory degree. There is, however, little point in going through the process of risk assessment if these precautions are not actually implemented in the workplace. The 'theory' of the risk assessment must become reality in a practical setting.

The importance of this is recognised in the *Management of Health and Safety at Work Regulations 1999* ('the Management Regulations') *(SI 1999 No 3242)* where *SI 1999 No 3242, Reg 3* requiring 'Risk assessment' is followed by *SI 1999 No 3242, Reg 5* entitled 'Health and safety arrangements'. *SI 1999 No 3242, Reg 5, para (1)* states that:

> 'Every employer shall make and give effect to such arrangements as are appropriate, having regard to the nature of his activities and the size of his undertaking for the effective planning, organisation, control, monitoring and review of the preventive and protective measures.'

SI 1999 No 3242, Reg 5, para (2) requires employers with five or more employees to record these arrangements.

Such a 'management cycle' has long been applied to other areas of business activity, such as finance, but relatively few employers have utilised it in relation

Legislation and Regulations (Health and Safety) **6.47**

to health and safety. Managers have often stated their good intentions but have not always set up the organisational structure and control to implement those intentions and have failed to monitor what is actually happening in the workplace.

The cycle can be applied to an employer's overall approach to health and safety:

- *plan* – through the statement of intent within the health and safety policy;
- *organise* – by allocating responsibilities for implementing the policy and making the necessary resources available;
- *control* – through application of relevant management systems and techniques and the use of performance standards;
- *monitor* – through health and safety audits and inspections; and
- *review* – in health and safety committee and management meetings.

At a different level the cycle can be applied to management systems or procedures, such as those for carrying out health and safety inspections:

- *plan* – through a statement of intent to conduct regular inspections;
- *organise* – by having a formal procedure relating to inspections – where, who by, how often;
- *control* – by specifying arrangements for reporting on inspections and implementing remedial actions, and by providing training in inspection techniques;
- *monitor* – through checking that inspections are being carried out as scheduled, reviewing the quality of inspection reports and making sure that remedial action is being implemented; and
- *review* – by investigating the reasons for any shortcomings in the system (and looping back to the start of the cycle to plan how to correct these).

British Standard 8800 '*Guide to Occupational Health and Safety Management Systems*' and many other commercially available management systems are based upon similar application of the management cycle. However, whether or not employers use such formal systems, the principles of the cycle should always be applied to all aspects of health and safety management. *SI 1999 No 3242* has made this a statutory obligation.

Within '*Successful Health and Safety Management*' the HSE emphasised the importance of establishing a positive health and safety culture within an organisation as a prerequisite of effective health and safety management. It referred to the 'four Cs' as key components in establishing such a culture:

- control;

6.48 *Legislation and Regulations (Health and Safety)*

- communication;
- competence; and
- co-operation.

While competence in health and safety matters is relevant throughout any workforce, it is particularly important at management levels. *SI 1999 No 3242* has taken this concept further by requiring (in *SI 1999 No 3242, Reg 7(1)*) that every employer must appoint one or more competent persons to assist them in undertaking the measures they need to take for the purposes of complying with the statutory health and safety requirements and prohibitions imposed upon them. *SI 1999 No 3242, Reg 7(8)* provides that, where there is a competent person in the employer's employment, that person must be appointed for the purposes of *SI 1999 No 3242, Reg 7(1)* in preference to a competent person not in their employment. A person is to be regarded as competent where they have sufficient training, experience or knowledge and other qualities to enable them to properly assist in undertaking the measures referred to in *SI 1999 No 3242, Reg 7(1)*. A revised Approved Code of Practice has now been published to accompany the new Regulations.

Due diligence

6.48 The term 'due diligence' in the legal context, is a term used to describe a defence where a person charged with an offence can show that they took all reasonable care, and exercised all due diligence to avoid commission of that offence by themselves or by a person under their control.

There are many circumstances where a defendant can demonstrate all good intentions in the course of their business, although a contravention may have emerged. In mitigation, therefore, the defendant can claim they exercised all due diligence, given that they did not commit any contravention knowingly and other than in good faith.

The areas of due diligence, though complex, are intended to help a defendant, however they must prove to the court that, on the balance of probabilities, they have taken all reasonable precautions and exercised all due diligence. Written records of their system and controls, specifications, training records, instructions, and corrective action are all important evidence.

Public liability

6.49 Under the *Health and Safety at Work etc Act 1974*, both employers and the self-employed have duties not only to their own workpeople but also to outside contractors, workers employed by them and to members of the public, whether within or outside the workplace, who may be affected by work activities. Business undertakings must be conducted in such a way as to ensure, so far as is reasonably practicable, that they do not expose people who are not their employees to risks to their health and safety.

The duty extends to, for example, risks to the public outside the workplace from fire or explosion, from falls of unsafe scaffolding or from the release of harmful substances into the atmosphere.

The general duties of protection owed to visitors apply equally to the emergency services who should be given a plan or map of the premises together with information on specific high risk areas where high voltage or dangerous chemicals may be present.

The young and the disabled

6.50 In general, the standard of protection required for visitors and others within business premises will be similar to that given to employees. There may, however, be a need to apply different criteria to achieve these standards when assessing the risks to members of the public. For example, it will be necessary to consider that certain people, such as the very young or disabled, may be more vulnerable than others and that people visiting or passing a workplace may have less knowledge of the potential hazards and of how to avoid them.

Unauthorised entry

6.51 The responsibilities of employers and the self-employed to non-employees will in certain circumstances extend to people entering workplaces without permission. The duty under the *Health and Safety at Work etc Act 1974* to conduct the business in such a way as not to expose people to risks to health and safety implies taking certain precautions to deter people from unlawfully entering the workplace, for example by the provision of fences, barriers and notices warning of the danger.

Duties of all people

6.52 It should be noted that the *Health and Safety at Work etc Act 1974*, imposes a duty on all people, both people at work and members of the public, including children. The duty is that they must not intentionally or recklessly interfere with or misuse anything that has been provided in the interests of health, safety or welfare, whether it has been provided for the protection of employees or other people. The purpose of the provision is clearly to protect things intended to ensure people's safety, including fire escapes and fire extinguishers, perimeter fencing, warning notices for particular hazards, protective clothing, guards on machinery and special containers for dangerous substances.

6.53   *Legislation and Regulations (Health and Safety)*

Sources of health and safety advice

6.53   Health and safety training is available from many different sources. The following organisations either provide training themselves or oversee training through accredited training centres.

- National Examination Board in Occupational Safety and Health ('NEBOSH')
  Tel: 0116 263 4700
- Institution of Occupational Safety and Health ('IOSH')
  Tel: 0116 257 3100
- Chartered Institution of Environmental Health ('CIEH')
  Tel: 0207 928 6006
- Royal Society for the Prevention of Accidents ('RoSPA')
  Tel: 0121 248 2000
- British Safety Council ('BSC')
  Tel: 0208 741 1231

There are many independent consultants who can provide advice and assistance on health and safety matters. Consultants are listed in the Yellow Pages and the Institution of Occupational Safety and Health (see above) maintains a consultants' register. Employers planning to use consultants' services can obtain a copy of a free *Health and Safety Executive* leaflet, '*Selecting a Health and Safety Consultancy*' (IND (G) 33).

The *Health and Safety Executive* itself can also be a valuable source of information and advice:

- *Health and Safety Executive* Books
  Tel: 01787 881165; fax: 01787 313995

The *Health and Safety Executive* has a huge range of priced publications and free leaflets, some of which are referred to elsewhere in this Introduction. '*The Essentials of Health and Safety at Work*' is a useful starting point for the small employer. There is further information on this aspect in **CHAPTER 13: SPECIALIST INFORMATION**.

- Health and Safety Executive Infoline
  Tel: 08701 545500

Open Monday to Friday 8.30am to 5pm to provide information on workplace health and safety.

- Health and Safety Executive Home Page on the Internet:
  http://www.hse.gov.uk

An online enquiry service can be accessed from the home page. HSE Books can be ordered online or by telephone (tel: 01787 881165).

- 'Escaping the Maze'

This video guide, designed to help companies through the maze of health and safety information and its different sources, is available from *Health and Safety Executive* Video (tel: 0845 741 9411).

7 How to carry out a fire risk assessment

In this chapter:	
Introduction	7.1
'Five steps' approach	7.4
Home Office guidance	7.10
Fire risk categories	7.11
General principle for escape routes	7.16
Means of fighting fire	7.23
Maintenance and testing of fire safety equipment	7.26

Introduction

7.1 An employer or someone other than the employer, who has 'control' of a workplace, is required by law to carry out a fire risk assessment of their workplace. This includes people, who by virtue of any contract or tenancy may have an obligation in relation to the maintenance, repair or safety of a workplace, as they are treated as having 'control' of that workplace.

The risk assessment is required to be carried out as a consequence of the legal requirements contained within both *Management of Health and Safety at Work Regulations 1999 (SI 1999 No 3242)* (see CHAPTER 3: LEGISLATION AND REGULATIONS (FIRE PRECAUTIONS (WORKPLACE) REGULATIONS 1997 AND 1999) above) and *Fire Precautions (Workplace) Regulations 1997 (SI 1997 No 1840)*, (as amended) (see also CHAPTER 3: LEGISLATION AND REGULATIONS (FIRE PRECAUTIONS (WORKPLACE) REGULATIONS 1997 AND 1999) above). These Regulations require employers to demonstrate that they have adopted a systematic and controlled approach in dealing with health and safety and risk assessment.

CHAPTER 3: LEGISLATION AND REGULATIONS (FIRE PRECAUTIONS (WORKPLACE) REGULATIONS 1997 AND 1999) *(SI 1997 No 1840 and SI 1999 No 1877)* should be read in conjunction with this chapter, as it provides detailed guidance on your legal obligations under both sets of Regulations.

However, the specific references to the fire precautions requirements contained within *SI 1997 No 1840, Pt II* are discussed later in this chapter (see **7.20**, **7.25** and **7.27** below).

Quite clearly there will be major differences in the manner and to the intensity and detail that a fire risk assessment will need to be conducted. This will vary considerably based upon the size of the workplace, the use to which the workplace is put, the complexity of the layout, the number of storeys, the age and construction of the building, the number of employees and use of flammable materials etc.

This chapter is designed to assist the employer or person designated to carry out a fire risk assessment of their workplace in order to establish the adequacy or otherwise of existing arrangements for the safety of employees and for the safety of all other people whom a fire in the workplace may affect.

Risk assessments – what are they?

7.2 The 'risk assessment' concept is used widely in general health and safety legislation for the formulation of safety procedures within the workplace. With the introduction of *Fire Precautions (Workplace) Regulations 1997 (SI 1997 No 1840)*, this concept has now been extended to specifically include fire safety in the workplace, in order to assess whether the existing fire precautions are adequate and in proportion to the perceived risk.

A fire risk assessment is in effect a formal audit inspection of your workplace and work activities aimed at establishing the degree of 'risk' of a fire occurring and the 'hazards', which would cause a fire. Following this assessment, you should be able to determine whether or not existing fire precautions are adequate. An employer will then need to decide whether they must change or improve existing arrangements.

At the outset, it is helpful to define the terms 'risk' and 'hazard'. Frequently the persons who may carry out fire risk assessments poorly understand the difference between the terms.

The British Standard BS 8800:1996 – *Guide to occupational health and safety management systems* provides guidance on general health and safety risk assessment, but the definitions of the terms 'risk' and 'hazard', can be applied equally to fire risk assessments. There is a clear distinction between the meanings of the two words.

British Standard BS 8800:1996 provides the following definitions.

- 'Risk assessment'

    The overall process of estimating the magnitude of risk and deciding whether or not the risk is tolerable or acceptable.

7.3 *How to carry out a fire risk assessment*

- 'Risk'

  The combination of the likelihood and consequence of a specified hazardous event occurring.

- 'Hazard'

  A source or a situation with a potential for harm in terms of human injury or ill health, damage to property, damage to the environment, or a combination of these.

- 'Hazard identification'

  The process of recognising that a hazard exists and defining its characteristics.

To summarise these terms it should be remembered that a 'hazard' is something that has the potential to cause damage or harm. A 'risk' is, in effect, the possibility 'high or low' of damage or harm occurring.

By having a clear understanding of this terminology, the task of carrying out a risk assessment, whether it is for general health and safety or for fire, is made easier.

Fire risk assessment

7.3   It is important to carry out a fire risk assessment appropriate to the workplace. It would, for example, be entirely inappropriate for an employer who may be the owner of a small modern lock-up ground floor flower shop, employing two people, to carry out the same level of risk assessment as an employer who is the owner of a multi-storey, converted Victorian mill building, employing two hundred people that is manufacturing plastic materials. Quite obviously, the potential risk of fires occurring and endangering staff, are completely different in the two premises.

A sensible policy and good practice is to involve staff in the process, as they may have identified a potential fire risk of which the employer is not aware. This also has the added benefit of raising the level of fire safety awareness amongst staff. It is in the interest of both an employer and employees to ensure that the premises are safe.

The two most important questions are:

- How likely is it for a fire to start in the workplace; and

- How easy is it for employees, and other people who may be affected, to escape to a place of safety in the event of a fire starting in the workplace?

If an employer is the owner of the flower shop, the answers to these two questions will be completely different to those of the employer who is the owner of the plastic manufacturing factory.

In larger workplaces, it is good policy to carry out a separate inspection for each significantly different section, area or department. For example the office area of the plastic manufacturing factory carries significantly less risk than the manufacturing parts of the workplace and could be dealt with as a separate entity. However, when conducting an assessment of an office section of a workplace, particularly if part of the same building is used as the factory, it would be necessary to take into account the fact that an incident occurring in the factory could potentially present a significant fire risk to the occupants of the office.

The whole of the workplace should be taken into account by the risk assessment, to include any outdoor areas and any rooms or parts of buildings, which are not currently in use.

When a workplace has been subject to previous approvals by the various enforcing authorities for other fire safety requirements, eg licensing or building legislation, an employer is still required to carry out an assessment of the fire precautions under *SI 1997 No 1840*. However, if there has been no significant change in the workplace, for example, in the number of employees or the activities undertaken by the business, it is unlikely that any significant additional fire precautions will have to be provided.

If an employer proposes to make changes to the fire precautions as a result of carrying out a fire risk assessment, the employer must ensure that these do not conflict with the controls imposed by other legislation. Where there is any doubt, an employer is strongly advised to consult a Fire safety officer from the local fire service.

Where a 'workplace' premises is shared by other employers, the employer has a responsibility to ensure that the other employers are made aware of any significant risks that may have been identified as part of an assessment and any remedial action have taken to reduce that risk.

Under *Management of Health and Safety at Work Regulations 1999 (SI 1999 No 3242)*, employers are also required to co-operate with other employers who share the workplace to enable both groups to discharge their responsibilities under the *Fire Regulations*. In addition, an employer should take all reasonable steps to co-ordinate fire safety measures with those of any other employers who may share your workplace.

There are obvious financial and common sense advantages in co-operating and mutually sharing any fire safety measures which may have been provided within the individual occupancies of a shared building.

'Five steps' approach

7.4 The five steps approach to risk assessment is the Health and Safety Executive's established assessment method for dealing with general health and safety legislation. As a consequence, the Home Office guidance contained within the publication *Fire safety – An employer's guide* also advocates this approach to carrying out fire risk assessments. This method is recommended, as by adopting this approach you will ensure that the task is carried out in a practical and systematic manner.

Step 1: Identify fire hazards

7.5 It is necessary to identify potential fire hazards in the workplace, which include potential sources of ignition, sources of fuel and any hazards associated with the processes carried out in the workplace.

It is important to remember that potential risks can very often be dealt with at minimum cost. For example an employer may well be storing larger quantities of flammable material in the premises than the employer actually needs. Simply removing or reducing the amount of material stored will significantly reduce the hazard.

For detailed advice and guidance on this subject, refer to **CHAPTER 8: PREVENTION OF FIRES**.

Step 2: Identify the location of people at significant risk in case of fire

7.6 In this step of the process it is necessary to determine whether employees, visitors or members of the public might be in danger in the event of a fire, or in escaping from a fire, and to note their location.

An assessment should try to identify anything that could result in a fire and if a fire did occur, could it hurt anyone? There may be parts of the premises where people are more at risk than others. Employers also need to take into account people, other than employees, who may be in the premises such as customers, members of the public, visitors, and contractors. Employers should also take into account the special needs of any disabled staff and visitors who may be in the premises.

Step 3: Evaluate the risks – are your fire precautions adequate?

7.7 This step requires an employer to evaluate the risks arising from the hazards in order to make decisions on whether existing fire precautions are adequate, or whether improvements are required to remove the hazard or to control the risk. As part of this evaluation it is necessary to look at any existing fire safety measures provided in terms of:

- control of ignition and fuel sources;
- fire detection and fire warning systems;
- means of escape;
- means of fighting fire;
- maintenance and testing of fire precautions; and
- fire safety training of employees.

The nature of the risk evaluation an employer carries out will depend very much on the nature of the workplace and your work activities. To illustrate this point, once again using the example of the flower shop, the employer or owner should be able to carry out a quick 'walk through' inspection of the shop and identify any potential hazards. It is unlikely, due to the nature of the business, that there will be any significant fire risk. Probably, in this particular example, the greatest risk in the shop will be the storage of paper and packaging materials.

However, to consider the use the other example, ie of the plastic manufacturing factory, the risk assessment will need to be far more detailed and searching, as there are obviously quite significant risks in the manufacture of plastic products and the standard of fire resistance in a relatively old building.

After having identified in Steps 1 and 2 both what the hazards are and who might be at potential risk from them, it is now necessary to analyse the potential risks, to establish whether the existing fire precautions are adequate or whether they need to be improved. In making an evaluation it is necessary to look in detail at the following aspects of the fire defence systems:

(a) Fire detection and fire warning systems

It is important to ensure that a fire can be detected reasonably quickly so that people can leave the workplace quickly. An employer will need to look at the particular nature of the workplace, to decide if the arrangements are adequate.

For example, as there is always someone present when a shop is open, anyone discovering a fire, in the flower shop example, could raise the alarm by shouting 'FIRE' and everyone in the premises would be immediately aware that there was a problem.

It would not, however, be possible to raise an alarm of fire by shouting and achieve the same result in the plastic manufacturing factory, which occupies many floors and probably contains noisy production machinery.

The factory will obviously need a different solution. This requirement would probably be met by the provision of an electrical fire alarm system which can be operated from anywhere in the building and would sound an alarm that can be heard throughout the building. It may also be necessary, given that a fire can develop very rapidly, to install a system of automatic fire detection throughout the workplace to ensure that an early warning of fire can be given. This type of system provides smoke or heat detectors, which will automatically detect a fire and sound an alarm. In this way an employer is not relying entirely on people to discover a fire.

For more information see **8.23** below.

(b) Means of escape in case of fire for people in the workplace

As an employer it is necessary to ensure that in the event of a fire both employees and other people who may be in the workplace can escape safely and quickly by their own unaided efforts. The term used to describe this is 'means of escape'.

In general, the normal routes in and out of a workplace will be sufficient for most means of escape purposes – particularly if an employer is confident that an early warning of a fire will be given allowing all employees, visitors and members of the public to vacate the premises quickly.

Routes within the premises that are used by occupants to reach a place of safety in the open air away from any danger should be short enough to enable everyone in the premises to escape quickly and as safely as possible. A general 'rule of thumb' in a workplace with more than one exit, should allow occupants to be able to reach a place of safety within two to three minutes.

However, if the workplace has only one entrance/exit or it contains a high risk of fire, the time taken to reach a place of safety should be reduced to as little as one minute.

This is a very approximate method, but it does serve to point an employer in the right direction and indicate whether the existing arrangements are inadequate. There are many other factors, discussed later in this chapter, that need to be taken into account as part of your assessment of the 'means of escape'.

Using the example of the flower shop, which may only have the one entrance or, more usually, may have an additional rear door to open air, it is obvious that the occupants could escape to a place of safety, very quickly, almost within seconds.

This would not be the case in the plastic manufacturing factory, in which the occupants would probably have to traverse large floor areas and negotiate staircases to reach safety in the open air. There may also be areas of the building, which have a high risk of a rapid spread of fire. This would require a very quick evacuation. To ensure that a speedy evacuation of this building can be achieved within the required time scales, sufficient escape routes and exits will need to be provided.

Step 4: Record findings and action taken

7.8 This step ensures that the findings of the fire risk assessment are recorded together with any action taken as a result of the assessment. It is also good practice to inform employees or their representatives about the fire risk assessment findings. In addition if there is a formal risk assessment report, an employer may wish to make it available to them, should they ask for it.

Where employers employ five or more people they are required under *Management of Health and Safety at Work Regulations 1999 (SI 1999 No 3242), Regulation 3(6)* to record the significant findings of the assessment and any group of their employees identified as being especially at risk. In addition under *SI 1999 No 3242, Reg 10*, an employer has a legal requirement to provide employees with 'comprehensive and relevant information'. For more information on this subject please refer to **CHAPTER 3: LEGISLATION AND REGULATIONS (THE FIRE PRECAUTIONS (WORKPLACE) REGULATIONS 1997 AND 1999)**.

Step 5: Keep assessment under review

7.9 As a fire risk assessment should be considered as an ongoing process, it is necessary for an employer to review, from time to time, the original assessment to ensure that it is still valid. After the initial inspection has been carried out, it is good practice to carry out an annual review of the workplace in order to ensure that no new risks have developed as a result of, for example, changes to the work processes, machinery, substances, or the number of people likely to be present in the workplace, an employer should also re-assess the workplace if they have carried out alterations or extensions, as they may have affected the fire precautions previously provided.

Home Office guidance

7.10 The following information on means of escape in case of fire, are extracts from the Home Office guidance publication *Fire Safety – An employer's guide*, which explains what is required to comply with *Fire Precautions (Workplace) Regulations 1997 (SI 1997 No 1840)*.

Means of escape in case of fire

'The principle on which means of escape provisions are based is that the *time available for escape* (an assessment of the length of time between the fire starting and it making the means of escape from

7.10 How to carry out a fire risk assessment

the workplace unsafe) is greater than the *time needed for escape* (the length of time it will take everyone to evacuate once a fire has been discovered and warning given).

Regardless of the location of a fire, once people are aware of it they should be able to proceed safely along a recognisable escape route, to a place of safety. In order to achieve this, it may be necessary to protect the route, ie by providing fire-resisting construction. A protected route may also be necessary in workplaces providing sleeping accommodation or care facilities.

The means of escape is likely to be satisfactory if the workplace is fairly modern and has had building regulation approval or if it has been found satisfactory following a recent inspection by the fire authority (and in each case an employer has not carried out any significant material or structural alterations or made any change to the use of the workplace). However, an employer should still carry out a risk assessment to ensure that the means of escape remains adequate.

If, as a result of the fire risk assessment, an employer proposes making any changes to the means of escape, he should consult the fire authority (in Scotland an employer, occupier or owner must seek the agreement of the building control authority) before making any changes.

When assessing the adequacy of the means of escape an employer will need to take into account:

- the findings of the fire risk assessment;
- the size of the workplace, its construction, layout, contents and the number and width of the available escape routes;
- the workplace activity, where people may be situated in the workplace and what they may be doing when a fire occurs;
- the number of people who may be present, and their familiarity with the workplace; and
- their ability to escape without assistance.

Assessing means of escape

The aim of the following paragraphs is to provide enough information for an employer to make a reasonable assessment of the escape routes from the workplace and decide whether they are adequate and can be safely used in the event of fire.

Because of the wide variation in the type of workplaces covered by the *Fire Regulations*, it is only possible to give a general guide to

> the level of precautions required to satisfy those *Regulations* in most workplaces. So this guide does not seek to give specific advice about each individual type of workplace. If a workplace is unusual, particularly if it is a large, complex premises or involved with specialised activities or risks, an employer may wish to seek specialist advice or refer to further specific guidance.
>
> In some cases, it may be necessary to provide additional means of escape or to improve the fire protection of existing escape routes. If, having carried out a fire risk assessment, an employer considers this may be the case in the workplace, they should consult the fire authority and, where necessary, your local building control officer before carrying out any alterations.
>
> It would be a time-consuming and complicated process, requiring specialist expertise, to establish the time needed in each individual case. So this guide uses an established method for assessing means of escape, which has been found to be generally acceptable in all but the most particular circumstances. This method is based upon limiting travel distances according to the category of potential fire risk the workplace falls into.
>
> These distances ensure that people should be able to escape within the appropriate period of time. An employer can use actual calculated escape times but should do so only after consulting a fire safety specialist with appropriate training and expertise in this field'.

Fire risk categories

7.11 In terms of fire risk, most workplaces can be categorised as being high, normal or low. The following brief descriptions give an indication of the types of workplaces falling within each category and can be used for assessing the means of escape. For more detailed guidance, the reader should refer to the guidance contained within the Home Office publication entitled *Fire safety: An employer's guide.*

High risk

7.12 A workplace where:

- there is a serious risk to life from fire;
- there are substantial quantities of combustible materials or highly flammable substances; or
- there exists a likelihood of the rapid spread of fire, heat or smoke.

7.13   *How to carry out a fire risk assessment*

Normal risk

7.13   A workplace where there are sufficient quantities of combustible materials and sources of heat to be at *greater* risk than a *low risk* property, but where a fire would be likely to remain confined, or to only spread slowly.

Low risk

7.14   A workplace where the risk of fire occurring is low, the potential for fire, heat and smoke spreading is also low, and where there is minimal risk to people's lives.

Further Home Office guidance

7.15   The following information on:

- general principles for escape routes (see **7.16** below);
- evacuation times and length of escape routes (see **7.17** below); and
- number and width of exits (see **7.18** below),

are further extracts from the Home Office guidance publication, entitled *Fire safety – An employer's guide*.

General principle for escape routes

7.16   Other than in small workplaces, or from some rooms of low or normal fire risk, there should normally be alternative means of escape from all parts of the workplace.

Routes that provide means of escape in one direction only (from a dead-end) should be avoided wherever possible as this could mean that people have to move towards a fire in order to escape.

Each escape route should be independent of any other and arranged so that people can move away from a fire in order to escape.

All escape routes should always lead to a place of safety. They should also be wide enough for the number of occupants and should not normally reduce in width.

Escape routes and exits should be available for use and kept clear of obstruction at all times.

Evacuation times and length of escape routes

7.17 The aim is, from the time the fire alarm is raised, for everyone to be able to reach a place of relative safety, ie a storey exit, within the time available for escape. (A storey exit is defined as an exit people can use so that, once through it, they are no longer at immediate risk). This includes a final exit, an exit to a protected lobby or stairway (including an exit to an external stairway) and an exit provided for means of escape through a compartment wall through which a final exit can be reached).

The time for people to reach a place of relative safety should include the time it takes them to react to a fire warning. This will depend on a number of factors including:

- what they are likely to be doing when the alarm is raised, eg sleeping, having a meal etc;
- what they may have to do before starting to escape, eg turn off machinery, help other people etc; and
- their knowledge of the building and the training they have received about the routine to be followed in the event of fire.

Where necessary, an employer, occupier or owner can check these by carrying out a practice drill.

To ensure that the time available for escape is reasonable, the length of the escape route from any occupied part of the workplace to the storey exit should not exceed the following.

- Where more than one route is provided:
    - 25 metres: high-fire-risk area;
    - 32 metres: normal-fire-risk (sleeping) area;
    - 45 metres: normal-fire-risk area; and
    - 60 metres: low-fire-risk area.
- Where only a single escape route is provided:
    - 12 metres: high-fire-risk area;
    - 16 metres: normal-fire-risk (sleeping) area;
    - 18 metres: normal-fire-risk area (except production areas in factories);
    - 25 metres: normal-fire-risk area (including production areas within factories); and
    - 45 metres: low-fire-risk area.

Where the route leading to a storey exit starts in a corridor with a dead-end, then continues via a route which has an alternative, the total distance should not exceed that given above for 'Where more than one route is provided'. However, the distances within the 'dead-end portion' should not exceed those given for 'Where only a single escape route is provided'.

Number and width of exits

7.18 There should be enough available exits, of adequate width, from every room, storey or building. The adequacy of the escape routes and doors can be assessed on the basis that:

- a doorway of no less than 750 millimetres in width is suitable for up to 40 people per minute (where doors are likely to be used by wheelchair users the doorway should be at least 800 millimetres wide); and

- a doorway of no less than 1 metre in width is suitable for up to 80 people per minute.

Where more than 80 people per minute are expected to use a door, the minimum doorway width should be increased by 75 millimetres for each additional group of 15 people.

For the purposes of calculating whether the existing exit doorways are suitable for the numbers using them, an employer should assume that the largest exit door from any part of the workplace might be unavailable for use. This means that the remaining doorways should be capable of providing a satisfactory means of escape for everyone present.

For further information on the subject of means of escape in case of fire, see **8.2** below.

Emergency lighting and exit signs

7.19 In order to ensure that the means of escape provided in a workplace can be used at all times, particularly if the workplace is in use during the dark hours, it is sometimes necessary to provide emergency lighting.

Generally, in premises which have a daytime occupancy only, emergency lighting will only be necessary when there is no (or insufficient) natural light for people to make their way out of a building safely, if the primary lighting should fail.

The need for escape lighting is greater in buildings, which are occupied by people who are unfamiliar with the building and its means of escape, such as in shops, hotels etc.

The provision of emergency escape and fire exit signs are necessary to indicate any emergency exit doors and routes which are not in common use.

For further information on the subject of emergency lighting and exit signs, see paragraphs **8.15** and **8.18** below respectively.

What the Fire Regulations say:

7.20

> *Fire Precautions (Workplace) Regulations 1997, Part II (Fire Regulations) (SI 1997 No 1840)*
>
> Regulation 5: Emergency routes and exits
>
> (1) Where necessary in order to safeguard the safety of employees in case of fire, routes to emergency exits from a workplace and the exits themselves shall be kept clear at all times.
>
> (2) The following requirements must be complied with in respect of a workplace where necessary (whether due to the features of the workplace, the activity carried on there, any hazard present there or any other relevant circumstances) in order to safeguard the safety of employees in case of fire–
>
> (a) emergency routes and exits shall lead as directly as possible to a place of safety;
>
> (b) in the event of danger, it must be possible for employees to evacuate the workplace quickly and as safely as possible;
>
> (c) the number, distribution and dimensions of emergency routes and exits shall be adequate having regard to the use, equipment and dimensions of the workplace and the maximum number of persons that may be present there at any one time;
>
> (d) emergency doors shall open in the direction of escape;
>
> (e) sliding or revolving doors shall not be used for exits specifically intended as emergency exits;
>
> (f) emergency doors shall not be so locked or fastened that they cannot be easily and immediately opened by any person who may require to use them in an emergency;
>
> (g) emergency routes and exits must be indicated by signs; and
>
> (h) emergency routes and exits requiring illumination shall be provided with emergency lighting of adequate intensity in the case of failure of their normal lighting.

7.21 *How to carry out a fire risk assessment*

Emergency plan

7.21 Employers are required to ensure that staff, know what to do in the event of a fire. To achieve this, the workplace should be provided with an emergency plan of action. Where an employer employs five or more people they must have a written emergency plan.

The plan should include what action is to be taken by an employer and staff in the event of fire. The plan should include an evacuation procedure and include arrangements to ensure that the fire brigade is called.

The essential element in any emergency action plan is 'simplicity' it should be *simple and easily understood*. The more complex an employer makes any plan, the less likely that people will remember what to do when an emergency occurs. If it is kept as simple as possible, there is every possibility that employees will remember what to do at the critical time when a fire breaks out.

Checklist

7.22

> The plan should include the following as a minimum:
>
> - what action employees should take if they discover a fire;
> - how to warn everyone else in the building; and
> - how to call the fire brigade.
>
> For a small workplace, such as the flower shop (mentioned previously at **7.3**), this information could be provided by a simple durable fire action notice on the wall in a position where staff are able see it and which should ensure that they become familiar with its content.
>
> Larger workplaces and those that are at a high fire risk obviously need more detailed plans, which should take into account the findings of the risk assessment. They should also identify any staff that may be at significant risk and their location within the premises should be noted. Notices giving clear and concise instructions on the action to be followed in case of fire should be prominently displayed throughout the workplace. It is also important to include a designated 'assembly point' in the notices.
>
> An assembly point is a place to which staff escaping from the workplace in an emergency should report, in order to ensure that they are accounted for. The assembly point should be a free, non-enclosed area away from the building, preferably indicated with a sign. An employer

> should also take into account any assistance that disabled or sensory-impaired employees may require in order to escape from fire. It may be necessary for staff to be trained in the correct procedures to cope with this eventuality.
>
> For further information on the subject of emergency action planning, see **8.7** below.

Means of fighting fire

7.23 Where necessary, in order to safeguard the safety of employees in case of fire, *Fire Precautions (Workplace) Regulations 1997 (SI 1997 No 1840)* require the provision and maintenance of appropriate fire-fighting equipment, fire detectors and alarms.

In determining what is appropriate fire-fighting equipment, account must be taken of:

- the dimensions and use of the buildings at the workplace;
- the equipment they contain;
- the physical and chemical properties of the substances likely to be present; and
- the maximum number of people that may be present at any one time.

In addition, any non-automatic fire-fighting equipment provided should be easily accessible, simple to use and indicated by signs.

In effect, this means that as an employer must ensure that the workplace has sufficient 'first aid' fire-fighting equipment meeting the criteria set out in SI 1997 No 1840, Reg 4, in short, it should be of the correct type, suitable for the potential risks within the premises and suitably located.

The term 'first aid' in this context means the fire-fighting equipment provided for use by staff to tackle a fire in its very early stages, ie before the arrival of the fire brigade. The types of fire-fighting equipment provided might be portable fire extinguishers, hose reels and fire blankets.

To assess which fire-fighting equipment an employer needs in the workplace, an employer needs to look at what type of fire is most likely to occur. To use the example of the flower shop, an employer may well consider that paper and packaging are most likely to be involved in fire and that water will extinguish fires involving paper and cardboard. An employer could reasonably assume that a water type extinguisher would be suitable for this risk. However, when an employer assesses the nature of the business, including the presence of numerous containers and buckets of water in which the flowers are kept, does an employer, in this case, actually need to buy an extinguisher.

7.24 *How to carry out a fire risk assessment*

The flower shop is an unusual example, but it is used to illustrate the need to look at the equipment and resources that you already have in your premises before you spend money unnecessarily.

Possibly, for example, the workplace may have a hose pipe permanently connected and used in connection with the 'day to day' business. It is reasonable, providing that the hose pipe is always available and is long enough to reach all parts of the workplace, to accept that this can be used for fires on which water can be used.

It is an altogether different and more complex process to assess the first aid fire-fighting needs of the plastic manufacturing factory. In this type of occupancy the risk of fire is more varied and therefore each potential risk should be analysed to ensure that the correct type of fire-fighting equipment is available and suitably located. There may be process, electrical and chemical risks, which will require the provision of specialist extinguishers or even specialised fixed fire-fighting installations.

These areas are discussed and explained in more detail at **11.31** and **11.33** below.

Fire safety training of employees

7.24 *Fire Precautions (Workplace) Regulations 1997 (SI 1997 No 1840)* also require the employer, where necessary, to ensure the safety of his employees in case of fire and to take appropriate fire-fighting measures in the workplace. In so doing the employer must take into account:

- the nature of his workplace activities;
- the size of the undertaking and of the workplace concerned; and
- persons other than his employees who may be present.

An employer must also:

- nominate employees to implement those measures; and
- ensure that the number of such employees, their training and the equipment available to them are adequate.

In ensuring that the equipment is 'adequate' the employer must take into account the size of the workplace and its specific hazards.

An employer is also required to arrange any necessary contacts with external emergency services, particularly as regards rescue work and fire-fighting.

SI 1997 No 1840 requires an employer to ensure that staff are aware of the risk of fire and that they know what do if a fire occurs. This means that staff should receive guidance, training or instruction on fire safety. Training should form part of the overall plan for dealing with emergencies within the workplace. It is important to ensure that staff fire training is relevant. To use the example of the low risk flower shop with only two employees, it would be entirely inappropriate and unnecessary to provide a formal fire safety training course for those two employees. In this instance it would be reasonable to perhaps provide a single sheet of paper to a new employee on induction, detailing the action to be taken in case of fire, and to discuss its content in order to ensure that the new employee is aware of what he or she has to do. Existing staff could be given the same information, which could be discussed informally, perhaps even over a cup of coffee.

However, the fire safety training of staff employed in the plastic manufacturing factory is an entirely different matter. This is a *high risk* type of business housed in an old complex building and employing two hundred people. Here it would be necessary to carry out more formal training. The employer's approach might be that certain designated members of staff such as foremen, supervisors and managers would be formally trained as floor or department fire marshals.

This type of specialised training could either be carried out in-house, if sufficient expertise was available, or by specialist fire safety training organisations. Once trained, the fire marshals could then be given the responsibility for ensuring that staff within their jurisdiction are given appropriate training relevant to the specific areas of the building in which they work and the specific work activities they carry out.

The point to note here is that it is not necessary for *all* staff to receive the same high level of training. This would be inappropriate and unmanageable. It is however necessary to ensure that *all* staff know:

- location and use of escape routes;
- location of a nominated Fire Assembly Point;
- how to use the fire equipment provided; and
- how to summon the fire brigade.

This subject is dealt with in more depth in **CHAPTER 9: MANAGING FIRE SAFETY (FIRE SAFETY PROCEDURES, TRAINING AND FIRE MANUAL)**.

What the Fire Regulations say:

7.25

> *Fire Precautions (Workplace) Regulations 1997, Part II (Fire Regulations) (SI 1997 No 1840)*
>
> Regulation 4: Fire-fighting and fire detection

7.26 *How to carry out a fire risk assessment*

> (1) Where necessary (whether due to the features of a workplace, the activity carried on there, any hazard present there or any other relevant circumstances) in order to safeguard the safety of employees in case of fire–
>
> (a) a workplace shall, to the extent that is appropriate, be equipped with appropriate fire-fighting equipment and with fire detectors and alarms; and
>
> (b) any non-automatic fire-fighting equipment so provided shall be easily accessible, simple to use and indicated by signs,
>
> and for the purposes of sub-paragraph (a) what is appropriate is to be determined having regard to the dimensions and use of the buildings at the workplace, the equipment they contain, the physical and chemical properties of the substances likely to be present and the maximum number of people that may be present at any one time.
>
> (2) An employer shall, where necessary in order to safeguard the safety of his employees in case of fire–
>
> (a) take measures for fire-fighting in the workplace, adapted to the nature of the activities carried on there and the size of his undertaking and of the workplace concerned and taking into account persons other than his employees who may be present;
>
> (b) nominate employees to implement those measures and ensure that the number of such employees, their training and the equipment available to them are adequate, taking into account the size of, and the specific hazards involved in, the workplace concerned; and
>
> (c) arrange any necessary contacts with external emergency services, particularly as regards rescue work and fire-fighting.

Maintenance and testing of fire safety equipment

7.26 An employer, is required to ensure that all the fire safety equipment provided in the workplace is in effective working order and in good repair. This includes any 'first aid' fire-fighting equipment, ie fire extinguishers and hose reels etc, together with the means of detecting and giving warning in case of fire, the means of escape, emergency signs and any emergency lighting that has been provided.

It is very much in an employer's own interest to ensure that the fire defence systems, which may represent a considerable financial investment, perform to expectation when an emergency arises.

How to carry out a fire risk assessment **7.26**

Obviously, it is important that equipment is fit for its purpose and is properly maintained and tested. All equipment provided to assist escape from the premises, such as fire detection and warning systems, emergency lighting, and all equipment provided to assist with fighting fire, should be regularly checked and maintained by a suitably competent person in accordance with the manufacturer's recommendations.

A 'competent person' is defined as someone who has the necessary training, experience and abilities to carry out the work.

The table below provides a simple guide to good practice and, is reproduced from the Home Office guidance entitled *'Fire safety – An employer's guide'*.

Equipment	*Frequency*	
Fire-detection and fire-warning systems including self-contained smoke alarms and manually operated devices.	Weekly	Check all systems for state of repair and operation. Repair or replace defective units. Test operation of systems, self-contained alarms and manually operated devices.
	Annually	Full check and test of system by competent service engineer. Clean self-contained smoke alarms and change batteries.
Emergency lighting including self-contained units and torches.	Weekly	Operate torches and replace batteries as required. Repair or replace any defective unit.
	Monthly	Check all systems, units and torches for state of repair and apparent working order.
	Annually	Full check and test of systems and units by competent service engineer. Replace batteries in torches.

7.27 *How to carry out a fire risk assessment*

Equipment	Frequency	
Fire-fighting equipment including hose reels.	Weekly	Check all extinguishers including hose reels for correct installation and apparent working order.
	Annually	Full check and test by competent service engineer.

Note: unless otherwise stated, the user can carry out the above actions. Manufacturers may recommend alternative or additional action. Further, more detailed information can be found in the relevant British Standards.

See also CHAPTER 5: LEGISLATION AND REGULATIONS (BRITISH AND EUROPEAN STANDARDS) and APPENDIX: SOURCES OF FURTHER INFORMATION. In addition, for more detailed information on the maintenance of fire precautions see CHAPTER 12: MANAGING FIRE SAFETY (MAINTENANCE, SERVICING ROUTINES AND RECORD KEEPING).

What the Fire Regulations say:

7.27

> *Fire Precautions (Workplace) Regulations 1997, Part II (Fire Regulations) (SI 1997 No 1840)*
>
> Regulation 6 Maintenance
>
> Where necessary in order to safeguard the safety of employees in case of fire, the workplace and any equipment and devices provided in respect of the workplace under regulations 4 and 5 shall be subject to a suitable system of maintenance and be maintained in an efficient state, in efficient working order and in good repair.

Fire risk assessment checklist

7.28 The following checklist is provided to assist an employer, owner or occupier to carry out a fire risk assessment. A number of the items detailed in the checklist may not apply to their workplaces, or there may be additional items to be included which are relevant to specific premises. Employers should develop checklists, using the suggestions provided and tailoring it so that it relates to *their specific workplace*.

Checklist

7.29

Fire safety features

Stairways/corridors

- Is the main stairway protected by self-closing fire doors?
- Is the emergency stairway protected by self-closing fire doors?
- Do all emergency stairway(s) discharge to open air via a 'risk free' area?
- Is the route to the emergency stairway clearly sign posted?
- Are there any 'dead-end' conditions where escape is possible in one direction only?
- Are the stairways clear of obstructions?

Escape routes/exits/doors/signs

- Are all escape routes (corridors, exit doors) free from obstructions?
- Is the route (within the emergency exit stairway) clearly sign posted?
- Are all exit doors unobstructed externally?
- Are the exits provided sufficient?
- Are exit doors free to open at all times (eg not locked)?

Fire/smoke stop doors

- Are fire doors fitted to risk rooms?
- Are fire doors fitted with 'Fire Door – Keep Shut' signs?
- Are fire doors effectively self-closing?

Fire alarm system

- Is there an electrical fire alarm system with call points and sounders?
- If *not* an electrical system – state type of fire alarm installed.
- Is it satisfactory for the risk? Will it meet the current legal requirements?
- Is the fire alarm maintained?

7.29 *How to carry out a fire risk assessment*

- Is the fire alarm tested on a regular basis? State date of last test.
- Does the fire alarm have any automatic fire detectors in the workplace?
- Does it cover the following areas:

(a) corridors   (state whether heat [   ] or smoke [   ]);
(b) stairways   (state whether heat [   ] or smoke [   ]); and
(c) risk rooms   (state location – State whether heat [   ] or smoke [   ])

- Is maintenance of the fire alarm recorded in a fire safety log book?

Fire equipment and protection

- Are fire extinguishers/hose reels available for use in the workplace?
- Are the extinguishers/hose reels maintained on a regular basis?
- Is fire extinguisher/ hose reel maintenance recorded in a fire safety log book?
- Is there an automatic water sprinkler system provided?
- If 'yes' to automatic water sprinkler – is it properly maintained?

Emergency lighting

- Is self-contained emergency lighting provided?
- Is the emergency lighting properly maintained?
- Is the maintenance of the emergency lighting system recorded in a fire safety log book?

Fire Safety Instructions

- Are Fire Action Notices posted throughout the workplace?
- Is there a fire safety manager?
- Are fire marshals appointed and trained in their duties?
- Give date of last training and who gave the instruction?
- Is there a designated Fire Assembly Point? If 'yes' – state where?
- Are arrangements made to evacuate staff, customers or visitors with disabilities?

Other fire and safety risks

7.30

- Is there a system for controlling the amount of combustible materials and flammable liquids and gases that are kept in the workplace?
- If the answer is 'yes' — is it working effectively?
- Are all combustible materials and flammable liquids and gases stored safely?
- Are all heaters fitted with suitable guards and fixed in position away from combustible materials?
- Are all items of portable electrical equipment inspected regularly and fitted with correctly rated fuses?
- Is the wiring of the electrical installations inspected periodically by a competent electrical engineer?
- Is the use of extension leads and multi-point adapters kept to a minimum?
- Are flexible electrical leads run in safe places where they will not be easily damaged?
- Is the upholstery of any furniture in good condition?
- Is the workplace free of rubbish and combustible waste materials?
- Have measures been taken to prevent against the risk of arson?
- Is there a NO SMOKING policy and is there a designated smoking area provided with adequate ashtrays?
- And finally, is there a prioritised action plan for remedial measures following an assessment?

8 Managing fire safety (awareness, prevention and fire safety policy)

In this chapter:	
Fire awareness and prevention	8.1
Formulating a fire safety policy for the workplace	8.2
Developing a 'fire-safe' environment	8.3
Who should be involved in the process?	8.4
Fire marshals	8.5
Raising staff awareness	8.6
Planning for emergencies: suggested fire safety checklists	8.7
What is risk management?	8.19
Policy for the disabled, the public and visitors	8.20
Protecting computer records	8.21
Bomb alerts and threats	8.22
Plan of action	8.23
Bomb call checklist	8.24
How the plan should work	8.26
Home Office guidance	8.33
Bombs: protecting people and property – A Guide for Small Business	8.34
Bombs – some useful information	8.35
Taking precautions	8.40
What to do if a telephone bomb threat is received	8.50

Fire awareness and prevention

8.1 Whenever premises, to conform with the requirements *Fire Precautions Act 1971*, require or are provided with a Fire Certificate, and/or are subject to the provisions of *Fire Precautions (Workplace) Regulations 1997 (SI 1997/1849)* and/or *Fire Precautions (Workplace) (Amendment) Regulations 1999 (SI 1999/1877)*, the employer, owner and tenants should have an awareness of fire and the need for fire prevention. This will be even more apparent when a fire risk assessment of the premises has, in accordance with the criteria contained in **CHAPTER 7: HOW TO CARRY OUT A FIRE RISK ASSESSMENT**, been undertaken.

It helps for an employer, owner or occupier to have a fire safety policy for the workplace, which promotes good housekeeping and reduces the possibility of a fire occurring. Carelessness and neglect not only make the outbreak of a fire more likely but also will inevitably create conditions, which may allow a fire to spread more rapidly (see further **8.2**).

Formulating a fire safety policy for the workplace

8.2 Fire insurance can provide reimbursement for damaged property and much of the income that may be lost as a result of a fire, however, it cannot protect the market share. Employers will need to know if their customers will remain loyal and wait for a business to rebuild after a fire has destroyed the premises, goods and facilities. Will they go elsewhere? Will the more experienced and skilled workers seek alternative employment after they have been laid off? Will the temporary loss be the competitors' permanent gain?

The answer is clear. An employer cannot afford to ignore the possibility that a fire could seriously affect their business premises and must take positive action to minimise harm to both their business and employees. The first step is to develop a fire safety policy for the workplace. This policy, if implemented effectively, will minimise the likelihood of a fire occurring at all and will protect employers and their employees from the catastrophic consequences of a fire.

The adoption of a proactive fire safety policy, dovetailed with the risk management concept (discussed below at **8.18**), can be regarded as insurance against loss of tangible assets, loss of market share and as insurance against a damaged reputation and business interruption.

Under *Fire Precautions (Workplace) Regulations 1997* (SI 1997 No 1840), an employer has a legal obligation to carry out a fire risk assessment of premises. In addition, *Management of Health and Safety at Work Regulations 1999* (SI 1999 No 3242) require every employer to make 'a suitable and sufficient assessment' of the risks to which employees are exposed while they are at work, and the risks to which persons not in his employment are exposed.

8.3 *Managing fire safety (awareness, prevention and fire safety policy)*

Given this onerous responsibility, which requires the investment of time and resources in carrying out fire and health and safety risk assessments, it makes good business sense to invest a little more into the process and develop an ongoing and long-term fire safety policy.

Many small to medium-sized companies have already seen the business advantages in this course of action and have appointed a designated individual within the organisation to manage fire and health and safety matters. Not only are such companies addressing their legal responsibilities competently in the eyes of the enforcing authorities, but they have also created a positive safety culture within their company, in which every employee is involved.

This chapter gives advice on protecting a business from fire (see **8.7–8.20** below) and offers recommendations for developing an effective fire safety policy (see **8.2–8.6**). Although the advice is primarily targeted at the small to medium sized business, the recommendations can be tailored to suit any size and type of business.

This chapter also contains a section on dealing with bomb alerts (see **8.21** below), which is a very real threat to many organisations. It is appreciated that this is not strictly a fire safety matter, but the response to a bomb alert is very similar to that of a fire alert. This is of particular importance in premises where 'phased evacuation' has to be adopted because the staircases may not have sufficient capacity to accommodate all the persons present within the building should, for any reason, immediate or 'simultaneous evacuation' of the premises be required. Under these circumstances it is sensible, to avoid confusion amongst employees, to combine similar policies for both fire and bomb incidents by using the same alert and evacuation procedures.

Developing a 'fire-safe' environment

8.3   No matter how small or large a business is, it can benefit from an environment in which the prevention of fire is considered by everyone in the organisation to be a high priority. The objective should be to provide a 'fire-safe' environment, which minimises the loss of life, personal injury and loss of property due to fire and associated hazards.

Developing a 'fire-safe' environment in which an employer and the employees can all work safely is not an instant process. It will take time to develop. However, through perseverance, the rewards can be considerable. Such efforts can minimise the potential for disastrous business disruption, which can inevitably result from a fire. As a consequence, through reducing the risks, a business can be safeguarded.

Who should be involved in the process?

8.4   The short answer is *everyone* in the organisation. It is known from experience that the most effective way of preventing fire is by ensuring that employees do things correctly. Incorrect actions are often the result of ignorance on the part of the people involved, in that they do not realise the consequences of their actions.

It is sensible for both the employer and the employees to work together, in a positive and proactive way, in aiming to prevent a fire. There is a vested interest in such prevention as, apart from protecting the business, such prevention protects the lives and jobs of the employees.

Depending on the size of the business and the number of employees, it is good practice to appoint someone with the responsibility of ensuring that the policy is implemented throughout the company. In larger organisations the role would probably be filled by a full time appointment. Often in large companies it is common to 'recruit' from elsewhere an individual who already has the necessary skills and expertise in this field. In smaller companies, where it does not warrant the appointment of a full time post holder, it is common for a designated individual to be given the fire and health and safety role, in addition to their normal duties within the company.

Fire marshals

8.5   An effective system for ensuring that the fire safety message is spread throughout an organisation is to 'delegate' the responsibility through all staff levels by appointing floor, department or section 'fire marshals'. The basis of this system is that the fire marshals are given a higher level of fire safety training than the average member of staff in order to look after a designated part of the premises.

Their role should be to check for hazards and potential fire risks within their area. Fire Marshals must check that first aid fire fighting equipment and other fire safety equipment is working and in place. They should also be responsible for organising the evacuation of staff, the public and visitors, in the event of a fire alarm. This would include a final sweep of their designated area, including toilet areas, to ensure that everyone is out and to report accordingly to the person in charge of the assembly point.

The level of suggested training for staff and fire marshals is dealt with in more detail within **CHAPTER 10: FIRE TRAINING MANUAL.**

Raising staff awareness

8.6   The fact that a business has formulated a fire safety policy and has raised the profile of the subject within the organisation inevitably means that staff will become more aware of fire safety issues. Improved awareness must be coupled with appropriate fire safety training for staff if the policy is to be effective. Fire safety training for all employees should begin on the first day of employment and should form an integral part of the policy.

8.7 *Managing fire safety (awareness, prevention and fire safety policy)*

Fire Safety is all about teamwork, which needs a combined approach in order to be successful. Raising the level of staff awareness of the potential effects of fire on both themselves and on the business encourages a climate of 'joint ownership' of the problem.

This 'hearts and minds' approach has been proved time and time again to be the most effective way of achieving the objective.

CHAPTER 9: MANAGING FIRE SAFETY (FIRE SAFETY PROCEDURES, TRAINING AND FIRE MANUAL) provides detailed recommendations for the training of staff.

Planning for emergencies: suggested fire safety checklists

8.7 The following suggested checklists are provided as a basis for developing a 'fire-safe' environment. Although the checklists are designed primarily for a manufacturing business, they can be adapted quite easily to suit any type of workplace environment.

Cleanliness and tidiness

8.8

- Are staff encouraged to tidy their personal work places?
- Are premises kept clear of combustible process waste and refuse?
- Are metal receptacles with closely fitting lids available for waste such as floor sweepings?
- Are separate, clearly labelled containers provided for waste and special hazards – eg flammable liquids, paint rags, oily rags?
- Are waste containers removed from the building at the end of each working day or more frequently if necessary?
- Is waste awaiting disposal put in a safe place, which is not accessible to the public?
- Is the burning of waste on site prohibited?
- Are cupboards, lift shafts, spaces under benches, gratings, conveyor belts and similar places kept free from dust and the accumulation of rubbish?
- Are pipes, beams, trusses, ledges, ducting and electrical fittings regularly cleaned?

- Are areas in and around the building kept free from accumulated packaging materials and pallets?
- Are metal lockers provided for employees' clothing?

Liquefied petroleum gas ('LPG') cylinders

8.9

- Are LPG cylinders stored safely, preferably in a fenced compound outdoors at least two meters away from any boundary fences?
- Is the store used only for gas cylinder storage?
- Are empty cylinders treated in the same manner, but kept separate and labelled empty?
- Are permanent warning notices prominently displayed prohibiting smoking and naked lights?
- Are cylinders stored with their valves uppermost?

Storage

8.10

- Are fire doors, exits, fire equipment and fire notices kept unobstructed?
- Are storage areas accessible to fire fighters?
- Are stack sizes kept as small as possible in the circumstances?
- Are there clear spaces around stacks of stored materials and adequate gangways between them?
- Are stacks stable and not liable to collapse easily?
- Are stocks of material arranged so that sprinkler heads and fire detectors are not impeded and are the required clearances beneath this equipment maintained?
- Are stocks maintained at appropriate levels, with the avoidance of excessive quantities in process areas?
- Is access to storage areas restricted to those who really need to be there?
- Are stocks kept well clear of light fixtures and hot service pipes?

8.11 *Managing fire safety (awareness, prevention and fire safety policy)*

- Are storage areas inspected regularly and at the end of the working day?

Maintenance of buildings

8.11

- Is every point of entry to the site and building secure against intruders?
- After close down of operations are all doors, windows and gates checked and secure?
- Is the building regularly inspected for damage to windows, roof and walls?
- Are the grounds surrounding the premises kept free of combustible vegetation by regular grass cutting and scrub clearance?
- Are all outside contractors supervised while on the premises and their work authorised by 'permit to work' and 'hot work permit' schemes?

Flammable liquids

8.12

- Are all stocks of flammable liquids kept in purpose-built flammable liquid stores?
- Is the flammable liquid store kept uncongested and tidy?
- Are flammable liquids carried in specially designed safety containers and not in open cans and buckets etc?
- Are quantities of flammable liquids in use kept to a minimum and when not required returned to safe storage?
- Are flammable liquids kept away from possible sources of ignition?
- Are suitable spark reducing tools provided for use in places where there may be flammable vapours?

Machinery

8.13

- Does all machinery and equipment receive regular scheduled maintenance?
- Is lubrication adequate?
- Are motors and all moving parts of machinery kept clean to prevent overheating?
- Is machinery located so as to prevent congestion among machines and materials?
- Are drip trays used where necessary and emptied regularly?
- Are oil leaks and drips absorbed with mineral absorbents, not sawdust?
- Is there adequate provision of cleaning materials – wipes, cloths and so on?
- Are vents on motors and other equipment kept free of blockages to prevent overheating?

Space heating and lighting

8.14

- Are there restrictions on using unauthorised heaters?
- Are combustible materials at a safe distance from appliances and flues?
- Is care taken that no materials are left on heaters?
- Are portable heaters securely guarded and placed where they cannot be knocked over or ignite combustibles?
- Are goods kept clear of lighting equipment?

Smoking

8.15

- Is smoking prohibited in all but designated 'smoking' areas?
- Are the non-smoking regulations strictly enforced?

8.16 *Managing fire safety (awareness, prevention and fire safety policy)*

- Where smoking is permitted is there an abundant supply of non-combustible receptacles for cigarette ends as distinct from containers for waste?
- Are these receptacles emptied at least once a day?

Damage control

8.16

- Where possible are goods stored clear of the floor?
- Are drains provided and are they kept clear of blockages by routine inspection and cleaning?
- Are duplicate copies of important records kept in another building?
- Have contingency plans been drawn up to enable production to resume with the minimum of delay in the event of a fire?

Fire defence equipment

8.17

- Are hydrants, fire extinguishers, fire alarms and sprinkler systems regularly maintained by suitably qualified people?
- Are fire doors kept closed?
- Are routine checks made to ensure equipment has not been obscured, moved or damaged?
- Are notices informing staff what to do in the event of fire prominently displayed?
- Is the fire alarm tested weekly?

Staff training

8.18

- Are new staff instructed in fire procedures and shown the fire escape routes on their first day at work?
- Have fire marshals been appointed and trained in their duties?

- Have all staff received training on what they must do in the event of fire?
- Have staff been instructed on what type of fire extinguisher to use for specific types of fires?
- Have staff had the opportunity to operate a fire extinguisher?
- Do staff know how to deal with the disabled, the public and visitors in the event of an evacuation?

What is risk management?

8.19 Risk management is a fairly recent concept that aims to protect the assets and profits of an organisation by reducing the potential for loss *before* it occurs. Examples of the types of losses, which can be anticipated, are:

- major fires;
- arson attacks;
- large-scale theft;
- terrorist attacks;
- industrial espionage;
- storm damage; and
- flooding and other natural disasters.

Although it is a specialist field, particularly when it comes to dealing with larger companies, the general concepts are explained briefly below. By being aware of these principles, an employer will be better placed to assess whether a risk assessment would be appropriate for their business and whether specialist outside advice is needed. Having said that, the principles of risk management can be applied very successfully to virtually any size business and they will assist with the formulation of a fire safety policy for the organisation.

The risk management process consists of four basic elements.

- *Risk assessment* – identifying and quantifying the potential exposures that threaten an organisation's assets and profitability.
- *Loss control* – reducing the frequency and severity of losses through preventive measures, such as sprinkler systems, improved housekeeping practices or preventive maintenance of key equipment.
- *Risk transfer* – shifting the financial burden of loss so that, in the event of a catastrophe, an organisation can continue to function without severe hardship to its financial stability.
- *Risk monitoring* – continually assessing existing and potential exposures.

8.20 *Managing fire safety (awareness, prevention and fire safety policy)*

Some other positive benefits of risk management are listed below.

- Effective risk management can be considered to be essential to an organisation's long-term health and stability.
- An important element of risk management strategy should be the communication of risk management benefits to an organisation. This will help you to develop a more effective, comprehensive risk management strategy.
- Having an effective written risk management policy statement can emphasise the importance of the risk management function throughout an organisation.

Policy for the disabled, the public and visitors

8.20 A fire safety evacuation policy should take into account the fact that disabled people, the public and visitors may well be on the premises when a fire occurs. As many of these people will be strangers to the building, they will rely entirely on staff to assist them to vacate the premises. Staff training should therefore include information on how to deal with the evacuation of these people.

With regard to the disabled and their evacuation in the event of fire, detailed information is provided in **CHAPTER 4: LEGISLATION AND REGULATIONS (THE DISABILITY DISCRIMINATION ACT 1995)**.

Protecting computer records

8.21 In today's electronic business environment a company will struggle to survive if it loses its computer data. Consider the consequences of a company completely losing its orders, invoice, finance and personnel records, as a result of a fire.

All businesses must have a rigid policy of at least making a daily back-up of their computer data. Any back-up of such data should be kept *off* the business premises. In smaller companies a secretary taking back-up discs or tapes home at the end of the day could achieve this. In larger companies operating network systems, the back-up can be stored electronically on a remote server located *off* the premises. New companies now exist who offer this back-up service via the internet.

The other important aspect of back-up procedures is to ensure that the back-up will actually restore the information completely in the event of data loss. Have you actually tested that you can read and restore your company computer and personal computer back-up files?

Although protecting computer records should form an important part of a fire safety policy, it is equally important for them to be protected for the normal day-to-day running of the business.

Bomb alerts and threats

8.22 It is true to say that the more sophisticated and advanced we become as a society, the more vulnerable we are to attacks from terrorists. The fact that the general public is able to come within close proximity of Government buildings, financial institutions and other prominent buildings and that explosives are relatively easy to obtain, allows terrorist organisations to achieve tremendous economic and psychological effect by their actions.

No matter how small or large a business is it makes sense to be aware that the premises could become a target for terrorism. It is important to remember that those responsible have three objectives, which are to:

- disrupt the normal flow of business and create an atmosphere of anxiety and panic;
- injure people and damage property; and
- obtain publicity for their cause.

In order to address this issue, it is important that all organisations develop a 'Bomb threat strategy'.

Plan of action

8.23 Every organisation and building within an organisation needs a plan of action to reduce or prevent injury to people and damage to property. It has been proved that a well rehearsed plan reduces anxiety and disruption to daily routine.

As most businesses have an emergency plan in the event of fire, a bomb threat plan of action can be developed and dovetailed within that existing framework quite successfully.

It is important that:

- a bomb call checklist is produced and distributed throughout the organisation;
- all personnel are aware of evacuation procedures; and
- emergency services are listed for contact in the event of a threat.

8.24 Managing fire safety (awareness, prevention and fire safety policy)

Bomb call checklist

8.24 Whilst bomb threats can be made in writing or in person, they are more commonly made by telephone. The caller wants to create anxiety, disruption and confusion and usually uses such words as 'There's a bomb in your building and its set to go off at . . .'

It is worth remembering that there are significantly more threats made than actual bombs 'planted' and the receipt of the threat call is therefore extremely important.

All personnel likely to receive such a call must treat the calls seriously and remain calm.

Remember, above all, to try to keep the line open. In all cases, such calls should be reported immediately to the police.

Checklist

8.25

In the event of a call.

- Do not hang up even if the caller does, the authorities may be able to trace the call.
- Attract the attention of another person and inform them of the threat.
- Keep calm and engage the caller in conversation about the threat.
- Note the exact wording used.
- Ask questions to clarify the threat.
  - When will it explode?
  - Where is it placed?
  - What does it look like?
  - What will make it explode?
  - Why did you place it?
  - What is your name?
  - What is your address?
- Take notes whilst the caller is talking and keep those notes for future reference.

- Take special note of mannerisms, accents and any background noise.
- Following the receipt of such a threat, it is important that management is informed immediately.

A checklist for dealing with telephone bomb threats is included at **8.29** below in the Home Office guidance leaflet.

How the plan should work

8.26 Familiarity with the checklist will ensure the prompt and accurate recording of the call and will ensure the most effective management of the information.

All staff should be made aware of what to do upon receipt of such a call by alerting the switchboard operator who will, in turn, implement their action plan.

It is important to remember that *evacuation* will *not automatically follow.* In fact – premature evacuation can unnecessarily endanger the lives of a greater number of people.

Search

8.27 The potential attackers will normally plant the device in an easy place and it is therefore essential that these places are searched before evacuation is ordered.

Obvious areas are:

- garden shrubbery near to buildings; and
- public reception areas or routes where large volumes of personnel or visitors have access.

What to look for

8.28 During a search it is helpful to know that you are attempting to locate something which:

- should not be there;
- cannot be accounted for; or
- is out of place.

8.29 *Managing fire safety (awareness, prevention and fire safety policy)*

It is important to remember that employees will know their own work areas and know if something is not right. The police and other emergency services will not have that special knowledge. The decision on whether to search or not must be taken by senior management.

The options for search are:

- search of immediate work areas;
- search of the entire premises; or
- if, because of the specific timed nature of the warning, there is not time for a search then partial or full evacuation must be considered.

Checklist

8.29

- if a package is found – do not touch or move it;
- clear all personnel from the immediate vicinity;
- notify senior management;
- describe the package accurately; and
- do *not* use a radio or mobile phone within 30 metres of the package.

Following such a discovery the police will arrange for examination and disposal by experts.

Evacuation procedure

8.30 Simple rules to remember if evacuating a building in event of a bomb threat:

- staff *must* take personal belongings with them – briefcases, handbags or any other item, which could be misconstrued by search teams; and
- car parks should not be generally used as evacuation assembly points.

Remember, it is essential to remain calm and do as you are told.

Preventive measures

8.31 As with fire safety, good housekeeping is essential and will reduce the risk of a placement.

You should:

- remove rubbish and unwanted packaging on a regular basis;
- ensure emergency doors, exits and walkways are free from obstruction; and
- plant rooms, cupboards and service ducts should be secured at all times when not in use.

Security measures

8.32

- develop inspection procedures for all incoming packages and parcels;
- determine likely locations for the planting of devices;
- control access to critical areas;
- ensure adequate interior and exterior security lighting; and
- report anything unusual and challenge all strangers.

Home Office guidance

8.33 To assist companies in formulating a policy to deal with bombs and bomb threats, the Home Office has produced a leaflet for small businesses entitled '*Bombs: protecting people and property – A Guide for Small Business*', which is reproduced below.

Bombs: protecting people and property – A Guide for Small Business

8.34 This leaflet has been written for the owners and managers of small businesses. But it may also be useful for people working in larger organisations who do not need the detailed advice contained within the Home Office handbook with the same title. A copy of this handbook may be obtained from the local police Crime Prevention Department.

The information in this leaflet can help to reduce the risk of a bomb attack against you, your staff, your customers and your property.

An assessment should be made of how likely a terrorist attack against the business is. If the organisation could be targeted by a particular terrorist group, this should be taken into account when planning how to protect the organisation. You should also think about getting more detailed advice than there is in this leaflet.

8.35 *Managing fire safety (awareness, prevention and fire safety policy)*

However, terrorists do not always target specific businesses when they plant bombs. They may plant them at random, so a business might be affected simply because of where it is located. The chances of suffering a bomb attack may also change over time. In some cases, those at high risk may receive information directly from the police. Reports in the media may provide a good idea of the level of terrorist activity.

Those that own or manage small businesses will find the advice in this leaflet helpful in terms of guidance on making staff feel safer and buildings more secure, as well as protecting the business against other forms of crime.

More information may be sought from the police Crime Prevention Department.

Bombs – some useful information

8.35   Bombs are easily disguised. They may be hidden in bags, cases, or other everyday containers and in out-of-the-way places. Vehicles can carry large bombs without showing any signs. Any object which is unusual or out of place should be treated with suspicion and dealt with in the way described below.

The four kinds of bomb which you should know about are:

- high-explosive;
- vehicle;
- incendiary; and
- postal.

High-explosive bomb

8.36   High-explosive bombs can kill or injure people by their blast or by causing flying debris, particularly glass. Bombs small enough to be hidden in a bag or holdall may be big enough to cause serious damage to property.

To protect against high-explosive bombs:

- stop people bringing them onto the property;
- reduce the chances of someone planting a bomb which will not be detected; and
- keep watch over the property.

Vehicle bomb

8.37 To protect against vehicle bombs:

- control access to the car park;
- if possible, make sure that people park their vehicles well away from buildings; and
- keep watch over the outside of the property.

If you do not know why a package or a vehicle is there and you are suspicious of it follow these instructions:

- do not touch or move it;
- clear people away from the area close by;
- dial 999 for the police;
- sound the alarm and ask everyone to leave the property;
- warn properties next to yours of the danger; and
- gather everyone together at a meeting point.

Everyone should know where this meeting point is. It must be at least 100 metres away from the place where the suspicious object was found and out of sight of it. Or, if the suspicious object is a vehicle or anything bigger than a suitcase, meet at least 400 metres away from it.

Incendiary bomb

8.38 The retail industry is particularly at risk from incendiary bombs. They are normally small and very difficult to detect. They may be hidden inside a cigarette packet or cassette box. Terrorists normally put them amongst goods on display in a shop eg in the pockets of clothes or inside furniture. Incendiary bombs are often designed to go off in the early hours of the morning when there is no one in the shop.

To protect against incendiary bombs.

- Look out for people who are acting suspiciously.
- Search the property regularly, especially at the end of the day. Pay particular attention to anything that might burn easily and spread the fire.

If an incendiary device detonates and you find it, sound the alarm. If you have been properly trained in using fire extinguishers make one quick attempt to

8.39 *Managing fire safety (awareness, prevention and fire safety policy)*

put the fire out. Then get to a safe place as quickly as possible. While you are doing this, everyone else should leave the building and someone should call the police and fire brigade.

Do not touch or move anything; it might be an incendiary device. It may kill or injure.

Postal bombs

8.39 Letter and parcel bombs are envelopes and packages designed to kill or injure people when they are opened. They may not come through the post. They may be delivered by hand.

Any of the following signs are warnings that a letter or package might contain a bomb:

- there may be grease marks on the envelope or wrapping;
- the envelope or package might smell like marzipan or machine oil;
- it might be possible to see wires or foil, especially if the letter or package is damaged;
- the envelope or package may feel very heavy for its size;
- it may be heavier in some places than others;
- the envelope may be soft but the contents will feel hard;
- the package may have been delivered by hand by somebody you do not know;
- the package may be wrapped more than normal;
- there may be poor handwriting, spelling or typing;
- the envelope or package may be wrongly addressed;
- it may come from somewhere unexpected; or
- there may be too many stamps for the weight of the package.

If suspicions are roused about a package and there is an address on it, try to contact the sender. It is sensible to ascertain whether anyone in the premises is expecting a package.

If there is any reason to suspect that a letter or package may contain a bomb:

- put it down gently and walk away from it;
- ask everyone to leave the area; and
- sound the alarm.

Do not put the letter or package into anything (including water) and do not put anything on top of it.

The local police station can provide posters about postal bombs. It may be good practice to display one in the area where a business deals with its post.

Taking precautions

8.40   A number of steps can be taken to protect staff and property against bombs. Some of these steps will also help prevent other kinds of crime.

Basic security

8.41   Keep terrorists out when the premises are closed by fitting good quality key-operated locks and bolts to doors and windows. For outside doors, it is best to use locks that meet British Standard 3621 (look for the 'Kitemark'). Care should be taken not to break the rules for fire exits.

Good security lighting will also put terrorists off. Lights over fencing, outside doors and windows are a good deterrent.

Glass

8.42   Protect people from flying glass by having a special thin polyester film fixed to the inside of your windows and hanging special net curtains. Alternatively, replace the glass in windows with laminated glass, which is at least 7.5 millimetres thick. The local police Crime Prevention Department are able to give names of companies in the area that can carry out such work.

Control who comes into your property

8.43   Make it as difficult as possible for people who should not be there to enter the property during business hours. If possible, separate the private areas of the business from the public areas. Then control who can get into the private areas by using locks or by checking people as they go in.

If the threat of a terrorist attack is high, search all handbags and luggage which people bring onto the property. An employer has the right to refuse entry to anybody who will not allow their bags to be searched.

Make sure that people do not leave personal belongings on the property without permission.

8.44 *Managing fire safety (awareness, prevention and fire safety policy)*

Keep the property tidy

8.44 If the inside and outside of the property is kept tidy, there is less chance of somebody planting a bomb where no one will see it.

Inside, try to reduce the number of places where a bomb could be hidden. Lock all cupboards and unused rooms, and decide whether all the furniture is needed. Pay particular attention to public areas (including toilets) and keep them tidy.

Outside, do not let rubbish pile up, especially on the pavement. A pile of rubbish can be used to hide a bomb. Hedges and bushes can also be good cover for a bomb, so do not let them become overgrown.

Be alert and on guard

8.45 Look out for suspicious or unusual behaviour and report anything which seems wrong or out of place.

Question people who are in an area where they should not be. In particular look out for the following suspicious behaviour:

- somebody leaving a package or other objects in an unlikely place (for example, a shop doorway or flowerbed);
- somebody *placing* (rather than *dropping*) something into a litter bin; or.
- in shops, somebody putting something in an unusual place (for example, among clothes or in furniture) – especially if somebody else is keeping a lookout for them.

Think about installing closed circuit television cameras outside the property and inside too. Ask the police Crime Prevention Department for advice on how to get a system, which will meet the needs for the premises. Liase with adjacent businesses to cover areas outside the workplaces.

Key holders

8.46 The police may urgently need to contact the person who keeps the keys to the property. Make sure that the local police station has the name and address of the key holder and that this information is kept up to date.

Dealing with telephone warnings

8.47 An employer or employee may receive a telephone warning that a bomb has been planted in the building or somewhere else. It is essential to obtain as much information as possible from the caller. There is a checklist (see **8.50** below) of what to do if a threatening call is received. Keep copies of this checklist by the telephone so that anyone who receives a bomb threat can fill one in. The information on the checklist may help the police to trace the caller and to find the bomb.

If the caller states that a bomb is on the property, it will be necessary to decide whether the threat is serious. There are no hard and fast rules. Employers should take into account whether their business is at risk from a particular terrorist group. They should also consider the call itself. If, for example, the caller is drunk, or a child, the employer may decide that the threat is not serious.

In all cases, whether or not it is considered that the call is serious:

- phone the police immediately; and
- decide whether to search or evacuate the area under threat.

Searching and evacuating the property

8.48 If a bomb threat is received, it will have to be decided whether the property should be evacuated immediately, or whether it should be searched first and then everybody moved out if something suspicious is found.

If there is reason to believe that there is a bomb, everyone should be asked to leave immediately. If not, a search may be made for the bomb first before considering evacuating the property.

The police will not normally search the property, but they will give advice on searching, evacuating and re-entering the property.

The search should be very thorough. Check the whole of the floor area, the furniture and the fittings right up to the ceiling. Do not forget cloakrooms, passageways and stairways.

Remember to include the car park and other areas around the property.

Searchers are looking for something that should not be there, something that is out of place, something that nobody can recognise or explain.

If something is found that may be a bomb, everybody should be moved away from it, move everybody out of the building to a meeting point at least 100 metres away from and out of sight of the place where the suspicious object was found. If the suspicious object is larger than a suitcase or is a vehicle, the meeting point should be at least 400 metres away. Call the police if you have not already done so. *Do not touch or move the suspicious object.*

It is sensible to designate a suitable assembly point and ensure everybody know the location before an emergency occurs. The meeting point could be shared with another business and should be at least 400 metres away from each business.

Some buildings have an area, which can be used, as a bomb shelter. But the area must be approved by experts first. If there is an approved bomb shelter, it may be safer to move people there than to move them outside. The police Crime Prevention Department can be contacted for advice on how to find out whether a building contains a suitable bomb shelter.

Re-entering the property

8.49   If the property is evacuated without being searched and there has not been an explosion, consideration will have to given about re-entering the premises. *Do not allow anyone to re-enter before the property has been thoroughly searched.* If a time for the explosion is given over the phone, allow at least one hour after that before a search is commenced.

What to do if a telephone bomb threat is received

8.50
- If possible, tell someone else immediately so that they can tell the owner or manager. But *do not put down the handset or stop the conversation.*
- Tell the caller which town or area you are speaking from.
- Try to keep the caller talking (apologise for a bad phone line, ask him or her to speak up, and so on). Get as much information as you can.
- Fill in the form as you go along.

Ask the questions below (exact words) if the caller does not give you the information that you need. Try and ask them in the order they appear so that you do not miss any out.

- What time is it now?
- Where is the bomb?
- What time will it go off?
- What does it look like?
- What kind of explosive is in the bomb?
- Why are you doing this?
- Who are you?

Managing fire safety (awareness, prevention and fire safety policy) **8.50**

When the call has finished give this form to the owner or manager. They will decide what to do next. The more information you can get, the easier it will be to decide whether the warning was serious or not.

The following details should be completed as soon as possible:

(i) The caller

- Man [ ]
- Woman [ ]
- Child [ ]
- Young [ ]
- Old [ ]
- Do not know [ ]

(ii) How they sounded

- Drunk [ ]
- Rational [ ]
- Rambling [ ]
- Laughing [ ]
- Serious [ ]
- Speech Impediment [ ]

Give details:

Was the caller reading the message?

Yes [ ]
No [ ]

(iii) Other noises during the call

Any noise on the phone line?

Yes [ ]
No [ ]

Give details:

Did the caller use a pay phone (pay tone or coins)?

Yes [ ]
No [ ]

8.50 *Managing fire safety (awareness, prevention and fire safety policy)*

Did you hear the operator?

Yes [ ]
No [ ]

Were there any interruptions to the call?

Yes [ ]
No [ ]

(iv) Other noises in the background

- Traffic [ ]
- Talk [ ]
- Typing [ ]
- Machinery [ ]
- Aircraft [ ]
- Music [ ]
- Children [ ]

(v) Other

- Your name:
- Number you received the call on:

9 Managing fire safety (fire safety procedures, training and fire manual)

In this chapter:	
Introduction	9.1
Legal responsibilities	9.2
Fire safety training policy	9.7
Fire training manual	9.11
Damage control/salvage	9.34
Where to obtain assistance with training?	9.35

Introduction

9.1 The introduction, of adequate and suitable fire safety procedures, is an essential element in 'managing fire safety'. Employers, owners or occupiers of all premises should ensure that the procedures are suitable for the use of the premises. The procedures will, of course, vary according to the use of the premises.

Many 'workplaces' will have a fairly static population who are fully familiar with the layout of the premises, eg offices and factories are probably prime examples. However, in other cases, eg shopping complexes, departmental stores and restaurants etc, the majority of the population will mainly consist of members of the public who are visiting the premises.

The physical condition of the members of the public, in these type of premises, may well cover a wide range of abilities ranging from small children, the elderly, the infirm through to wheelchair disabled persons. These aspects will have to fully taken into account during the fire risk assessment and fully reflected in the 'fire safety' procedures and accompanying staff training.

A high standard of staff training is a vitally important weapon in any company's fight against fire. An employer should ensure that all of the staff have an understanding of how to react during an emergency. When the training to complement this understanding is prepared to a high standard, if a

9.2 *Managing fire safety (fire safety procedures, training and fire manual)*

fire occurs, the collective reactions of trained staff should ensure that the effects of the emergency are minimised.

Once an organisation has formulated a good fire safety training policy this will raise the profile of fire safety within that organisation and will inevitably mean that staff become more 'fire aware'. Fire safety training for all employees should begin on their first day of employment and should form an integral part of the induction programme.

Teamwork is critical, as by raising the level of awareness amongst staff about the dangers of fire and its devastating effects an employer can encourage a climate of 'joint responsibility' to the problem. All staff must receive adequate training based around a prepared emergency plan of action for the place of work.

An employer, owner or occupier of premises that have been issued with a fire certificate under the *Fire Precautions Act 1971*, has a legal responsibility to ensure that all staff are aware of the risk of fire and that they know what to do if a one occurs. To discharge that legal responsibility an employer needs to ensure that all staff receives guidance, training or instruction on fire safety. Training should form part of the overall plan for dealing with emergencies within the place of work. It is also important to get staff fire training into perspective and relate it to your premises.

The plan should include:

- action that should be taken by the employer, owner or occupier of the premises in the event of fire;
- action that should be taken by staff in the event of fire;
- an evacuation procedure; and
- detailed arrangements to ensure that the fire service is called to any fire or suspicion of fire.

It is also important to provide refresher training for staff on a regular basis to ensure that employees remain familiar with the fire precautions in the workplace and are reminded about what to do in an emergency.

See **9.11** below for guidance and information on how to formulate a staff fire safety training policy. All recommendations are intended to be flexible enough to be tailored to suit any business regardless of the size.

Legal responsibilities

9.2 The following paragraphs deal with the legislation on the training of employees, explaining the legal responsibilities of an employer, occupier or owner of premises.

The Health and Safety at Work etc Act 1974

9.3 *Section 2, HSWA 1974: Duties of employers*

Section 2(1), HSWA 1974 (a catch-all provision) states:

> 'I shall be the duty of every employer to ensure, so far as is reasonably practicable, the health, safety and welfare at work of all his employees.'

Section 2(2), HSWA 1974 goes on to detail more specific requirements relating to:

- provision and maintenance of plant and systems of work;
- use, handling, storage and transport of articles and substances; and
- provision of information, instruction, training and supervision.

Fire Precautions Act 1971

9.4 *Section 5(2A), HSWA*

> 'Where an application is made for a fire certificate with respect to any premises it is the duty of the occupier to secure that, when the application is made and pending its disposal–
>
> (a) the means of escape in case of fire with which the premises are provided can be safely and effectively used at all material times;
>
> (b) the means for fighting fire with which the premises are provided are maintained in efficient working order; and
>
> (c) any persons employed to work in the premises receive instruction or training in what to do in case of fire.'

In addition, a fire certificate may place requirements on the occupier or owner of the premises in respect of:

- maintenance of the means of escape provided and to ensure that it is free from obstruction;
- maintenance of any other fire precautions set out in the certificate;
- training of employees on the premises as to what to do in the event of fire and to ensure that suitable records of training are kept;
- any limitation of the number of persons who at any one time may be on the premises; and
- any other relevant fire precautions.

9.5   *Managing fire safety (fire safety procedures, training and fire manual)*

Fire Precautions (Workplace) Regulations 1997 (SI 1997 No 1840)

9.5   Regulation 4(2)

'An employer shall, where necessary in order to safeguard the safety of his employees in case of fire–

(a) take measures for fire-fighting in the workplace, adapted to the nature of the activities carried on there and the size of his undertaking and of the workplace concerned and taking into account persons other than his employees who may be present;

(b) nominate employees to implement those measures and ensure that the number of such employees, their training and the equipment available to them are adequate, taking into account the size of, and the specific hazards involved in, the workplace concerned; and

(c) arrange any necessary contacts with external emergency services, particularly as regards rescue work and fire-fighting.'

Legal note

9.6   There have been a number of cases where a person using an extinguisher has been seriously injured, due to a lack of training (for further information see **CHAPTER 11: MANAGING FIRE SAFETY (FIRE DEFENCE INCLUDING ACTIVE AND PASSIVE FIRE PRECAUTIONS**).

Fire safety training policy

9.7   It is good practice to formulate a written policy for the training of staff that can also incorporate other aspects of health and safety training. Such a document will also serve to demonstrate to the enforcing authorities that the company has a responsible attitude and a commitment to discharging its legal responsibilities.

The production of a written fire safety training policy document, if given enough weight and importance by senior management, will quickly become a safety 'bible' to everyone employed within the organisation.

As a guide to the production of a policy, the following suggestions for inclusion are made, which can be tailored to suit the particular circumstances within the business.

Managing fire safety (fire safety procedures, training and fire manual) 9.7A

Checklist

9.7A

- Training should be based upon the plan of action to take in the event of fire.
- Training should be based upon the written fire procedures.
- Training should include general fire preventative measures, good housekeeping practices and hazard spotting, specific to the staff being trained and the premises.
- The evacuation procedure for the building that should include:
  - avoiding the use of lifts;
  - any special arrangements for the physically disabled and sensory-impaired staff and members of the public;
  - the checking of all areas of the buildings;
  - informing and reassuring members of the public;
  - directing or escorting members of the public to the exits; and
  - carrying out of a roll call of staff at a fire assembly point.
- All of the following should be decided in advance:
  - the level of training that various grades of employees will receive;
  - how often training is to be carried out;
  - who will carry out the training; and
  - who will be responsible for ensuring that training is carried out.
- The arrangements for recording that the training has been carried out which should include:
  - date of the instruction or exercise;
  - duration;
  - name of person giving the instruction;
  - names of persons receiving the instruction; and
  - the nature of the instruction, training, drills or exercises.

It should be remembered that the fire training policy should take into account that the disabled, the public and visitors may well be in the premises when a fire occurs. As many of these people will be strangers to the building, they will need to rely entirely on staff to assist them to ➤

9.8 *Managing fire safety (fire safety procedures, training and fire manual)*

> vacate the premises. The staff training should therefore include information on how to deal with the evacuation of these people.
>
> With regard to the disabled and their evacuation in the event of fire, detailed information is provided in the **CHAPTER 4: LEGISLATION AND REGULATIONS (DISABILITY DISCRIMINATION ACT 1995).**

What level of staff training should a firm have?

9.8 It is important to get fire safety training into perspective and to relate it to the business premises and the circumstances. For example, it would be entirely inappropriate and unnecessary to provide a formal fire safety training course for just two employees working in a small shop. In this instance it would be reasonable to provide a single sheet of paper for a new employee on induction, detailing their action in case of fire. The contents of the instructions can be discussed in order to ensure that the new employee is aware of what they have to do in the event of a fire. Existing staff should be given the same information, which can be discussed informally.

On the other hand, the fire safety training of a large number of staff employed in larger and more complex premises, is an entirely different matter. In this latter instance it would be necessary to carry out more formal training, taking into account complexities and risks identified within the premises.

The employer's approach in these circumstances may be to employ a fire safety manager in order to ensure that the company's legal responsibilities are carried out effectively. In addition, certain designated members of staff such as foremen, supervisors and managers might be formally trained as floor or department fire marshals. This type of specialised training could either be carried out in-house, if sufficient expertise was available, or by specialist fire safety training organisations.

Once trained, the fire safety manager and fire marshals could then be given the responsibility of 'cascading' training down through the organisation, by ensuring that staff within their jurisdiction are given appropriate training relevant to the specific areas of the building in which they work.

It is not necessary for all staff to receive the same high level of training. This would be inappropriate and unmanageable. However it is necessary to ensure that:

- all staff know the location and use of escape routes;
- the location of a nominated fire assembly point;
- how to use the fire equipment provided; and
- how to summon the fire service.

It is particularly important that all new staff, including casual employees, should be shown the means of escape, and told about the fire procedures as soon as possible after they start work.

There is also a need to ensure that part time staff, occasional workers and others who may work in the premises outside normal hours, such as cleaners, IT and security staff are similarly instructed. Any staff who may have a limited understanding of the English language should be provided with training in a manner that they can understand.

Fire Training Manual (general)

9.9 The Fire Training Manual sets out ideas and suggestions for the production of a printed manual designed to cover all aspects of staff fire safety training (see **9.11** below).

In organisations with a large number of employees, it is sound practice to produce a Fire Training Manual for use by trainers within the organisation, in order to ensure the continuity and consistency of the training given. Ideally, the manual should set out the various levels of responsibility and include a syllabus for each of those levels. The suggested format of the manual is to use training 'modules', which reflect the various levels of fire safety responsibility within the organisation.

Practice your systems

9.10 The purpose of fire drills and evacuation exercises are to ensure that all staff are trained in the role they would play if a fire should occur. It is therefore good practice for a fire drill and evacuation exercise to be carried out at least once every six months, simulating fire conditions in which one or more of the normal escape routes from the building may be obstructed. During these exercises, a member of staff should operate the fire alarm and the fire routine should be fully rehearsed. Initially, until staff are fully familiar with the evacuation drill, advance notice should be given of the date and time of the drill so that staff are fully aware. As time progresses, and staff become more familiar with the evacuation procedures, they should be advised that in future unannounced evacuation drills will be carried out. In smaller properties, similar arrangements should be made to ensure that all staff know their roles and what they must do in the event of an emergency.

Fire Training Manual

9.11 This clause and most of the remainder of this chapter contains various ideas and suggestions for the production of a printed manual designed to cover all aspects of staff fire safety training within an organisation.

9.12 *Managing fire safety (fire safety procedures, training and fire manual)*

It is particularly suitable for owners and managers of larger organisations who have a responsibility to ensure that all of their employees receive adequate training in what they must do in the event of a fire.

The production of a Fire Training Manual for use by fire safety managers and trainers within an organisation will considerably improve the effectiveness of training by ensuring that the training material is always available and that the level and quality of the training is consistent. As time progresses and staff changes take place amongst the trainers, the provision of a manual will also ensure the continuity of the training given.

The format set out later is based on the concept that the manual is divided into a set of predetermined training 'modules'. These modules provide a basic syllabus for courses of varying duration, which reflect the variety of training necessary for those individuals given a designated fire safety role within the organisation.

To assist with the formulation of the appropriate level of training for those post holders with a higher level of fire safety responsibility, guidance is also provided on what their roles should ideally encompass, as part of the overall management of fire safety within the organisation.

In addition, it should be noted that the majority of the information recommended for inclusion within the various training modules, is contained within the chapters of this handbook.

Fire safety manager's role

9.12 In today's business environment, the employment of dedicated fire safety staff is becoming rarer. It is much more likely for companies to 'out source' their fire safety training requirements to specialist consultancy firms in common with many other services that companies need on an infrequent basis.

Companies consider that the employment of dedicated personnel is a sound, cost-effective method of ensuring the health and safety of their employees. Where it is the policy that all staff fire safety training is to be carried out in-house, it is likely that the task of carrying out the training will be given to a fire safety manager. He may have additional responsibilities, perhaps as part of a larger, overall health and safety brief. Or alternatively, he may well be given the task of ensuring that the training is 'cascaded' down through all levels of the organisation.

If the particular post holder has not been recruited from a fire safety background, it is advisable for the individual concerned to attend a recognised external course. There are many courses on fire safety available throughout the UK from various organisations that may include your local fire service.

However, if the local fire service cannot provide the necessary training, they will be able to advise an employer where it can be obtained.

Job specification

9.13 To assist companies who may wish to employ a fire safety manager, to determine the post holder's key responsibilities, the following suggestions are made for inclusion in a job specification or job description.

- Depending on the overall size of the workplace, the fire safety manger should appoint an appropriate number of fire marshals (sometimes also known as fire wardens, floor wardens or floor marshals) who, in turn, should each have deputies.
- Designate an area of the workplace for which each fire marshal will be responsible.
- Specify the fire marshal's duties in writing.
- Establish procedures to enable the fire marshal's reports to be actioned.
- Train the marshals and their deputies in their duties.
- Maintain continuity when fire marshals are on holiday, leave the company or are moved to another area.
- Control the work of contractors, which includes the issuing of any 'hot work' permits that may be necessary.
- Liase with the local fire service.
- Prepare a written fire precautions policy, which includes:
  - a fire safety management structure;
  - how to eliminate or reduce hazards;
  - ways to monitor all unavoidable hazards;
  - ways to prevent the spread of smoke and flames;
  - ways to ensure the safety of all staff and visitors;
  - the level and type of training staff should receive to ensure they know what to do in a fire;
  - measures to be taken to protect against disruption of business; and
  - any other measures which might attract savings in insurance premiums.

Fire safety manager – general responsibilities

9.14 The fire safety manager's general responsibilities are set out below.

9.14 *Managing fire safety (fire safety procedures, training and fire manual)*

1. To ensure that all escape routes from the premises are not blocked or obstructed and that emergency exit doors are maintained available for immediate use whenever people are in the workplace.

2. To take immediate action if, at any time when the property is occupied, emergency exit doors are found to be locked or obstructed. Self-closing 'fire doors' provided within the property, must not be held in the open position, except by an approved electromagnetic 'hold-open' device, operated from automatic smoke detectors linked into the fire alarm system.

3. To ensure that the fire alarm system is tested on a regular (weekly) basis and that the results of such tests are recorded in a fire safety log book maintained for the purpose. Any faults with the system should immediately be reported to a reputable company of competent fire alarm engineers. A comparable fire engineering company must properly maintain the fire alarm system, ie on an annual basis. All tests, faults and maintenance occurrences should be entered in the fire safety log book. Smoke and heat detectors as well as electromagnetic door 'hold-open' devices should also all be checked and the details similarly entered in the log book.

4. To ensure that the first aid fire-fighting equipment (fire extinguishers, blankets and hose reels etc) are kept readily available in allotted positions and always ready for use in case of fire. All fire extinguishing equipment must be checked at least once a year by a competent firm of extinguisher maintenance engineers and the details recorded in the fire safety log book.

5. To ensure that all emergency lighting units (where fitted) are operating correctly and that a monthly check is made to ensure that all units will illuminate under mains or circuit failure conditions. The results of all checks should also be recorded in the fire safety log book.

6. To ensure that all safety and emergency signs and notices are in good condition and, where necessary, either repaired or replaced. Fire instruction notices detailing the action to be taken in case of fire must be prominently posted throughout the premises. A designated fire assembly point must be nominated in case of a fire evacuation.

7. To ensure that the arrangements for calling the fire brigade and/or other emergency services are clear, precise and unambiguous.

8. To ensure that appropriate training is given on the operation of first aid fire extinguishing equipment and on the action to be taken in case of fire. A fire evacuation drill should be carried out at least once a year and the details of this, including attendees, should be recorded in the fire safety log book.

9. To ensure, as far as practicable, that all practices and procedures in the workplace are safe and take account of good fire safety advice and precautions and that the 'prevention of fire' is the aim of everyone in the

organisation. Any potentially dangerous practices identified should to be given urgent attention so that they can be immediately dealt with.

10. To take account of the ever-present risk of arson or wilful fire raising, taking appropriate steps to safeguard the workplace at all times.

General notes

9.15

- Anyone nominated to undertake the duties of a fire safety manager must have sufficient training, experience or knowledge to enable them to undertake these responsibilities.
- Where necessary, a deputy should be nominated in order to ensure continuity in the safety arrangements in the event of holidays, sickness and other absences.
- The duties detailed above are not listed in any particular order of priority and it will be up to the person concerned to determine the best method of ensuring they are carried out.

Fire marshals

9.16 A tried and tested method of ensuring that the fire safety message is spread throughout the organisation is to delegate some of the responsibility for fire safety down through all staff levels, by appointing floor, department or section fire marshals. The obvious candidates for this role are those members of staff such as foremen, supervisors and managers who already have a supervisory responsibility for a section of the organisation. In addition, because of their managerial roles in the organisation, they are in a better position to resolve any fire safety issues, as they occur. Their supervisory capacity will also ensure that employees will obey any instructions that are given by them.

The role of the fire marshal

9.17 The basis of this system is that the fire marshals are given a higher level of fire safety training than the average member of staff in order for them to be responsible for the fire safety within a designated part of the premises.

Their role should be to check for hazards and potential fire risks within their area and to check that first aid fire-fighting equipment and other fire safety equipment is working and in place.

They must also organise the evacuation of staff, the public and visitors in the event of a fire alarm. This would include a final sweep of each marshals'

9.18 *Managing fire safety (fire safety procedures, training and fire manual)*

designated area to ensure that everyone is out and to report accordingly to the person in charge of the assembly point.

To summarise this aspect, their role should be:

- to take appropriate and effective action if fire occurs;
- to ensure that escape routes are kept available for immediate use; and
- to identify and report fire hazards in the workplace.

Checklist

9.18

> If a fire is discovered, the fire marshal should:
>
> - ensure the fire alarm has been activated;
> - check that all equipment and processes are made safe;
> - evacuate staff from the building to the designated fire assembly point;
> - check that any staff or visitors with disabilities are assisted out of the building;
> - call the designated reporting centre or control room and give details of the location, severity and cause of the fire, if known; and
> - fight the fire if it is safe to do so.

Training module for fire marshals

9.19 It is suggested that preparing a one-day course, which should include practical sessions on first aid fire fighting, would be an appropriate way to train individuals for this level of responsibility.

Syllabus

9.20 The following suggestions are made as a basis for inclusion in the syllabus.

- Fire safety management within the company:
  - company fire safety policy;
  - fire marshal's role in the management of fire safety;
  - developing a 'fire-safe' environment for the firm; and

Managing fire safety (fire safety procedures, training and fire manual) **9.22**

- ○ company fire procedures.
- The nature of fire:
  - ○ the triangle of fire.
- Means of escape:
  - ○ fire warning;
  - ○ escape routes;
  - ○ emergency exits;
  - ○ signage;
  - ○ fire evacuation; and
  - ○ fire drills.
- Fire prevention:
  - ○ good housekeeping;
  - ○ fire safety checklist;
  - ○ hazard spotting; and
  - ○ arson prevention.
- Fire-fighting:
  - ○ first aid fire-fighting equipment;
  - ○ classification of fires;
  - ○ colour coding of extinguishers; and
  - ○ practical fire-fighting.

It will also be appropriate to include training in the syllabus on any special risks or hazards, which relate specifically to the department or premises.

Security staff

9.21 In larger premises, it is not uncommon for security staff to be employed on a 24-hour basis. Their primary role is generally to ensure the security of the premises round the clock.

Their fire safety role

9.22 Security staff can provide a vitally important contribution to the fire safety defence of any organisation. Employers often overlook this aspect, with the greater emphasis being placed on their primary role in the prevention of theft.

9.23 *Managing fire safety (fire safety procedures, training and fire manual)*

As security staff, normally patrol the protected premises at regular intervals, with additional training they can be a significant asset to the fire protection of a building. In their specialised role it is appropriate for them to be given a higher level of fire safety training than the average employee, which should include practical first aid fire-fighting. During normal business hours it makes good sense for the fire safety role of security staff to be integrated with that of the fire marshals.

Training module for security staff

9.23 *Course duration* – A one-day course which includes practical sessions on first aid fire-fighting would be sufficient to ensure that security staff can undertake fire safety responsibilities. However, in some major buildings, eg shopping complexes, they will have a vital role in assisting in the evacuation of the premises with particular regard to the wheelchair disabled, elderly and infirm persons etc.

Syllabus

9.24 The following is a suggested syllabus to train security staff on fire safety responsibilities and is essentially the same as the one used for training fire marshals.

- Fire safety management within the company – the same as on the syllabus for fire marshals (see **9.20** above) earlier in this chapter, but also including:
  - night and weekend patrols;
- The nature of fire (see **9.20** above);
- Means of escape (see **9.20** above);
- Fire prevention (see **9.20** above);
- Fire-fighting (see **9.20** above);
- Fixed fire-fighting and suppression systems;
  - purpose; and
  - method of operation.

General staff training

9.25 The policy objective of all businesses should be to ensure that all staff receive formal training within three months of their employment with the company. They should however receive a basic briefing on fire safety during their first day of employment.

Staff training should be formulated around the following basic criteria but should be tailored to suit the individual circumstances within an employer's business:

- a description and explanation of emergency procedures;
- particular hazards associated with the type of work carried on at your business;
- the duties and responsibilities of staff;
- the findings of any risk assessment carried out at your business under the *Fire Precautions (Workplace) Regulations 1997 (SI 1997 No 1840)*; and
- all instruction should be clear and easy to understand.

Training module – all staff

9.26 *Course duration* – A two-hour course would be appropriate to brief all staff on basic fire safety. However if it is considered necessary for all staff to receive practical sessions on first aid fire-fighting then the course should be extend to one half-day.

Syllabus (all staff)

9.27
- Fire safety management within the company:
  - company fire safety policy;
  - staff role in the management of fire safety;
  - developing a 'fire-safe' environment for the firm; and
  - company fire procedures.
- Fire warning:
  - action to take on discovering a fire;
  - how to raise the alarm and what happens next;
  - action to take upon hearing the fire alarm; and
  - the procedures for alerting members of the public and visitors including, where appropriate, directing them to exits.
- Means of escape:
  - evacuation procedures;
  - location of the escape routes, especially those not in regular use;
  - how to open all escape doors, including the use of any emergency fastenings;

9.28 *Managing fire safety (fire safety procedures, training and fire manual)*

- ○ the reason for not using lifts (except those that have been specifically installed or adapted for the evacuation of disabled people);
- ○ the fire assembly point and its purpose; and
- ○ fire drills.
- Fire prevention:
  - ○ importance of general fire safety and good housekeeping; and
  - ○ importance of keeping fire doors closed to prevent the spread of fire, heat and smoke.
- Fire-fighting:
  - ○ the location and the use of fire-fighting equipment; and
  - ○ practical fire-fighting (if considered appropriate).
- General:
  - ○ arrangements for calling the fire service; and
  - ○ where appropriate, how to stop machines, processes and how to isolate power supplies in the event of fire.

Staff induction briefing

9.28 It is recommended that newly inducted staff, which should include all casual and temporary staff, be provided with a 30-minute briefing session on fire safety, during their first day of employment. It may be that this briefing can be incorporated as a segment of the general induction training given to all staff when they join the company. In smaller companies, who may not have formal induction training, the new employee should be told what to do in the event of fire and shown where all the escape routes are. A simple but effective way of ensuring that this information is conveyed to the employee is for someone to physically walk with the employee through all of the emergency escape routes in the premises, pointing out the fire safety equipment en route.

Contractors briefing

9.29 Accidental fires are more likely to occur when contractors are working in a building.

It is important whenever construction work is carried out within premises (particularly if the premises are in normal use during the course of that construction work) that the fire safety policy is consistent with any policy set up by the contractors.

An employer should ensure that the fire warning and evacuation procedures for both the business employees and the contractors are dovetailed. This will ensure that if a fire occurs in either the occupied part of the premises or in the area of responsibility of the contractor, both sets of employees are made immediately aware so that a total evacuation takes place. These procedures should be established before the construction work commences.

It is essential that contractors follow clear rules for 'hot work' such as cutting, welding, brazing, grinding, sawing, soldering, thawing frozen pipe, and applying roof covering or sealing plastic shrink-wrap. This applies even in instances where the works do not form part of major construction work in the premises. It is important to include the contractors in any established formal permit-to-work systems in use within the company. This system should be established to ensure that there is no 'hot work' is commenced, by the contractors, without written authorisation from a competent person within the company.

A contractor's briefing should be carried out prior to any work being carried out on the premises. It is advisable to have a written briefing note, which can be given to contractors upon their arrival at the premises.

Trainers

9.30 Some of the information suggested for inclusion in the various syllabuses above is contained within other chapters of this handbook. Therefore it is a relatively simple process for trainers to obtain information from the relevant chapter and prepare 'ready-made' lecture notes. The information must, of course, be expanded and tailored so that it relates specifically to the premises concerned.

Continuity and refresher training

9.31 The nature, content and frequency of any staff refresher or continuity training will necessarily vary according to many factors, eg the nature of the premises concerned and the degree of fire risk contained within it. Other factors, which need to be taken into account are whether the premises are in multi-occupation, staff turnover, including temporary employees, part-time employees, structural alterations which may significantly alter the fire safety evacuation procedures, or changes in the company's fire safety policy that will need to be communicated to all employees and managerial staff.

The important point to remember, is that it will be necessary to carry out pre-programmed refresher training periodically. A good rule of thumb is that for average-risk premises refresher training should be carried out annually. For high-risk and sleeping-risk premises refresher training should be carried out twice a year. A one-hour session will normally be adequate for refresher training.

Maintaining a record of training

9.32 It is of vital importance to keep a permanent record of all fire safety training carried out by an employer's staff. There is a legal obligation for all companies with five or more employees, under the *Management of Health and Safety at Work Regulations 1992 (SI 1992 No 2051)*, to record this information.

For management purposes, it may also be appropriate to note that a particular employee has received training on their personal record file.

For smaller businesses all the staff training and instruction should be recorded in a fire safety log book.

In any event, whatever method is used to note the information the record should include the following details:

- date of the instruction or exercise;
- duration;
- name of the person giving the instruction;
- names of the persons receiving the instruction; and
- the nature of the instruction, training or drill/exercise.

Should the firm have fire teams?

9.33 Where a company's premises are located a long way from a fire station, the site is extensive, or is in a rural location it is possible to experience a wait of up to twenty minutes before the arrival of the fire service in an emergency.

In these instances an employer may consider it advantageous to set up an in-house 'fire team' from amongst the employees. A properly trained fire team can be a very effective way of extinguishing or controlling a fire in its early stages prior to the arrival of the fire service. It should be stressed, however, that it is extremely important that all members of any fire team receive a high standard of training. The local fire service should be consulted before fire teams are to be formed as they will normally be prepared to give advice on basic training, or they may even provide the training. However, for this service, they will probably charge a fee. Remember even if a 'fire team' is to be provided the fire brigade should always be called to any fire or suspicion of fire.

Damage control/salvage

9.34 By way of preparing a business for the worst possible scenario, a survival plan for recovery after a fire is vital. Any survival plan should seek to initiate salvage and damage control measures at a very early stage immediately after a fire. In some cases large quantities of water may be used during fire-fighting operations. This will inevitably cause additional damage to the building and its contents if no action is taken to mitigate the effects of water damage.

An employer may wish to incorporate salvage and damage control measures into staff training, perhaps as a segment on the course for fire marshals. Such a module should contain the following elements:

Salvage operations:

- initiate operations as soon as it is safe to do so;
- separate damaged from undamaged equipment, stock, etc;
- pump out any standing water;
- check electrical systems before starting up equipment;
- wipe down and cover equipment and stock;
- dry out, clean and test equipment;
- retrieve building plans, equipment specifications, selective information eg shop layouts;
- dehumidify damaged areas if needed;
- document the damage as recovery work proceeds;
- maintain or re-establish security during the recovery stage;
- establish surveillance to control looting and theft;
- set up physical access barriers as needed; and
- secure software and vital written records.

More information on the survival of a business after a fire is contained within **CHAPTER 14: FIRE (BEFORE, DURING AND AFTER)**.

Where to obtain assistance with training

9.35 A good starting point if you require information on the availability of local fire safety training is the local fire service. Many fire brigades now offer, on normal business terms, appropriate training for most types of commercial premises. Where the local fire service is unable to provide specific training themselves the fire brigade will normally be able to advise where the relevant training support can be obtained.

Should this approach be unsuccessful, there are many consultancy firms in the United Kingdom that can provide this service. Many of these companies will be listed in the Yellow Pages.

10 Managing fire safety (prevention of fire and liaison with authorities)

In this chapter:	
Introduction	10.1
Statistics	10.2
Legal obligation	10.3
Main causes of fire	10.4
A proactive approach	10.5
The 'triangle of fire'	10.6
Fuel	10.7
Health and Safety Executive (HSE)	10.39
Health and Safety Executive – what they do	10.40
Where to get help and advice	10.41

Introduction

10.1 Preventing fires should be considered as an essential facet of any business's overall health and safety risk management strategy. The intention of this chapter is to discuss the most common causes of fires in business and to offer advice on how to prevent them occurring in your business.

By investing the necessary time and resources needed to implement the advice, it is possible to eliminate many potential causes of fire. In doing so, the exposure of the business to the risk of a devastating fire, with all its consequential 'knock-on' complications, will be minimised.

Unfortunately, statistics indicate that businesses, which have not invested in fire prevention, may eventually have to pay the price and suffer the consequences.

In business today, it is inevitable that an employer will need to interact and liase with many different organisations and enforcing authorities. With regard to fire safety matters, the three primary organisations with whom an employer, owner or occupier will need to liase with are the fire service, local authority and the Health and Safety Executive ('HSE').

Statistics

10.2 In 2000, fire brigades in the UK attended more than 33,000 fires in premises, which could be regarded as falling into the category of workplaces, even where this was not the main use of the premises. These fires were responsible for the deaths of about 40 people and about 1,900 injuries.

Not only do fires take lives, they also cost money and jobs since many businesses do not fully recover after a fire and many never reopen.

The statistics clearly indicate the need for businesses to have a planned fire prevention programme.

In fairness it is necessary to state that these figures are actually now showing a slight downward trend. However, this improvement appears to be directly related to the fact that there are now significantly fewer fires that are, or were, caused by smoking materials. This would seem to indicate that the introduction of the 'non-smoking' workplace or the designation of restricted areas for smoking appears to be a major 'breakthrough' in fire prevention and consequently fire safety in workplaces.

Legal obligation

10.3 Under *Fire Precautions (Workplace) Regulations 1997 (SI 1997 No 1840)* and *Management of Health and Safety at Work Regulations 1999 (SI 1999 No 3242)*, if an employer employs one or more people, there is a legal obligation to ensure that employees are not exposed to any risk, in the workplace. The required 'risk assessment' of the premises, carried out under the Regulations, should include an assessment of the fire risk, any measures necessary to protect the employees against that risk, as well as others who may be affected by fire.

For further information about carrying out risk assessments, see **CHAPTER 7: HOW TO CARRY OUT A FIRE RISK ASSESSMENT**.

Main causes of fire

10.4 It is often said, albeit quite often in a light-hearted manner, that the three main causes of fire are men, women and children! The truth in this statement lies in the fact that fires generally start as a result of people's actions or conversely their lack of action.

10.5 *Managing fire safety (prevention of fire and liaison with authorities)*

When the list of the commonest causes of fires are considered it indicates that directly or indirectly they are all caused by people doing something that they should not do or by not doing something that they should have done.

The most common causes of fires in workplaces, ie, both industrial and non-industrial business premises, which are in no particular order of significance, are:

- fire raising and arson;
- careless disposal of cigarettes or matches;
- combustible material left near a heat source;
- accumulation of easily ignitable rubbish or paper;
- carelessness on the part of contractors, maintenance workers;
- electrical equipment left switched on when not in use;
- misuse of portable heaters;
- obstructing ventilation of heaters, machinery or office equipment;
- inadequate cleaning of work areas; and
- inadequate supervision of cooking activities.

The most effective way for an employer to prevent fires is to ensure that people who are employed or are involved in the business do things in a correct manner. Frequently it is ignorance on the part of employees involved that is the cause as they do not realise the potential consequences of their actions.

A proactive approach

10.5 It is sensible for an employer and employees to work together in a proactive way, in order to try to prevent fires from occurring. All have a vested interest in preventing fires. Protecting the workplace and instituting a sensible fire prevention regime regarding the risk from fire are also protecting employees' lives and jobs.

An employer can achieve this 'proactive approach' to the problem by advising, informing and training employees about the dangers of fire. These aspects are dealt with more fully in **CHAPTER 9: MANAGING FIRE SAFETY (FIRE SAFETY PROCEDURES, TRAINING AND FIRE MANUAL)**.

The 'triangle of fire'

10.6 Fire is a chemical reaction which requires three fundamental elements to be in place. This is referred to as the 'triangle of fire'. It may appear basic but remembering this can aid in fire prevention.

The three prerequisites for fire are:

- oxygen – which is therefore always present;
- fuel – the combustible substance, which can be a solid, a liquid or a gas; and
- ignition source or other source of heat energy.

As all three 'points' of the 'triangle' must be in place to cause a fire, it follows that an employer should attempt to prevent all three coming together. Employers and their employees should therefore strive to prevent sources of fuel and sources of ignition from coming together.

Fuel

10.7 Virtually anything that burns can be regarded as 'fuel'. An employer has only to inspect the premises to determine that there are substantial quantities of materials that can be regarded as 'fuel'.

In fact, the structure of the premises may well be made of materials which will readily burn when exposed to fire or flame, if for example the building is an older one which has been constructed with timber as opposed to a more modern one made from brick and concrete which does not burn.

Checklist

10.8

> The following list will assist an employer in understanding and identifying some of the potential sources of 'fuel', which may be found within the premises:
>
> - upholstered furniture, carpets and curtains;
> - office storage, filing and other storage cabinets;
> - wood and paper;
> - packaging materials, such as cardboard, expanded polystyrene foam, bubble wrap, wood and wool;
> - paints and adhesives;
> - white spirit, turpentine, thinners, solvents, petrol, paraffin and other flammable liquids; and
> - liquefied petroleum gas ('LPG'), acetylene and other flammable gases.

10.9 *Managing fire safety (prevention of fire and liaison with authorities)*

> The adoption of the good 'housekeeping' regime, detailed below at 10.20 will assist in reducing the risk to an acceptable level, by not allowing the accumulation of potentially flammable 'fuel'.

Ignition sources

10.9 Employers should inspect their premises to try to identify any potential sources of ignition that may be present. Some sources of ignition will be immediately apparent, but others may not be quite so obvious.

The following paragraphs should assist in identifying some of the potential sources of ignition that may be found within premises and offers solutions to minimise the risk.

Arson

10.10 This is becoming an increasingly serious problem and a number of suggested solutions can be found in **CHAPTER 1: CAUSES OF FIRE**.

Cigarettes and matches

10.11 Wherever possible a 'no smoking' regime should be established. However, this cannot always be achieved and the adoption of a sensible smoking policy for employees, which recognises the need for people to smoke, will eliminate some of the risk. It is important to provide a designated 'smoking' area or room that is 'set aside' for the purpose and is provided with suitable deep metal ashtrays. If an employer does not make suitable arrangements, it is inevitable that smokers may still smoke. Should this happen employees may well use unsupervised areas or storage cupboards that may contain flammable materials.

Faulty electrical and lighting equipment

10.12 The adoption of a planned ongoing maintenance policy will do much to prevent problems with electrical equipment. It is also important that when fuses are replaced, the correct fuse rating for the appliance is used.

Any naked flames

10.13 As these are normally used as part of a manufacturing process, management will be more than aware of these obvious potential sources of ignition.

Heating appliances

10.14 Stoves, boilers and portable heaters can all be potential sources of ignition. Employers should try to use only fixed heating installation if possible, rather than portable equipment. Where the use of portable heaters is unavoidable, they should stand on a non-combustible surface, be properly guarded and kept well clear of combustible materials. In this connection if portable LPG heaters are used, extreme care should be taken when changing and handling the fuel cylinders.

Hot processes

10.15 Welding, cutting and grinding work which can generate sparks, are always potential ignition sources. Where this is part of the normal manufacturing process, care should be taken to ensure that as a matter of policy, no combustible materials are allowed in the vicinity of the work. When this work is carried out on an infrequent basis, perhaps by maintenance staff or contractors, an employer should implement a 'hot work' permit-to-work system. This system is described more fully in **CHAPTER 13: SPECIALIST INFORMATION**.

Cooking

10.16 This is invariably a potential source of ignition, whether cooking is part of the manufacturing process in a food production plant, or carried out in order to prepare meals in a restaurant or a staff canteen. An employer should ensure that all cooking equipment is in a good state of repair and that all grease filters and flues are regularly cleaned. Staff should also exercise great care in using frying pans and deep-fat fryers. In addition, as part of your planned maintenance programme, all thermostats and other heat controls should be regularly checked for correct temperature operation.

Static electricity

10.17 This is a potential source of ignition that is often overlooked and not considered. In certain manufacturing processes, particularly when dealing with flammable liquids and finely divided 'dust like' materials, static electricity can be generated by friction and could be a potential source of ignition. For expert advice on this subject you are advised to consult the Health and Safety Executive (HSE).

Hot processes

10.18 When used continuously, some machines can generate high levels of heat that could be a potential source of ignition. If this is the case an employer

10.19 *Managing fire safety (prevention of fire and liaison with authorities)*

should ensure good air circulation around the machines and keep combustible materials at a safe distance. In addition, friction caused by damaged or loose belt drives and bearings could be another source of ignition. A planned maintenance and replacement programme will eliminate many of these potential problems.

Office equipment

10.19 There is an increasing amount of electrical and computer equipment being introduced into commercial and manufacturing premises. Equipment such as computers are very often left running on a continuous basis. As they generate heat, they are provided with cooling fans in order to dissipate the heat generated. However, if the ventilation openings become obstructed, either by materials being placed next to them, or more commonly by the build up of dust and fluff, the equipment may overheat and become a potential source of ignition. Good housekeeping practice and planned maintenance programmes should eliminate these problems.

Good 'housekeeping'

10.20 Poor housekeeping standards are the greatest single cause of fire. A carelessly discarded cigarette end, especially into a container of combustible waste or amongst combustible storage, often results in fire. The risk is higher in areas that are rarely in use as there is less likelihood of early detection of a fire. The fire develops and therefore becomes more difficult to extinguish.

Checklist

10.21 The following 'best practice' guidelines are recommended in order to minimise the risk.

- Where smoking is permitted, suitable deep metal ashtrays should be provided. Ashtrays should not be emptied into combustible waste unless the waste is to be removed from the building immediately. It is recommended that smoking should cease one hour before close of work, so that if smouldering occurs this will be detected before staff leave the premises.
- Combustible waste and contaminated rags should be kept in separate metal bins with close fitting metal lids.
- Cleaners should, preferably, be employed in the evenings when work ceases. This will ensure that combustible rubbish is removed from the building to a place of safety before the premises are left unoccupied.

Managing fire safety (prevention of fire and liaison with authorities) **10.23**

- Rubbish should not be kept in the building overnight, or stored in close proximity to the building.

- 'No smoking' areas should be strictly enforced, especially in places which are infrequently used, eg stationery stores, oil stores, or telephone equipment intake room. Suitable 'No Smoking' notices should be displayed throughout such areas.

- Where 'no smoking' is enforced due to legal requirements (for example, areas where flammable liquids are used or stored) or in areas of high risk or high loss effect, it is recommended that the notice should read 'Smoking Prohibited – Dismissal Offence'.

- Materials should not be stored on cupboard tops, and all filing cabinets should be properly closed, and locked if possible, at the end of the day.

- Ensure that the grease filters and flues on cooking equipment are regularly cleaned and kept in a good state of repair.

Liaison with authorities

10.22 The enforcing authorities have exacting standards and it is essential for all employers, owners or occupiers to ensure that as far as practicable their premises comply with all current standards.

Why liase?

10.23 Professionals working in the various organisations that deal with fire and health and safety matters have a wealth of knowledge and expertise behind them.

Although these agencies' primary role is to enforce legislation, they are generally only too pleased to share their knowledge and will freely give advice and information to help you solve your particular problem. All an employer has to do is ask. In many instances, seeking their advice can help an employer and the company save money.

Liaison is particularly relevant in today's legislative climate where the emphasis is currently on self-compliance with regard to fire safety legislation. The fire service, and, to some extent, the other enforcing authorities no longer police premises in the same way. Prior to current legislation, the enforcing authority would have inspected the premises and produced a detailed report on what an employer was required to change or improve in order to comply with the law.

An employer is now required to determine how to comply with legislation. It is entirely their responsibility.

10.24 *Managing fire safety (prevention of fire and liaison with authorities)*

As a consequence, it is all too easy for salespeople to sell systems and equipment that are unnecessary or inappropriate. The recommendation is to seek advice from an unbiased source such as the local fire service's fire safety department.

Fire Service

10.24 The *Fire Service Act 1947* places a duty on fire authorities to set up free advisory services on all matters relating to fire safety and prevention.

As a consequence, the fire service has traditionally been to the forefront in protecting life and property, together with ongoing campaigns to educate the public. The provision of advice and information on preventing fires with a view to educating business communities and the general public has always been a high priority to the fire service and one in which they have invested heavily.

At the same time, the fire service has an enforcement function. An increasing number of statutory enforcement duties have been imposed upon it in recent years.

The primary legislation for which the fire service has enforcement responsibility is the *Fire Precautions Act 1971*, which requires certain hotels and boarding houses, factories, offices and shops to hold a fire certificate. An extension of this legislation, *Fire Precautions (Workplace) Regulations 1997 (SI 1997 No 1840)*, covers virtually all premises with the exception of private dwellings in single occupation.

In addition, there is a mandatory requirement placed upon the fire service to give advice to those agencies responsible for the fire provision under many other pieces of legislation.

The fire service in most areas also administers legislation dealing with the storage of explosives and petroleum.

What is the legal responsibility?

10.25 If an employer is the occupier of premises with a fire certificate issued under the *Fire Precautions Act 1971* they are obliged to consult the fire authority before doing any of the following:

Checklist

10.26

- making any material extensions or structural alteration to the premises;

- making a material alteration in the internal layout of the premises;
- changing the layout of furniture or equipment within your premises; or
- keeping explosive or highly flammable materials in, on or under the premises.

When or where an employer is unsure whether or not the proposals constitute 'material' change, it is good practice to consult the local fire safety officer for advice on the matter.

In addition, should an employer have either petroleum or explosives licences they should consult and discuss any changes that are to be made with regard to the storage of petroleum and explosives with the fire authority.

Local authorities

10.27 The main contact with a local authority in relation to fire safety is likely to be as a result of proposed alterations, extensions or changes to buildings or premises. When an employer intends to make changes it may be necessary to make an application to either (or both) the Building Control or the Planning departments of the local authority.

What is Building Control?

10.28 The Building Control system exists to make sure that buildings are properly designed and constructed so as to ensure the health, safety, welfare and convenience of people using them. See also **CHAPTER 13: SPECIALIST INFORMATION**.

The principal areas covered by *Building Regulations 1991 (SI 1991 No 2768)* that were introduced in 1991 (which have subsequently been amended several times) are:

- structural stability – will the building safely carry anticipated loadings;
- fire precautions – both controlling the materials used according to the degree of risk and making sure that buildings could be evacuated without loss of life in the event of a fire;
- prevention of damp and condensation in buildings;
- sound resistance of walls and floors between dwellings;
- ventilation of habitable rooms and unheated voids;
- hygiene, sanitary appliances and drainage;
- heating appliances including the safe discharge of flue gasses;

- stairways, ramps and vehicle barriers;
- conservation of fuel and power, and properly insulated building etc;
- access to buildings and facilities for disabled people; and
- the safe positioning of glazing in windows and doors etc.

The above points relating to fire precautions, stairways etc and access to buildings are areas of particular relevance to fire safety of the workplace.

How does building control work?

10.29 Building control generally operates as a two-stage process, referred to as the full plans application. The first stage, known as the plan stage, involves the applicant submitting detailed plans for approval. These plans are checked by a building control officer to ensure that all necessary information is shown, and that it complies fully with *Building Regulations 1991 (SI 1991 No 2768)*. Wherever possible, applicants are given the opportunity to make any required amendments, before either an approval, conditional approval or rejection is given.

The second stage, the inspection stage, starts when work commences on site after which a series of site visits are made to check that the work proceeds in accordance with the plan, and therefore complies with the Regulations.

For more simple works an alternative type of application known as the building notice application is also available together with more specialised application types to meet specific needs. For more information about these types of application you are advised to contact your local building control department.

What is the planning department's role?

10.30 The following information will help you understand the role of your local authority's planning department and how it might relate to any building proposals that you may have for your business premises.

Town and country planning

10.31 Town and country planning in England and Wales is principally based on the provisions of the *Town and Country Planning Act 1990* as amended by the *Planning and Compensation Act 1991*. The main purpose of the system is to regulate the use and development of land and buildings in the public interest.

This is a complex task involving the consideration of many issues and balancing many competing interests.

It involves ensuring that:

- sufficient land is available to meet the community's needs for housing, industry, and business, commercial and recreational purposes;
- adjacent land uses are not incompatible;
- proposed land uses do not involve people in additional travel by car thus increasing CO_2 emissions;
- development occurs in locations that are capable of being served by existing infrastructure and services such as roads, sewers, mains etc;
- high quality landscapes and townscapes are protected from poor or unnecessary development;
- development which is permitted is of a good environmental and design standard; and
- providing a policy framework in which development decisions can be made.

The Office of the Deputy Prime Minister (ODPM) through the Planning Inspectorate, also determines appeals where planning permission has been refused and has the power to 'call in' major applications for determination by himself.

When is planning permission required?

10.32 Planning permission is required for any 'development'. Development is defined in the *Town and Country Planning Act 1990, section 55* as 'the carrying out of building, engineering, mining or other operations in, on, over or under land or the making of any material change in the use of any buildings or other land'.

Planning permission is normally not required for internal alterations unless the building is listed as being of architectural or historic interest or unless the work involves sub-division of residential accommodation or a change of use is involved.

However, permission may still be required under the *Building Regulations*.

Permitted development

10.33 The legislation does allow some house extensions and other minor development to be carried out without the express permission of the local planning authority. These works are referred to as 'permitted development' and are defined in the *Town and Country Planning (General Permitted Development) Order 1995 (SI 1995 No 418)* and various amendments. It is advisable to check with your local Planning department, as it is important that expert advice is obtained before proceeding with any development.

Even where planning permission is not required, consent may again be required under *Building Regulations 1991 (SI 1991 No 2768)*. Formal confirmation as to whether planning permission is required for a particular development should be sought in writing.

Types of applications

10.34 There are two main types of planning application.

Outline application

10.35 Outline applications are submitted where approval of the principle of the proposed development is required. This allows the applicant to get a decision on the type of development being proposed without having to go to the expense of having detailed plans prepared. If permission is granted, conditions may be applied relating to 'reserved matters' that are the details of siting, design, external appearance, means of access and landscaping. These must be submitted for approval within three years of the decision and the development should commence within five years of the granting of 'an outline' permission, or within two years of the date when the reserved matters were approved.

Full application

10.36 Full planning applications are submitted where all the details and information are provided. When planning permission is granted, it is generally valid for five years and development must commence within that time. Applications for change of use should be made as if they are full applications.

How to make a planning application

10.37 Anyone can make a planning application in relation to any piece of land, even land outside their ownership, provided that they have completed an appropriate certificate of ownership and, if necessary, have notified the owner of the land that they are making a planning application.

Application forms, notes for guidance and a scale of application fees can be obtained from your local planning department.

In order to avoid delays, it is important that the forms are correctly and fully completed, the correct fee is paid, and the application drawings are accurately drawn to a metric scale and are well presented. In many cases, it is sound advice to consult a member of the planning control staff before submitting a planning application. They will outline any possible conflict with local council's policies and suggest modifications to the proposal (where required). This will greatly assist the progress of the application once it is submitted and, in the long run, will save time and money.

Other areas where planning permission is required

10.38 An applicant is advised to seek advice from the planning department of the planning authority when considering carrying out any of the works listed below.

- Listed building applications are required for internal or external alterations or any development within the curtilage of any building listed as being of architectural or historic interest, or for demolition of all or part of a listed building.

- Conservation area – with certain exceptions, consent is required for the demolition of a building within a conservation area.

- Advertisements – the display of certain types of advertisement requires consent under the *Town and Country Planning (Control of Advertisements) Regulations 1992 (SI 1992 No 666)*.

Health and Safety Executive (HSE)

10.39 Where a business is in the manufacturing sector it will probably be necessary to consult the Health and Safety Executive on matters of fire safety, particularly where production involves hazardous or specialist processes.

In addition, the Health and Safety Executive has an enforcement role on all matters of health and safety across the whole spectrum of business and commercial organisations.

Health and Safety Executive – what they do

10.40 The role of the Health and Safety Executive is clearly set out in their own statement. According to the Health and Safety Executive an employer must:

> 'Ensure that risks to people's health and safety from work activities are properly controlled.
>
> The law says:
> - employers have to look after the health and safety of their employees;

10.41 *Managing fire safety (prevention of fire and liaison with authorities)*

- employees and the self-employed have to look after their own health and safety; and
- all have to take care of the health and safety of others, for example, members of the public who may be affected by their work activity.'

The Health and Safety Executive then add:

'Our job is to see that everyone does this.

The Health and Safety Executive is interested in the health and safety of people at work – that includes people who may be harmed by the way work is done (for example because they live near a factory, or are passengers on a train). In some situations, we are also concerned with the way work affects the environment.

The Health and Safety Executive develop new health and safety laws and standards, and play a full part in international developments, especially in the European Union and:

- inspect workplaces;
- investigate accidents and cases of ill health;
- enforce good standards, usually by advising people how to comply with the law, but sometimes by ordering them to make improvements and, if necessary, by prosecuting them;
- publish guidance and advice;
- provide an information service;
- carry out research;
- carry out various activities such as nuclear site licensing and accepting off shore installation safety cases.

This describes what we do in general terms.'

Where to get help and advice

10.41 For more detailed information relating to some specific activities within the HSE contact the following.

Health and Safety Commission, HSE Infoline, Caerphilly Business Park, Caerphilly, CF83 3GG. Tel: 08701 545500; fax: 02920 859260. Website: www.hse.gov.uk; e-mail:hseinformationservices@natbrit.com

Health and Safety Executive, HSE Infoline, Caerphilly Business Park, Caerphilly, CF83 3GG. Tel: 08701 545500; fax: 02920 859260. Website: http://www.hse.gov.uk; e-mail:hseinformationservices@natbrit.com

Health and Safety Executive office addresses and telephone details (open 9am to 5pm – Monday to Friday).

Health and Safety Executive contact points for specific activities.

- Domestic Gas Safety

  Where an employer wants advice about domestic gas safety, phone (free): Health and Safety Executive Gas Safety Advice Line. Tel: 0800 300363.

- Mining

  Mines Inspectorate Room 611, Daniel House, Trinity Road, Bootle, Merseyside, L20 7HE. Tel: 0151 951 4133.

- Railways

  Railway Inspectorate, 2nd floor SW, Rose Court, 2 Southwark Bridge, London, SE1 9HS. Tel: 020 7717 6533.

- The Nuclear Industry

  Nuclear Safety Directorate, Information Centre, St Peter's House, Balliol Road, Bootle, Merseyside, L20 3LZ. Tel: 0151 951 4103.

- The Offshore Oil and Gas Industry

  Offshore Safety Division, Information Centre, Lord Cullen House, Fraser Place, Aberdeen, AB25 3UB. Tel: 01224 252500; fax: 01224 252662.

- The Manufacture, Transport, Handling and Security of Explosives

  Explosives Inspectorate, St Anne's House, Stanley Precinct, Bootle, Merseyside, L20 3RA. Tel: 0151 951 4025.

- The Manufacture, Processing and Storage of Chemicals and other Onshore Major Hazards including Gas Transmission and Distribution, Pipelines, and the Road Transport of Dangerous Substances.

  Contact the Health and Safety Executive offices or enquiries may also be directed to:

  Hazardous Installations Directorate Secretariat, 6th Floor, St Anne's House, Stanley Precenct, Bootle, Merseyside, L20 3RA. Tel: HSE Infoline (08701 545500).

- Health and Safety Executive's Research Facilities

  Health and Safety Laboratory, Information Manager, Broad Lane, Sheffield, S3 7HQ. Tel: 0114 289 2920; fax: 0114 289 2830. Website: http://www.hsl.gov.uk; e-mail: hslinfo@hsl.gov.uk

10.41 *Managing fire safety (prevention of fire and liaison with authorities)*

- Public Enquiries

    Note that the Health and Safety Executive Infoline Tel: 08701 545500 Public Enquiry service is available between 8am to 6pm Monday to Friday. Website: www.hse.gov.uk; e-mail: hseinformationservices@natbrit.com

    Or write to:

    HSE Infoline, Caerphilly Business Park, Caerphilly, CF83 3GG

11 Managing fire safety (fire defence including active and passive fire precautions)

In this chapter:

Introduction	11.1
Means of escape – the principles	11.2
Means of escape – structural protection	11.5
Means of escape – measures to facilitate	11.10
Fire warning	11.24
Fire alarms	11.25
Automatic fire detection systems	11.26
Fire-fighting extinguishers – types of fire	11.31
Types of fire extinguisher	11.33
Hose reels	11.45
Automatic sprinklers	11.46
Fixed fire suppression systems	11.51
Foam inlets	11.54
Dry risers	11.55
Wet risers	11.56
Emergency procedures – staff training	11.57
Evacuation drills	11.61
Fire action notices	11.62
Preventative measures – planned maintenance programmes	11.66
Portable fire-fighting equipment	11.69
Fixed fire-fighting equipment	11.70

11.1 *Managing fire safety (fire defence including active and passive fire precautions)*

Fire alarm systems	11.71
Signs and notices	11.72
Emergency lighting	11.73
Security staff and systems	11.74
Insurance	11.75

Introduction

11.1 Over the last forty years, or so, there has been a considerable increase in the volume of fire safety related legislation introduced by successive governments, which impacts directly on businesses and their premises. Most of the legislation has been introduced primarily in response to a series of fatal fires, but additionally as a result of the increasingly complex society in which we live and work.

The net result is that the fire protection industry has grown in proportion to the growth in fire safety legislation. This is driven by the need to provide tailored solutions to the problems of protecting buildings and their occupants from fire. The industry now produces a very wide range of sophisticated fire defence products designed to assist architects and the owners of buildings in complying with their legislative responsibilities.

This chapter is designed to be a ready source of reference, in which each system of fire defence is briefly described. A fuller description has been included in respect of certain fire defence measures that are of growing importance (such as automatic sprinklers) but where there is little information readily available. The aims are to provide the reader with a basic understanding of the terminology involved and, where appropriate, sources of further information have been included. To gain a more in-depth appreciation and knowledge of any particular aspect of fire defence it will of course be necessary for the reader to consult the relevant British Standard, Codes of Practice or other guidance document.

There are two basic systems of fire defence in buildings, ie active and passive measures. Active measures in general terms relate to any system where some form of positive action is used in fire defence, eg an automatic sprinkler system. Passive measures are of a constructional type, eg compartment walls and floor. These aspects are developed later in this chapter.

For ease of reference the various systems of fire defence have been divided into the type of defence and protection that they are designed to provide, as indicated below.

Means of escape – structural protection

See **11.2–11.8** below.

11.1A These are the measures used to provide physical structural protection to enable people to escape from a building. Some of these measures, in addition to providing protection for people escaping from a fire within a building, also serve to contain and restrict any fire, which may occur within that building. They generally provide structural fire protection to a building and its components and will usually have been installed to comply with the *Building Regulations 1991 (SI 1991 No 2768)* (in England and Wales) ('Building Regulations'), (Approved Document B 2000 Edition), the *Building Standards (Scotland) Regulations 1991 (SI 1991 No 158) (as amended)* or the *Building Regulations (Northern Ireland) 1994 (SI 1994 No 243)*.

Means of escape – measures to facilitate

See **11.9** below.

11.1B These are measures that are not necessarily of a structural nature but designed to assist to enable the occupants of a building to escape safely.

Fire warning

See **11.23** below.

11.1C This consists of the measures used for detecting and/or giving warning of a fire to the occupants of a building.

Fire-fighting

See **11.30** below.

11.1D This section includes measures used for 'first aid' fire-fighting, such as portable fire extinguishers and hose reels. It includes both automatic fixed fire-fighting equipment, such as sprinklers, fixed installations and equipment to be used by the fire service to fight fires.

Emergency procedures

See **11.64** below.

11.1E Measures which form part of an overall strategy of preparing and training the occupants of a building on how to react to an emergency situation.

11.1F *Managing fire safety (fire defence including active and passive fire precautions)*

Preventative measures

See **11.73** below.

11.1F Measures, which are either designed to prevent fires occurring or to minimise their effects.

Means of escape – the principles

11.2 The term '*means of escape*' in a fire safety context refers to the means that are provided in a building to enable people to escape safely and quickly from a fire, by their own unaided efforts.

It generally refers to the routes within premises that occupants must use in order for them to reach a place of safety, in the open air, away from any danger. Ideally, these routes should be short enough to enable everyone in the premises to escape quickly and as safely as possible. The primary objective with any means of escape scheme will be to provide for escape in more than one direction, unless the distances are very short, from any point in a building. The reasoning behind this is that if one route is unavailable due to a fire, or the location of a fire, people within the building will be able to make their escape via alternative routes, leading to separate exits.

Unsatisfactory means of escape

11.3 The following are generally considered to be unsatisfactory for means of escape purposes and therefore should *not* be used in the event of a fire:

- lifts;
- portable ladders;
- long spiral staircases;
- escalators; or
- lowering lines or other 'throw out' devices.

Travel distances

11.4 In order to ensure that buildings are provided with adequate means of escape the design of buildings should take into account the distance that a person must travel before reaching a place of safety, or relative safety. A place of relative safety can be regarded as a protected route, which leads ultimately to a place of safety. The term '*protected route*' refers to a route within a building which is physically protected from the remainder of the building by fire resisting construction and fire resisting self-closing doors which ultimately leads to a place of safety outside of the building in the open air.

Managing fire safety (fire defence including active and passive fire precautions) 11.4

In calculating the maximum travel distances allowed for the various types of building occupancy, the following three stages, in one form or another, are normally taken in to account:

(i) travel within rooms;

(ii) travel from rooms to a stairway or storey exit; and

(iii) travel within stairways to a final exit.

The maximum amount of permitted travel distance allowed will vary according to the use and occupancy of a building and the level of fire risk contained within it.

Further guidance on the recommended travel distances for specific types of premises can be found in the following publications.

Type of building:	Publication:	
New buildings.	*'Building Regulations 1991: Approved Document B: Fire Safety'* (2000 Edition)	Published by the Department of the Environment, Transport and the Regions ISBN 1 851123512
Existing factories, offices, shops and railway premises that require a fire certificate under the Fire Precautions Act 1971.	*'Guide to Fire Precautions in Existing Places of Work that require a Fire Certificate; Factories, Offices, Shops and Railway Premises'*	Published by the Stationery Office ISBN 0 11 341079 4
Existing hotels, guesthouses and bed and breakfast type premises.	*'Guide to Fire Precautions in Premises used as Hotels and Boarding Houses which require a Fire Certificate'*	Published by the Stationery Office ISBN 0 11 341005 0
Existing places of entertainment and assembly, such as nightclubs, dance halls and village halls used for public entertainment.	*'Guide to Fire Precautions in Existing Places of Entertainment and Like Premises'*	Published by the Stationery Office ISBN 0 11 340907 9

11.5 *Managing fire safety (fire defence including active and passive fire precautions)*

Means of escape – structural protection

11.5 Virtually all escape routes within a building, which include stairways are enclosed and provided with appropriate levels of structural fire protection. The generally accepted minimum of fire resistance is 30 minutes.

Compartmentation

11.6 Passive fire protection is provided in buildings where part of the structure of a building is constructed of inherently fire resistant materials. Most buildings are divided into fire-resisting compartments, which are usually built of brick, block or concrete. They serve one of two functions: either to protect property or to save lives. Compartments are designed to:

- contain and limit the spread of fire; or
- to provide physical fire separation within a property in order to protect escape routes by enclosing the escape route within a fire-resisting compartment.

With the exception of small buildings, internal fire escape stairways are normally contained within fire-resisting compartments. Persons are not always aware of the existence of these compartments until they appreciate that access from one compartment to another is only possible by means of fire resisting doors. The doors in these openings are nearly always self-closing fire resisting door sets. It is essential that these doors are not obstructed and are allowed to self-close freely at all times, in order to preserve the fire resisting integrity of the compartment. Any glazing in the doors must also be fire resisting and, if damaged, must be repaired to the appropriate standard.

Fire resisting doors

11.7 Fire resisting self-closing doors, which, in this section are termed 'fire doors' are a vital and essential form of fire defence in both the protection of the structure of a building and the protection of the occupants. Fire doors are almost universally found in business premises, with perhaps the exception of very small single storey buildings. They are normally provided to carry out one of two functions, either to:

- protect the integrity of a structural fire compartment; or
- to protect the means of escape for the occupants of a building for a sufficient period of time for them to escape safely from the building.

Fire doors will have been specified for a building either to comply with the requirements of the relevant Building Regulations when the premises were built or, in older buildings which were not built to modern Building Regulations standards, to comply with fire safety legislation, as a requirement of the appropriate enforcing authority.

For new buildings or buildings being altered or extended, the requirements relating to fire safety are set out within '*Building Regulations 1991 (SI 1991 No 2768): Approved Document B*' (2000 Edition). This document requires that fire resisting doors must be fitted so as to protect escape routes or to preserve structural compartmentation. The exact requirements are as follows:

- Approved Documents B1 and B3 give the locations where fire doors are required;
- Appendix B gives the specific requirements for fire doors; and
- Table B1 defines performance requirements.

For *SI 1991 No 2768* purposes doors are classified in standard periods of fire resistance, the most common being:

- FD30: which will provide up to 30 minutes structural integrity plus 30 minutes insulation; and
- FD60: which will provide up to 60 minutes structural integrity plus 60 minutes insulation.

The suffix '**S**' (for example in 'FD30S') indicates a door which is also required to resist the passage of 'cold smoke' which is the smoke produced in the early stages of a fire. Cold smoke presents a significant risk as it reduces visibility and constitutes a toxic hazard for the occupants. There are various types of smoke seals available from manufacturers that are also available for upgrading existing fire doors to provide or improve their smoke-stopping capability.

All manufactured fire doors must have been tested in accordance with the requirements of the British Standard 'BS 476: Fire tests on building materials and structures'. The manufacturer will have obtained a test certificate from an independent testing body stating that a door meets the particular requirements for which it has been tested. The tests are very stringently controlled, they must be carried out on the whole door assembly – which includes the door itself, the door frame, any permitted glazing and any additional fixtures and fittings provided with the doorset.

All modern manufactured fire doors are fitted with '*intumescent*' edge strips around either the door, or the doorframe lining. The term *intumescent*, in this context, refers to a proprietary material which when exposed to heat or flame will expand and provide a protective heat insulation layer. This protective formed layer will then insulate the material beneath it from the fire. When exposed to heat the intumescent strips in fire doors will also expand and fill the gap between the door and the frame to provide physical, smoke and fire-stopping protection.

11.8 *Managing fire safety (fire defence including active and passive fire precautions)*

Although FD30 and FD60 are by far the most commonly used fire doors within commercial premises, doors with a higher time rating may be found in certain circumstances, perhaps to provide suitable protection to a specific risk within a building or as a requirement under *SI 1991 No 2768* for compartmentation purposes. Fire doors that provide this higher level of time protection are normally of metal construction.

In certain difficult situations, perhaps in a historically sensitive building that requires the provision of a fire door, there are a number of proprietary products available, which use a combination of fire resisting materials and intumescent paints, to achieve a measure of fire resistance. This type of treatment is suitable for adapting existing architecturally attractive doors that may be situated in non-standard door openings. However, it must be stressed that before work of this nature is carried out full consultation should take place with the appropriate enforcing authority.

Automatic fire door releases

11.8 Fire doors are fitted throughout commercial premises to prevent the spread of smoke and fire. To be effective they must be kept shut, except when they are actually being used, which is why they are always provided with a self-closing device. This requirement is very often compromised in busy commercial environments where there is a constant need for people to pass through the doors as part of their daily business. When doors are in constant use, perhaps by employees carrying goods and materials, or to facilitate the movement of disabled people, the continual opening and closing of the doors becomes irritating or impractical.

This frequently results in the fire doors being wedged or propped open. It is not uncommon for the nearest fire extinguisher to be used for this purpose, which, from a fire safety point of view, adds insult to injury.

The solution to this problem, and one that is becoming increasingly used is for fire doors to be held in the open position by automatic releases, eg electro-magnetic devices. These devices are then linked either to smoke detectors installed in the areas on both sides of the door or are linked to the fire alarm system and, where installed, the automatic fire detection system provided for the building. On the operation of the smoke detectors, a manual fire alarm call point or the automatic fire detection system the doors are released and close to fulfil their role as 'fire doors'.

There are a number of different types of electro-magnetic devices available from manufacturers, some of which are contained within the actual self-closing mechanism for the doors. In addition, there are a number of self-closing devices available which will allow a delayed closing of the door to allow a disabled person in a wheelchair or the elderly sufficient time to negotiate the door.

Where premises have not been provided with these devices and the occupants of the building are inclined to wedge fire doors open, it is perhaps worth considering fitting them with automatic releases. This will not only protect the original investment in the fire doors but should ensure that if fire breaks out an employer will know that the doors will be closed and that the spread of fire and smoke will be contained. It is important to remember and to 'highlight' during staff fire safety training sessions that a fire door can only fulfil its function when *it is closed*.

The British Standard BS 5839: 'Fire detection and alarm systems for buildings Part 3: 1988 – Specification' for automatic release mechanisms for certain fire protection equipment', provides more information and detail about these types of devices.

Staircases

11.9 Buildings are obviously provided with staircases for normal 'day-to-day' access to the upper or lower floors. However, they may also be fire protected and designated as emergency staircases to provide means of escape in case of fire. In addition, some buildings are provided with staircases purely for emergency purposes, which may be internal or external. It is fair to say that the vast majority of buildings will require at least two staircases.

Internal emergency staircases will normally be as follows.

- Totally enclosed and physically separated from the remainder of the building by fire resisting construction and fire resisting self-closing doors. The fire resistance provided will normally be to a minimum standard of 30 minutes.

- Part of a continuous route that ultimately leads to a place of safety outside the building, ie in the open air.

- At least 800 mm in width, but wider, when necessary, to cope with the maximum number of people likely to use the staircase at any one time.

- Provided with a protected by-pass arrangement if, without the by-pass, people would have to pass through the staircase enclosure in order to reach an alternative escape route.

External emergency staircases will normally be as follows.

- Enclosed, in order to provide weather protection. However there are a large number of external emergency staircases in use that are not enclosed as they were built prior to the introduction of this requirement.

- Provided with access from the building only by means of a thirty-minute standard, fire resisting, self-closing door. The only exception being the access door leading into the staircase from the top floor.

11.10 *Managing fire safety (fire defence including active and passive fire precautions)*

- Provided with a thirty-minute standard of fire resistance to any glazing, or any other openings, beneath or within 1.8 metres horizontally of the external emergency staircase. Any glazing within the specified area should be fixed permanently shut.

Means of escape – measures to facilitate

11.10 There are a range of measures that have to be applied to ensure that the means of escape routes within a building are available and there location can be identified at all times.

Escape routes and exits

11.11 The emergency escape routes and exits provided in businesses throughout the United Kingdom are not always obvious or straightforward. Fire signage for 'way finding' is vital. In some premises, eg shopping complexes precise detail relating to persons always being within a defined distance of a fire exit sign is of great assistance. In this particular example intermediate signs should be provided so that no part of a public common area is more than 25 metres from an exit sign or directional exit sign. This guidance for shopping complexes is particularly helpful, as it is an example of premises where the number of persons can be extremely large.

Ensuring that escape routes and exits are always available in an emergency can be a formidable task for those responsible for the safety of employees. However, the following checklist should assist the owners and occupiers of business premises to discharge their statutory duty by ensuring that the escape routes and exits within their premises are always available to provide speedy and safe evacuation.

Checklist

11.12

Escape routes must:

- be kept clear of obstruction, goods and storage at all times;
- lead as directly as possible to a place of safety, in the open air, outside of the building;
- be clearly indicated by pictorial exit signs and any necessary directional exit signs;
- be provided with emergency lighting where required;
- not use revolving or sliding doors as part of the escape route;

Managing fire safety (fire defence including active and passive fire precautions) **11.14**

> - only have doors within the escape route that open in the direction of exit from the building;
>
> - have the equipment and contents of any occupied room organised and arranged so that there is unhindered access to the general exit or to the exits from the room.
>
> If, in the normal course of the 'day-to-day' business, the constant movement of goods, materials and equipment causes obstructions then clearly defined painted (hatched) or guarded pathways to the exits should be provided across the room.

Emergency exits

11.13 Emergency exits that are normally, but not necessarily, the final exit from a building, have been provided in virtually all commercial premises. They will have been provided as a requirement of either the *Building Regulations 1991 (SI 1991 No 2768)* (when the building was built or altered) or by the enforcing authority of the particular fire safety legislation relevant to the premises.

The following checklist will assist the owners and occupiers of business premises to ensure that they discharge their statutory duty with regard to the emergency exits provided within their premises;

Checklist

11.14

> - Emergency exit doors should never be locked whilst persons are in the building. Where for security reasons, the doors have to be locked, panic bolts or panic latches should be fitted. The essential requirement is that the doors should be capable of being opened from the inside (without the aid of a key) at all times that the building is occupied. Many enforcing authorities will not normally accept keys in glass boxes. When emergency exit doors are locked for security reasons, outside of normal business hours, account must be taken of any cleaners or security staff who may be in the premises at night and at weekends.
>
> - They must be kept free from obstruction and storage at all times.
>
> - The minimum width of an emergency exit door should be 750 mm and if it is likely that the disabled in wheel-chairs will use them they should be at least 800 mm in width.
>
> - Emergency exit doors should open outwards, particularly if the door leads from an area of high fire risk or if the door is likely to

11.15 *Managing fire safety (fire defence including active and passive fire precautions)*

> be used by more than 50 people. Inward opening doors may be acceptable for a small number of people, subject to the approval of the enforcing authority.
>
> - They must be clearly visible by the provision of pictorial exit signs.
>
> - They should not open directly onto steps or stairs without the provision of a stepping flat area immediately in front of the open door, as defined in the *Building Regulations 1991 (SI 1991 No 2768)*.
>
> - If large numbers of people are likely to use the exit, there must be ample space immediately outside of a final exit door to allow people to disperse quickly from the area.

Inner rooms

11.15 Occupants of a room within another room may not realise that a fire has broken out in the outer room and they may therefore become easily trapped by the fire. This arrangement is quite common in office buildings and offices within factories. To overcome this problem one or more of the following solutions is normally acceptable.

- The door or wall of the inner room is provided with a clear glazed vision panel.

- The outer room is provided with an automatic smoke detector linked to the building's fire alarm system. In some instances, where the fire risk in the outer room is low, such as an office, it may be acceptable for a domestic type battery operated smoke alarm to be provided in the outer room.

- The provision of a gap of at least half a metre between the wall of the inner room and the ceiling of the outer room.

As a reminder of an employer's, occupier's or owner's statutory obligations, if the premises or building have been issued with a fire certificate under the *Fire Precautions Act 1971*, the following aspects will be essential requirements of the certificate:

Checklist

11.16

> - all doors affording means of escape in case of fire should be maintained and readily available at all times that people are on the premises; ➤

Managing fire safety (fire defence including active and passive fire precautions) **11.17**

> - all doors not in continuous use affording a means of escape in case of fire should be clearly indicated; and
> - sliding doors should clearly indicate the direction of opening.
>
> Doors should be adequately maintained and should not be locked or fastened in such a way that persons leaving the premises cannot easily and immediately open them. Moreover, all gangways and escape routes should be kept clear at all times.

Emergency lighting

11.17 Emergency lighting that is sometimes referred to as escape lighting is provided in buildings in addition to the ordinary normal or primary lighting, which is supplied by an electricity company.

In the event of a fire affecting local sub-circuits within escape routes, those circuits will fail and people using the escape routes will find themselves in darkness. Many buildings are provided with automatic generators to ensure lighting is provided during a 'power cut' which enables the business to carry on as normal. However because a building has a generator does not automatically mean that emergency lighting will work if there is a fire. Regardless of the source of power (whether it be from an electricity company or from a generator) the lighting is provided using the *same circuits* so if the circuits are fire-damaged then there will be no emergency lighting. A failure of this type would be a relatively rare event.

It is possible to install an emergency lighting system that is either run by an automatic generator or by a central battery system. The system must provide the lighting points via an entirely separate circuit from that used by the primary lighting. In addition, the wiring circuit must use fire resistant cabling systems. This therefore makes such a system expensive to install.

For this reason, when emergency lighting is required for means of escape purposes the most common type used is the relatively inexpensive, self-contained unit type that has its own internal battery. These units are wired into the local lighting circuit, within the escape routes, which provides a constant trickle charge to the unit's internal battery. The units are normally provided with a small neon charging light located within the unit that indicates that charging is taking place. If for any reason the normal electricity supply is interrupted each individual lighting unit will automatically be illuminated. The units are provided with batteries that are capable of providing continuous illumination for anything from one to three hours.

11.18 *Managing fire safety (fire defence including active and passive fire precautions)*

When should it be provided?

11.18 Generally, in premises that have only a day time occupancy, emergency lighting will only be necessary when there is not, or may not, be sufficient natural light for people to make their way out of a building safely should the primary lighting fail.

The need for escape lighting is greater in buildings that are occupied by people who are unfamiliar with the building and its means of escape – such as in shops, hotels etc.

The *Building Regulations 1991 (SI 1991 No 2768): Approved Document B* (see also **11.6** above) suggests that the following situations should be considered for the provision of emergency lighting in daytime occupancy type premises where:

- there is underground or windowless accommodation;
- stairways are situated in the centre of buildings;
- there are long internal corridors greater than 30 metres in length; or
- there are open plan offices exceeding 60 square metres.

Siting

11.19 Emergency lighting units are normally sited throughout escape routes and spaced according to the manufacturer's instructions, or the appropriate British Standard. As sudden darkness can potentially cause panic it is essential that adequate units are provided to ensure sufficient illumination for the whole of the escape route. Where possible, lighting units should illuminate fire alarm call points and fire-fighting equipment and emergency exits.

In places of public assembly or entertainment such as nightclubs, discotheques, cinemas and theatres, 'sustained' emergency lighting units are generally provided above each of the fire exit doors. These 'sustained' types of unit are provided with two internal bulbs, one that is illuminated at all times by the normal primary lighting circuit and, the other being powered from the internal trickle charged battery. This system ensures that there is uninterrupted illumination in areas where there are large numbers of people and where there is a possibility people will panic.

Exit signs

11.20 The provision of exit signs to indicate doors and to indicate escape routes are an important aspect of fire defence in any building. They are particularly important in premises in which people are not familiar with the means of escape provided, such as shops, public buildings, hotels and places of

Managing fire safety (fire defence including active and passive fire precautions) **11.22**

public assembly and entertainment. Their primary role is to guide and help people to find a pre-determined safe escape route from a building.

Directional exit signs are also used on longer more complex escape routes which also incorporate directional arrows – 'up' for straight on and 'left, right or down' – according to the direction of the route to be taken.

Advice on the use of all emergency signs, including exit signs, can be found in the Health and Safety Executive (HSE) publication, '*Safety Signs and Signals*', ISBN 0 7176 0870 0, which gives guidance on the *Health and Safety (Safety Signs and Signals) Regulations 1996 (SI 1996 No 341).*'

These Regulations bring into force the *European Union Safety Signs Directive (92/58/EEC)* that sets out the requirements for the provision and use of safety signs at work. The purpose of the Directive is to encourage the standardisation of safety signs throughout the member states of the European Union, in order that safety signs, wherever they are seen in Europe, have the same meaning.

SI 1996 No 341 applies to all places and activities where people are employed, but exclude signs used in connection with the supply of dangerous substances, products and equipment, or the transport of dangerous goods.

SI 1996 No 341 states that from 1 April 1996 all emergency exit signs must include a pictorial running man symbol, with or without directional arrows, as appropriate. Text-only signs will be deemed not to comply with *SI 1996 No 341* after the 24 December 1991.

Other emergency signs

11.21 As well as specifying signs for emergency exit purposes, *Health and Safety (Safety Signs and Signals) Regulations 1996 (SI 1996 No 341)* specify minimum requirements for safety signs and signals in general which are designed for communicating health and safety information to employees and members of the public. These Regulations cover the use of illuminated signs, signs for hand and acoustic fire alarms, spoken communication alarms and the marking of pipework containing dangerous substances. Fire safety signs for fire-fighting equipment are also covered by these regulations.

Many of these signs are new concepts for the United Kingdom, although *SI 1996 No 341* does specify the traditional hazard signs that are commonly used in the United Kingdom, such as prohibition and warning signs.

What do the Regulations require?

11.22 They require employers to provide specific safety signs whenever there is a risk that has not been avoided or controlled by other means, for

11.23   *Managing fire safety (fire defence including active and passive fire precautions)*

example by engineering controls and safe systems of work. Where a safety sign would not help to reduce that risk, or where the risk is not significant, there is no need to provide a sign.

They also require employers to maintain the safety signs, provided by them, and to ensure that they provide instruction on the meanings of appropriate safety signs to their employees and tell them what they need to do when they see a safety sign.

Do existing signs need to be changed?

11.23   In the case of fire safety signs, where employers decide that a previously acceptable sign is not of a type referred to in the *Health and Safety (Safety Signs and Signals) Regulations 1996 (SI 1996 No 341)* they must have replaced it by the 24 December 1991. All other fire safety signs, including any new signs provided, must now meet the requirements of the Regulations.

Advice on the use of fire safety signs can be obtained from the fire safety officer of your local fire service. In general, *SI 1996 No 341* will not require any changes where existing fire safety signs containing symbols comply with British Standard 5499: Part 1: 2002 'Graphical symbols and signs – Safety signs, including fire safety signs', which may have been provided in order to comply with the requirements of a fire certificate, issued under the *Fire Precautions Act 1971*. This is because the signs in BS 5499, although different in detail to those specified in *SI 1996 No 341*, follow the same basic principle and are therefore considered to comply with the Regulations.

Fire warning

11.24   Fire warning can range enormously from the very simple single point alarms found in very small commercial premises to the very sophisticated computerised, automatic, intelligent, addressable systems found in larger premises, which enable you to pin-point the exact fire or false alarm location.

As a consequence of modern day fire safety legislation, virtually all commercial and industrial premises are almost certainly now required to have a fire alarm system installed, of one type or another.

Fire alarms

11.25   The simplest type of electrical fire alarm system is a manual system, which must be operated manually by someone, either by breaking a glass panel on a call point or in older systems by pressing a button. Another manual method of raising an alarm is by the use of a public address system.

Public address systems are also being increasingly used to broadcast taped evacuation messages in larger premises, however, these are normally linked to more sophisticated automatic detection systems. These systems are considered appropriate as an alternative to electronic sounders or bells, where a phased or staged evacuation is required. In addition these systems are likely to be installed where large numbers of employees or members of the public are likely to be involved in order to reduce the possibility of panic. Examples of these buildings are shopping centres, departmental stores, high rise office buildings, larger hotels, nightclubs, theatres, cinemas, conference and exhibition centres.

Automatic fire detection systems

11.26 Automatic fire detection systems are becoming more common in business premises as technology progresses and as they become more reliable and less prone to false alarms. These types of system are more accurately described as a fire detection and alarm system, because the system's primary function is firstly to detect a fire or fire condition, and secondly to sound audible alarms in all of parts of the building. These systems can be fairly complex and incorporate a variety of different types of smoke, heat, flame and other types of detector, which may be provided throughout the building. In addition, on certain larger sites containing many separate buildings, the systems are often inter-linked with other buildings on the site. The systems are normally linked in order to provide an immediate warning of a fire anywhere within a complex, with the termination point usually at a twenty-four hour security manned central control point.

The main components of an automatic detection system.

- Detectors – these can be of various types, with the most common as follows.
  - Smoke detectors – these can be either of the 'optical' or 'ionisation' type and are devices that detect the actual presence of smoke in the area of the detector.
  - Heat detectors – these can be either of the 'fixed' temperature or 'rate of rise' type. These detectors constantly monitor the temperature of the air within the area in which they are located and will give an alarm if either the temperature rises above a predetermined fixed temperature or, alternatively, if it rises too quickly.
  - Flame detectors – these devices, as their name suggests, are designed to detect the presence of a flame and to sound an alarm when flame is detected.
  - Beam detectors – these detectors transmit an infrared beam across a room, area or space, at a high level, to a receiver sensor. If the beam is broken, perhaps by the presence of smoke, an alarm is

11.27 *Managing fire safety (fire defence including active and passive fire precautions)*

sounded. These detectors are generally used when other types of detector are unsuitable or inappropriate, perhaps for architectural reasons in an historic building.

- Control panel – in a modern, 'intelligent' system the fire alarm control or indicator panel can be regarded as the heart of the system as it constantly monitors the detectors, the wiring and any other devices installed. It will monitor for faults on the system or a fire alert. If the system identifies an alarm the control panel will operate the fire alarm sounders located throughout the premises, so as to warn the occupants. In some instances the fire alarm signal can be transmitted direct to the local fire service. However, this is becoming less common as any fire alarm signal is nowadays likely to be sent direct to independent security call-handling centres that will, in turn, call the fire service, if necessary. The control panel will also give an indication of which 'zone' within a building a detector has been activated. A 'zone' is normally a specific area or floor level within a building.

  With more sophisticated addressable systems, the control panel will give the specific 'address' of the detector or other device which has activated. In order to ensure a constant power supply in the event of a fire the control panel will be provided with standby batteries, which are constantly being trickle-charged from the mains electricity supply. The panel is normally located in a position where the fire service can easily find it and is normally located near the front entrance of a building. They can also be found in locations at main entrance where they are visible from outside of the building. This is normally the case in buildings which are unoccupied at night, as it allows the fire service to assess whether the alarm is false, to save having to break into the building unnecessarily. In larger complexes where considerable distances can be involved, premises are sometimes provided with secondary or 'repeater' panels, which give the same information as the main control panel.

- Wiring and cabling – fire-resisting cabling is generally used throughout the installation of a fire alarm system.

  The type, design and installation of a fire alarm system will normally have been agreed with the enforcing authority for the premises concerned. All modern systems will have been installed using the universally specified British Standard BS 5839: 'Fire detection and alarm systems for buildings'. This Standard is a very comprehensive document, which is divided into several separate parts (publications), which cover, in detail, the installation of every type of fire warning systems for the majority of situations.

Types of protection

11.27 Details of the specific types of systems required for the various types of premises, which are required to have fire alarm systems installed, are detailed below. They fall essentially into two categories, which are, systems for protecting *life* and systems for protecting *property*.

The types of system specified in British Standard BS 5839: 'Fire detection and alarm systems for buildings' are contained in the following clauses:

Systems for protecting life

11.28

- Type M–Manual System – This is a system that relies upon a manual break glass call point being operated by the occupants of a building. When a break glass call point is broken the alarm will sound.

- Type L1–Life 1 – This system provides total coverage of the building, with automatic detectors installed throughout. Additionally, break glass call points are installed on all exits and between zones.

- Type L2–Life 2 – This system has automatic detectors installed along all escape routes as well as in high risk areas, where a fire would cause a high risk to life. Additionally, break glass call points are installed on all exits and between zones.

- Type L3–Life 3 – This system has automatic detectors installed only along escape routes and in areas where free passage is essential to protect life.

Additionally, break glass call points are installed on all exits and between zones.

Systems for protecting property

11.29

- Type P1–Property 1 – This system provides total coverage of the building, with automatic detectors installed throughout. Additionally, break glass call points are installed on all exits and between zones. To all intents and purposes a type P1 system can be regarded as having the same cover as an Type L1 system.

- Type P2–Property 2 – This system has automatic detectors installed only in high risk areas, such as plant rooms, storage facilities or any other area where there is a high risk of fire. Additionally, break glass call points are installed on all exits and between zones.

Systems for various types of buildings

11.30 As a general guide, British Standard 5839 gives examples of the types of systems to be installed in various types of buildings, as follows.

11.31 *Managing fire safety (fire defence including active and passive fire precautions)*

- Type M – A small office building with clear escape routes and occupants who know the building.
- Type L1 – A residential care home or hotel. Or a building with specific access or structural risks.
- Type L2 – A large complex office building where many people work. An older style building with many corridors and small rooms.
- Type L3 – A medium sized office building, retail premises or factory. Where large numbers of people are present, but escape is relatively easy.
- Type P1 – A large complex office building where a high risk exists throughout or where a small fire could easily spread and cause extensive damage.
- Type P2 – A listed building or older style premises where fire damage could be expensive.

However, it must be emphasised that if an employer, owner or occupier is being required to install a system as a result of a legislative requirement they must consult the enforcing authority to determine their exact requirements before commencing any installation.

More information on the various British Standards, relating to fire alarm systems, can be found in **CHAPTER 5: LEGISLATION AND REGULATIONS (BRITISH AND EUROPEAN STANDARDS)**

Fire-fighting extinguishers – types of fire

11.31 In order to determine the most appropriate type of fire extinguisher needed for a particular type of risk within a premises it is necessary to look at the 'fire classification'.

There are four common categories of fire which are related to the fuel involved and the method of extinction, as follows.

- Class A: fires generally involving solid organic materials, such as coal, wood, paper and natural fibres, in which the combustion takes place with the formation of glowing embers.

  Extinction is achieved through the application of water in jet or spray form.

- Class B: fires involving:

  (a) liquids, which can be separated into those liquids which mix with water eg acetone, acetic acid and methanol; and those which do not mix with water eg waxes, fats, petrol and solvents; and

  (b) liquefiable solids eg animal fats, solid waxes, certain plastics.

Foam, vaporising liquids, carbon dioxide and dry powder can be used on Class A and B type fires. Water spray can be used on liquids that mix with water, but not on fats, petrol, etc. In all cases, extinction is principally achieved by smothering, with a certain degree of cooling in some cases.

- Class C – fires involving gases or liquefied gases eg butane, propane.

  Both foam and dry chemicals can be used on small fires following spillage, preferably supported by water to cool a leaking container or spillage collector. Extinguishers used on liquid gas fires work by smothering or inhibiting air for further combustion.

- Class D – fires involving certain flammable metals, such as aluminium or magnesium.

  In this case extinction is achieved by the use of dry powders that include soda ash, dry sand, limestone and talc. Such powders have a smothering effect.

Electrical fires

11.32 This classification is no longer used. Fires involving electrical apparatus must always be tackled by first isolating the electricity supply and then by the use of carbon dioxide, vaporising liquid or dry powder.

The table below classifies fires, which can be controlled by portable fire appliances. Further guidance can be found in British Standards BS EN 3 and BS 6643.

Class of fire	Description	Appropriate extinguisher
A	Solid materials, usually organic, with glowing embers	Water, foam, dry powder, vaporising liquid, CO_2
B	Liquids and liquefiable solids: miscible with water eg acetone, methanol	Water, foam (but must be stable on miscible solvents), CO_2, dry powder
	Immiscible with water eg petrol, benzene, fats, waxes	Foam, dry powder, CO_2, vaporising liquid

Types of fire extinguisher

11.33 The fire-fighting extinguishing medium in portable extinguishers is expelled by internal pressure that is either permanently stored or pressurised when operated by means of an internally stored gas cartridge.

11.34 *Managing fire safety (fire defence including active and passive fire precautions)*

The type of extinguisher provided should be suitable for the risk involved, adequately maintained and appropriate records should be kept of all inspections, tests etc. Generally speaking, portable fire extinguishers can be divided into five categories according to the extinguishing medium they contain, which are:

- water;
- foam;
- powder;
- carbon dioxide; and
- vaporising liquids, including halons.

Water

11.34 This type of extinguisher is suitable for ordinary combustible fires, for example wood and paper, but is not suitable for flammable liquid fires. Such extinguishers should also be labelled '*not to be used on fires involving live electricity*'. Water spray extinguishers are recommended.

Foam

11.35 These are suitable for small liquid spill fires or small oil tank fires where it is possible for the foam to form a blanket over the surface of the flammable liquids involved. Foam extinguishers may not extinguish a flammable liquid fire on a vertical plane. Where foam is required for hydrocarbon fires, light water is recommended – preferably by spray applicator. As alcohol is miscible with water, when on fire it will break down ordinary foam and should be considered a special risk.

Dry powder

11.36 This type will deal effectively with flammable liquid fires and is recommended, as it is capable of quick knockdown of a fire. The size of the extinguisher is important and it must be capable of dealing effectively with the possible size of the spill fire, which may occur with some of the extinguishing medium in reserve. The recommended minimum size is a 9 kg trigger-controlled extinguisher with carbon dioxide (CO_2) discharge.

(Dry powder extinguishers will also deal with fires involving electrical equipment.)

BCF

11.37 A BCF extinguisher (Bromochlorodifluoromethane: a lightweight, efficient vaporising extinguisher) is suitable for fires where electrical or electronic equipment may be involved. This type of extinguisher can also be used on flammable liquid fires, although such use may produce large quantities of toxic irritant gases. The hotter the fire, the more toxic the vapours produced. Therefore, a quick knockdown is essential. BCF extinguishers should not be used on high temperature, metal or deep fat fires especially in confined areas.

It should be noted that these extinguishers are more commonly known as halon extinguishers. Due to the risk posed by their use in confined spaces, and the international agreement to discontinue the use of halons (known as the 'Montreal Protocol') this type of extinguisher is being phased out. This is being achieved by a stop on production and no refilling by manufacturers, once existing extinguishers are discharged.

Halon extinguishers will be banned, except for defined essential uses, after 31 December 2003. For special risks one of the other liquefied gas-type extinguishers may be used.

Substitutes for halon are carbon dioxide or dry powder extinguishers, depending on the nature of the risk.

Carbon dioxide (CO_2)

11.38 For fires involving electrical equipment, carbon dioxide extinguishers are recommended. carbon dioxide extinguishers are quite heavy and may be at high pressure. A minimum size of 4.5 kg is recommended. carbon dioxide is not recommended for flammable liquid fires, except for small fires. Training in the use of carbon dioxide extinguishers is essential as they are extremely noisy when operated and can be disconcerting when used for the first time.

Dry powder or carbon dioxide extinguishers which are too small, can be hazardous due to the danger of re-ignition or flashback.

A triple F ('AFFF')

11.39 This type of extinguisher uses aqueous film forming foam and is suitable for both Class A and B fires. It is particularly suitable for flammable liquids such as petrol, oil, solvents and paints, as it provides excellent 'burnback' resistance. With burning liquids, this extinguisher operates by using the foam to provide a protective film, which excludes oxygen and therefore extinguishes the fire in the liquid. The AFFF (aqueous film forming foam) protective film does not break down as readily as other foam compounds and consequently gives good protection against the possibility of the re-ignition, or 'burnback' of the liquid.

This type of extinguisher can, if fitted with an appropriate spray nozzle which has passed the electrical conductivity test of British Standard EN3, be

11.40 *Managing fire safety (fire defence including active and passive fire precautions)*

specified for Class A and B fires within the proximity of live electrical equipment in accordance with British Standard BS 5306: Part 3: 1985.

Colour coding of portable fire extinguishers

11.40 All new certified fire extinguishers for use throughout the EU have been coloured red, since 1 January 1997, following the introduction of British Standard BS EN 3: Part 5 and the removal of British Standard BS 5423. Manufacturers will be allowed, under British Standard BS 7863 (new), to affix different coloured panels on or above the operation instruction label. A zone of colour of up to 5% of the external area, positioned immediately above or within the section used to provide the operating instructions, may be used to identify the type of extinguisher. This zone should be positioned so that it is visible through a horizontal arc of 180 degrees when the extinguisher is correctly mounted. The colour coding should follow the recommendations of British Standard BS 7863.

Prior to this, the entire body of the fire extinguisher had been colour-coded with various colours and consequently large numbers of them will still be in use for many years to come.

It is worth noting that existing extinguishers need not be replaced until they have served their useful life.

Fire extinguishers, if properly maintained and serviced, may be in service for at least 20 years. There may be situations where a building will have a mixture of new and old fire extinguishers with the same type of extinguishing medium but with different colour coded markings. In these cases and to avoid any confusion, it is advisable to ensure that extinguishers of the same type, but with different colour-coded markings, are not mixed either at the same location in single-storey buildings or on the same floor level in multi-storey buildings.

Older style fire extinguishers must not be painted red to try to comply with the new standard, as this would contravene British Standard BS EN 3.

Standards to use

11.41 Fire extinguishers should conform to a recognised standard such as British Standard BS EN 3 for new and, British Standard BS 5423 for existing, extinguishers. For additional assurance you should look for the British Standard Kite mark, the British Approvals for Fire Equipment ('BAFE') mark or the Loss Prevention Council Certification Board ('LPCB') mark.

Siting of extinguishers

11.42 All fire extinguishers should be fitted on wall brackets as it has been found that if this is not done extinguishers get lost or are knocked over and damaged.

To assist in lifting, the carrying handle of larger, heavier extinguishers should be about one metre from the floor. However it is acceptable for smaller, lighter extinguishers to be mounted at a higher level. An employer should ensure that the weight of the equipment falls below the guidelines recommended in the *Manual Handling Operations Regulations 1992 (SI 1992 No 2793)*. This will ensure that extinguishers are easy to handle and use, by both male and female employees.

Fire extinguishers should normally be located in conspicuous positions on escape routes, preferably near exit doors or on the line of exit. Wherever possible, in larger premises, fire-fighting equipment should be grouped to form fire points. These fire points should be clearly visible or their location clearly and conspicuously indicated so that they can be readily identified. Where premises are of a uniform layout, extinguishers should normally be located at similar positions on each floor.

If for any reason extinguishers are placed in positions hidden from direct view, perhaps to prevent vandalism or theft, the *Health and Safety (Safety Signs and Signals) Regulations 1996 (SI 1996 No 341)* require that their location should be indicated by signs and, where appropriate, directional arrows. Suitable signs are described in the HSE Guidance on the Regulations. For further information, see **CHAPTER 10: MANAGING FIRE SAFETY (PREVENTION OF FIRE AND LIAISON WITH AUTHORITIES)**.

The most commonly found and most useful form of fire-fighting equipment for general fire risks, is the water-type extinguisher or a hose reel. A general rule is that one water extinguisher should be provided for approximately each 200 square metres of floor space, with a minimum of one per floor. If each floor has a hose reel that is known to be in working order and of sufficient length to cover the whole floor area that it serves, there may be no need for water-type extinguishers to be provided. However, some insurance companies may still insist upon the provision of extinguishers in addition to hose reels.

Legal note

11.43 There have been a number of cases where a person using an extinguisher has been seriously injured. Investigations have shown that either the wrong type of extinguisher was supplied or the operator had no training in the correct use of the extinguisher. The latter should not need to be overemphasised, especially in areas of special risk, eg oil dipping tanks, furnace areas, highly flammable liquids, gas or cylinder fires etc.

11.44 *Managing fire safety (fire defence including active and passive fire precautions)*

If the building concerned requires a fire certificate under the *Fire Precautions Act 1971*, the fire authority is normally responsible for determining the correct type of extinguisher for the building and the siting of extinguishers will be entered on the fire certificate.

The following recommendations are given in order to allow evaluation of an existing problem and may need to be related to process risks.

- It is essential that persons be trained in the use of extinguishers, especially in areas where special risks require a specific type of extinguisher to be provided.

- Any person employed to work who is requested to deal with a fire, should be clearly instructed that at no time should they jeopardise their own safety or the safety of others.

- Employees who may be wearing overalls contaminated with oil, grease, paint or solvents should not be instructed to attack a fire. Such contaminated materials may vaporise due to heat from the fire and ignite.

Fire blankets

11.44 Fire blankets should be located in the vicinity of the fire hazard they are to be used to extinguish, but in a position that can be readily and safely accessed in the event of a fire. They are classified as either light-duty or heavy-duty. Light-duty fire blankets are suitable for dealing with small fires in containers of cooking oils and fires involving clothing. Heavy-duty fire blankets are for industrial use where there is a need for the blanket to resist penetration by molten metal or other materials.

Hose reels

11.45 It is common in larger premises for fire-fighting hose reels to be provided as an alternative or in addition to portable water extinguishers. The main advantages of hose reels over extinguishers are that they give an unlimited supply of water and are easy to use by employees who are generally more familiar with the operation of a hose pipe than they are with extinguishers. In addition, they can be speedily used by the fire service to fight fires, which can save valuable time whilst their own equipment is being brought into the building.

Hose reels can either be brought automatically into operation by the use of an automatic valve mechanism, which turns itself on as the reel is unwound, or provided with a manually operated on/off valve.

Hose reels should be approved to British Standard BS EN 671-1:1995 and can be of various lengths with hose diameters of either 19 millimetres (three-quarters of an inch) or 25 millimetres (one inch).

Where hose reels are provided, they should be located where they are conspicuous and always accessible, such as in corridors and on escape routes.

When an employer, owner or occupier is considering the installation of hose reels and are unsure about the number or type required, they should seek advice from the local fire service's fire safety officer, before purchasing the equipment.

Automatic sprinklers

11.46 Automatic sprinklers are becoming increasingly more common, particularly in buildings with a large cubic capacity and in high rise modern buildings. They were originally conceived as a means to protect the contents, fabric and structure of a building from fire by ensuring that any outbreak was detected and extinguished before the fire became out of control. Sprinkler systems were very often installed on the insistence of the building's insurers as a pre-condition of insuring the property. However, in recent years they have been increasingly specified and installed for life safety purposes, as they are now acknowledged to be an effective means of reducing the risk to life from fire. As a consequence, it is anticipated that in the future, fire safety legislation is more and more likely to specify their use for the protection of life.

Sprinkler systems must be specifically designed and installed to the appropriate hazard category in accordance with an approved code of practice such as British Standard BS 5306: Part 2.

The British Automatic Sprinkler Association ('BASA'), which is the national trade association for the industry, supplies the following information on sprinklers.

Further information can be obtained from them at Richmond House, Broad Street, Ely CB7 4AH. Tel 01353 659187; fax 01353 666619. E-mail info@basa.org.uk

Sprinkler systems explained

11.47 All areas of the building to be protected should be covered by a range of pipes with sprinkler heads fitted into them at regular intervals. Water from a tank via pumps or from the town mains (if it can give adequate flow and pressure) fills the pipes.

Each sprinkler head will open when it reaches a specific temperature and spray water onto a fire. The hot gases from a fire are usually enough to make it operate. Only the sprinklers over the fire open and the others remain closed. This limits any damage to areas where there is no fire and reduces the amount of water needed.

11.48 *Managing fire safety (fire defence including active and passive fire precautions)*

The sprinkler heads are spaced, generally on the ceiling, so that if one or more operate there is always sufficient flow of water. The flow is calculated so that there is always enough to control a fire taking into account the size and construction of the building and the goods stored in it or its use.

Sprinkler heads can be placed in enclosed roof spaces and into floor ducts to protect areas where a fire can start without being noticed. In a large warehouse, sprinklers may be placed in the storage racks as well as the roof.

At the point where the water enters the sprinkler system there is a valve. This can be used to shut off the system for maintenance. For safety reasons it is kept locked open and only authorised persons should be able to close it. When a sprinkler opens and water flows through the valve it lets water into another pipe that causes a bell to ring. In this way the sprinkler system both controls the fire and gives an alarm using water, not electricity.

Types of sprinkler systems

11.48 There are sprinkler systems to cater for all types of buildings.

Wet pipe

11.48A These are the most common systems and are used in buildings where there is no risk of freezing. They are fast to react because water is always in the pipes above the sprinkler heads.

Wet systems are required for multi-storey or high rise buildings and for life safety.

Alternate systems

11.48B As the name suggests, alternate systems can have the pipes full of water for the summer and be drained down and filled with air (under pressure) for the winter. This is important for buildings that are not heated.

Dry pipe

11.48C The pipes are filled with air under pressure at all times and the control valve holds the water back. When a sprinkler head opens the drop in air pressure opens the valve and water flows into the pipework and on to the fire. Dry pipe systems are used where wet or alternate systems cannot be used.

Pre-action

11.48D In a similar manner to dry pipe systems, the pipes are filled with air but water is only let into the pipes when the detector operates (eg smoke detectors etc). Pre-action systems are used where it is not acceptable to have the pipes full of water unless there is a fire.

Deluge and re-cycling installations

11.48E These are not strictly sprinkler systems and are only used in special cases for industrial risks.

What standards are there for sprinkler systems?

11.49 To make sure the system will work it must be properly designed and fitted. The independently accredited Loss Prevention Certification Board ('LPCB') publishes standards for sprinkler systems based on British Standards. When a system is installed by a company approved by the LPCB they will issue a *certificate of conformity* to prove it has been installed satisfactorily.

BASA can provide a list of their members, all of whom are approved by the LPCB.

The LPCB issues certificates of conformity as proof that a system meets their professional standards of performance and quality. Only installers recognised by the LPCB can install systems that carry their certificate of conformity.

The certificate is proof to fire brigades, local authorities and insurance companies that the system meets the sprinkler rules. It is also third party verification for the owner that his system is to the correct standard.

Legislation

11.50 Approved documents accompanying the *Building Regulations Building Regulations 1991 (SI 1991 No 2768* in England and Wales make specific reference to the use of sprinklers (*Building Standards (Scotland) Regulations 1991 (SI 1991 No 158) (as amended)* and *Building Regulations (Northern Ireland) 1994 (SI 1994 No 243)* differ slightly).

For life safety, multi-storey buildings over 30 meters high must be fitted with sprinklers to meet *SI 1991 No 2768 Approved Document B* standards. Similarly an unpartitioned area in a shop or in commercial premises over 2,000 square meters requires sprinkler protection. There are corresponding Regulations applying to buildings for industrial or storage use. In addition, from 1 July 2000, new single storey shops in excess of 2,000 square meters were required to install sprinkler protection.

The installation of sprinklers can allow buildings to be built closer (half the normal space separation is required) to adjoining premises. This is a major benefit where there is limited space on a site.

233

11.51 *Managing fire safety (fire defence including active and passive fire precautions)*

In shops sprinklers can be taken into account when calculating fire growth and smoke volume. This in turn may allow the approval of longer distances of travel to exits.

The guidance issued to interpret *SI 1991 No 2768* now recognises the use of sprinklers for life safety and it is clear that future legislation will call for the increased use of sprinklers.

Fixed fire suppression systems

11.51 Because of the specific fire risks that can exist with certain hazardous processes or the need to protect delicate or irreplaceable equipment and data, it is not uncommon, in many industrial and commercial premises, for fixed fire suppression systems to be installed. The types of equipment which are normally covered by this type of equipment are:

- electrical switchgear and transformers;
- control, computer and data-processing equipment;
- flammable materials storage; and
- commercial cooking ranges.

Experience has shown that fires involving these types of risks can effectively be dealt with by the installation of fixed fire-fighting systems that may be either automatically or manually operated.

Examples of these are found protecting process equipment and machinery that handles flammable substances, such as printing machines, rolling mills and oil-filled switchgear. Various types of extinguishing media are used which range from dry powder, foam, carbon dioxide or other inert gases to more recent innovations which use water mist technology. This technology is being increasingly used to protect food-processing equipment.

Due to the high cost and value of control and data-processing equipment, many larger and more sensitive installations have been protected by various systems designed to fill the room totally, space or control equipment cabinets, with a total flood extinguishing medium. These systems use a range of various types of gas extinguishing media. When there is a possible situation where these total flood systems may fill a space or room that is occupied, suitable safety systems are put in place to ensure that a distinctive warning is given to the occupants, to enable them to vacate the area before the flood system operates. This is an obvious precaution as the flood systems are designed to smother fire by the exclusion of oxygen and would also smother people in the vicinity.

In addition it is quite common for large bulk storage tanks of flammable materials and similar facilities to be protected by fixed deluge water or foam systems.

Maintenance of fixed fire suppression systems

11.52 Having made a considerable financial investment in providing a fixed system to protect a particularly vital part of a business, it makes sense to ensure its continued effectiveness by ensuring that there is an adequate maintenance programme in place. In the main, it is good policy to follow the manufacturer's advice, with regard to maintenance. It is common practice for some of the routine maintenance to be carried out by the employer's own personnel, with major maintenance being carried out by the manufacturer.

Design and installation

11.53 The design and installation of fixed fire-fighting and suppression systems requires a high degree of technical expertise in order to select the appropriate system and fire-fighting medium for any particular given set of circumstances. These systems are usually custom designed and therefore can be expensive. In considering the installation of a fixed system, it is sound policy to consult the relevant enforcing authority and discuss your requirements with a reputable company at the earliest possible opportunity.

Further information

11.53A The British Fire Protection Systems Association ('BFPSA') can supply you with a list of companies in your area that are prepared to carry out this type of work.

The address and contact numbers for the BFPSA is Neville House, 55 Eden Street, Kingston-upon-Thames, Surrey KT1 1BW. Tel: 020 8549 5855; fax: 020 8547 1564. E-mail: info@abft.org.uk Website: http://www.bfpsa.org.uk

The British Standard BS 5306 – 'Fire extinguishing installations and equipment on premises' should be consulted for further guidance on this subject.

Foam inlets

11.54 Foam inlets are simple fixed installations provided within certain buildings for use by the fire service. Their purpose is to provide ready means for the fire service to use their foam-making equipment on fires involving flammable liquid risks. The installation comprises an inlet at street level that is normally enclosed within a reinforced glass-fronted metal box, connected to an empty pipe that leads to the area within the building that contains the protected risk. The foam inlet is therefore merely a simple means by which the fire service can speedily apply fire-fighting foam to a burning flammable liquid fire.

Dry risers

11.55 Dry risers are simple installations installed within high rise buildings for use by the fire service to speedily tackle fires on the upper floors of the building. The installation comprises of an inlet at ground or street level that is normally enclosed within a reinforced glass-fronted metal box connected to an empty pipe that rises through all floor levels within the building. At each floor level there are landing valve cabinets that may also be contained within reinforced glass-fronted metal boxes. On arrival at a building on fire the fire service simply connect their pumps and hoses to the inlet at ground level that fills the dry riser with water. They can then connect their own hoses to the landing valves at any of the floor levels to extract as much water as they need for fire-fighting.

This system enables the fire service to quickly get to grips with tackling fires on the upper floors of buildings, with the minimum of delay and without having to waste time whilst they bring their own equipment into the building.

Wet risers

11.56 Wet risers are essentially the same type of system as dry risers, the difference being that they are kept permanently charged with water for fire-fighting. Wet risers in general are less common than dry, but they do have the advantage of providing even quicker access to fire-fighting water by the fire service, as there is no delay whilst waiting for the riser to charge with water.

Emergency procedures – staff training

11.57 An important facet of a business's fire defence is to ensure that all staff have an awareness and an understanding of how to react in an emergency. To achieve this objective it is necessary to ensure that all staff receive adequate training based around a prepared emergency plan of action for your place of work.

The training of staff should be formulated around the following basic criteria.

Managing fire safety (fire defence including active and passive fire precautions) **11.60**

Checklist

11.58

> Training should:
>
> - describe and explain emergency procedures;
> - take account of the particular types of work carried on in the business, to include any specific hazards associated with it;
> - explain the duties and responsibilities of staff;
> - take account of the findings of a risk assessment carried out under the *Fire Precautions (Workplace) Regulations 1997 (SI 1997 No 1840)* (as amended 1999); and
> - be easily understood by staff.

Staff training in large premises

11.59 In larger premises, with consequently larger numbers of staff, it is a sensible policy to include an evacuation exercise in the training. Although fire drills and evacuation are generally treated quite light heartedly by staff, if they are carried out regularly an employer can be assured that in a real emergency their staff will know exactly what to do.

Checklist

11.60

> Training should aim to include and cover the following areas of fire safety, which should be tailored to suit the individual circumstances within the business:
>
> - action to take on discovering a fire;
> - how to raise the alarm and what happens next;
> - action to take upon hearing the fire alarm;
> - procedures for alerting members of the public and visitors including, where appropriate, directing them to exits;
> - arrangements for calling the fire service;
> - evacuation procedures for everyone in the workplace to reach an assembly point at a safe location;

11.61 *Managing fire safety (fire defence including active and passive fire precautions)*

- location and, when appropriate, the use of fire-fighting equipment;
- location of the escape routes, especially those not in regular use;
- how to open all escape doors, including the use of any emergency fastenings;
- importance of keeping fire doors closed to prevent the spread of fire, heat and smoke;
- where appropriate, how to stop machines, processes and how to isolate power supplies in the event of fire;
- the reason for not using lifts (except those that have been specifically installed or adapted for the evacuation of disabled people);
- importance of general fire safety and good housekeeping.

An employer, owner or occupier can obtain further guidance on training from the information contained within the *'Approved Code of Practice to the Management of Health and Safety at Work Regulations 1999'*, obtainable from the Health and Safety Executive (HSE).

In addition, this subject is dealt with in more depth in **CHAPTER 9: MANAGING FIRE SAFETY (FIRE SAFETY PROCEDURES, TRAINING AND FIRE MANUAL).**

Evacuation drills

11.61 An employer is required to ensure that all their staff know what to do in the event of a fire. To achieve this, the workplace should be provided with an emergency plan of action.

The plan should include what action is to be taken by an employer and the staff in the event of fire or other similar emergency. The plan should include an evacuation procedure and include arrangements to ensure that the fire service is called.

This is both a statutory requirement in procedures relating to *'events of serious and imminent danger'* detailed in *Management of Health and Safety at Work Regulations 1999 (SI 1999 No 3242) Regulation 8(1)* and possibly the requirement of a fire certificate issued under the *Fire Precautions Act 1971*. Employers should acquaint their workforce with the arrangements for fire evacuation. This normally consists of putting up notices in prominent locations stating the action employees should take upon:

- hearing the alarm; or
- discovering a fire.

Managing fire safety (fire defence including active and passive fire precautions) **11.63**

Ideally, employees should receive regular fire evacuation drills, which should be carried out at least annually, even though this may mean that normal working is interrupted. As part of the process of raising staff awareness, fire alarms should be tested weekly so that your employees may familiarise themselves with the sound and its meaning. This also has the added benefit of ensuring that the fire alarm system is functioning correctly.

Trained employees should be designated as fire wardens or marshals and carry out head counts following an evacuation. In addition their role should be to act as the last person out, following a total sweep of their area of responsibility, ensuring that everyone has been evacuated. Their role should also include generally advising and shepherding the public out of the building.

This subject is dealt with in more detail in **CHAPTER 9: MANAGING FIRE SAFETY (FIRE SAFETY PROCEDURES, TRAINING AND FIRE MANUAL)**.

Fire action notices

11.62 The essential element in any emergency action notice is that it should be *simple*. The more complex it is the less likely that people will remember what to do when an emergency occurs. An employer, owner or occupier should keep the notice as simple as possible so that employees will remember what to do, at the vital time, when a fire breaks out.

The following guidelines are suggested as a basis for an emergency action notice, which can be tailored to suit your individual circumstances.

Typical fire action notice

11.63

When the fire alarm sounds:
- switch off electrical equipment and leave room, closing doors behind you;
- walk quickly along the escape route to open air;
- do not stop to collect personal belongings;
- do not use lifts;
- report to fire marshal at the assembly point that is ....; and
- do not re-enter building.

239

11.64 *Managing fire safety (fire defence including active and passive fire precautions)*

When you discover a fire:

11.64

- raise alarm by breaking the nearest available fire alarm call point or by telephone, stating name and location;
- leave the room, or area making sure that you close doors behind you;
- leave the building by the escape route to open air;
- do not stop to collect personal belongings;
- do not use lifts;
- report to fire marshal at Assembly Point that is . . . .; and
- do not re-enter building.

Assembly points

11.65 An assembly point is an area designated as a meeting point for persons who have evacuated from an area that is on fire. An assembly point for the safe evacuation of employees and members of the public should be designated and included in the fire action notices provided throughout the premises. The assembly point should be a free unenclosed area outside and away from the building and preferably indicated with a sign. It must also be capable of holding the number of people who might reasonably be expected to arrive there in an emergency.

The assembly point is where the fire marshals meet after they have carried out their 'last person out' sweep of their area of responsibility, in order to report whether everyone is accounted for, or not. This is the point where all information about the location of employees or members of the public is collated, including when necessary, a roll call. It is essential that the fire service is informed at the earliest opportunity, if anyone is missing.

Preventative measures – planned maintenance programmes

11.66 Having made a considerable financial investment in providing sophisticated fire protection measures for the premises and its occupants, the investment will have been in vain if regular maintenance is not carried out on the various systems installed, to ensure their ongoing effectiveness. More significantly, an employer will require the systems to work if the need should arise.

Managing fire safety (fire defence including active and passive fire precautions) **11.67**

To ensure this happens, it is perhaps necessary to formulate a written maintenance policy that includes a pre-programmed maintenance timetable, for the fire safety equipment provided.

The maintenance programme should include:

- what equipment is to be maintained;
- who is responsible for ensuring that maintenance is carried out;
- how often it is to be carried out; and
- a record of *all* routine maintenance is kept.

The following guidance is provided in order to help an employer determine what equipment should be maintained and how frequently it should be carried out.

Means of escape – structural and integral

11.67   Each year a programme of planned maintenance should be drawn up to ensure that the means of escape are readily available in case of fire, as required by legislation and codes of practice.

This planned maintenance programme should cover the structural and integral parts of the means of escape that includes walls, ceilings and floors. The fire protection of openings, such as doors, glazed areas, hatches and shafts, should also be included in the programme.

- Walls, ceilings and floors should be maintained in good condition so as not to hinder persons escaping from fire or allow the passage of flame and smoke. Approved fire stopping units and fire barriers should be employed wherever gaps cause a reduction in fire resistance or a breach of the integrity of the structure.

- Fire and emergency doors should be inspected and maintained to ensure that they fit their frames correctly and that they do not jam in the closed position. Smoke seals and intumescent strips, where fitted, should be checked to ensure that they are functioning satisfactorily. The hinges should be checked to ensure that they do not allow the door to drop and catch the floor and that the automatic self-closing devices function correctly.

- Final fire exit doors that need a degree of security to make them inaccessible from outside the building should be fitted with panic latches or other 'hands free' operation devices of the 'push pad to open' or 'push bar to open' types that assist speedy and safe evacuation. This equipment should be inspected, tested and serviced on a regular basis as recommended by the equipment manufacturer. Signs giving instructions on how to use the door fastenings should also be maintained in a position where they can clearly be seen in an emergency.

11.68 *Managing fire safety (fire defence including active and passive fire precautions)*

- Fire resisting glazing should be inspected and maintained to ensure that it is installed correctly in secure frames. The glazed area should not exceed the recommended sizes for the task it is to perform and only approved fire resisting glass is used. Glazing should be in good condition without defect, cracks, or chips.
- Hatches and shafts should be inspected and maintained to ensure that they perform the function for which they were designed. Doors and hatch covers should be checked to ensure that they have the correct level of fire resistance for the risk. Fire dampers where fitted in ducting should be clearly marked and capable of easy operation from outside the protected area.

Means of escape – planned maintenance

11.68 A programme of planned maintenance for measures to facilitate means of escape should be drawn up at the same time as the programme of planned maintenance for the structural and integral components of a building. This is to ensure that those features not covered in the clause (see **11.69** below) are inspected and maintained.

This programme should cover paths leading away from the building to the place of safety, emergency lighting, emergency signs and assembly points.

- Paths leading to a place of safety should be inspected at various times of day and at night (if the premises are in use during the night). They should also be checked at various times of the year to ensure that they are maintained in a good condition for those who might be reasonably expected to use them.
- Paths should, wherever possible, be level and should have the least number of steps as possible. They should also be wide enough to accommodate the number of people likely to be escaping from the premises.
- Paths should be well lit at all times and clearly signed.
- Emergency lighting should be installed and maintained wherever necessary to ensure that persons evacuating the building may do so in safety should there be a mains power failure (see **11.17** above).
- Signs will be in lettering and pictograms that conform to the current *Health and Safety (Safety Signs and Signals) Regulations 1996 (SI 1996 No 341)*.

Portable fire-fighting equipment

11.69 Portable fire-fighting equipment should be provided at strategic locations, in accordance with and as required by legislation and codes of practice. Within larger premises, 'fire points' indicated by pictogram and lettered signs should be provided. Equipment that is considered suitable for the risk should be provided in the immediate vicinity and located at designated fire points.

- All extinguishers should be serviced and inspected annually, additionally they should be discharge tested every five years. Qualified engineers should carry out inspections and tests to the current standards, and the results should be recorded on each item of equipment. Test certificates should be obtained from the servicing contractor.

- Fire blankets should be positioned in those areas where food is cooked and where there is a possibility of persons becoming involved in fire. Fire blankets should be inspected annually and the results should be recorded on each fire blanket carrier.

- Hose reels should be checked for corrosion of the connections. Hoses should be checked annually by a qualified service contractor for any signs of deterioration and damage.

Fixed fire-fighting equipment

11.70 Where private on-site fire hydrants are provided, suitable fire hydrant marker posts should indicate their position. Hydrants should be tested annually. This type of service is normally available from your local fire service.

Dry rising mains should be subjected to an annual inspection and flow test by contractors and, a satisfactory test certificate obtained.

Automatic fire suppression and extinguishing systems should be maintained in accordance with the manufacturer's instructions.

Fire alarm systems

11.71 Both manual and automatic fire alarm systems should be subject to a rigorous testing regime. The planned maintenance programme should ensure that each zone, sector, detector, call point and alarm bell is tested on a regular basis as well as any other equipment linked to the system, such as fire door electro-magnetic automatic closing devices. The master control and any repeater indicator panels should be tested at the same time. Full service records should be maintained, which are normally supplied by a service contractor.

Signs and notices

11.72 Exit signs and emergency notices provided throughout the premises should be inspected annually. In addition they should be checked to ensure that they are replaced following any building work, refurbishment or redecoration.

11.73 *Managing fire safety (fire defence including active and passive fire precautions)*

Emergency lighting

11.73 Emergency, or escape, lighting, where fitted, should be inspected at least monthly to ensure that the bulbs are not defective. The system should be subject to annual maintenance checks. A qualified person or a contractor should carry out these checks on the system. Service records should be obtained on completion.

Security staff and systems

11.74 In larger premises, it is not uncommon for security staff to be employed on a twenty-four hour basis. Their primary role is generally to ensure the security of the premises round the clock. Security staff can provide a vitally important aspect of fire safety defence that is often overlooked by employers, with the greater emphasis being placed on their role in the prevention of theft. Their very presence of security staff 'round the clock' can mean the difference between early detection of a fire and one that is allowed to develop and grow undetected. As their role is normally to patrol the protected premises at regular intervals, with additional training security can be a significant asset to the fire protection of a building. In their specialised role it is appropriate for them to be given a higher level of fire safety training than the average employee.

As a guide, the training of security staff should encompass the following aspects of fire prevention and fire-fighting.

- Training and guidance should be given on the reasons and importance of carrying out hourly patrols on all parts of the premises should be given.

- It should be emphasised that the patrols should be carried out during the hours of darkness or at other times when the buildings are unoccupied.

- The patrols should include all risk rooms, process areas and kitchens.

- In addition, during the first patrol, all fire doors should be checked to ensure that they are closed. It is also good policy to ensure that *all* internal doors are closed, as even the flimsiest of doors will be capable of containing a fire for a limited amount of time.

- Each person carrying out the patrol should record and sign to state that each patrol has been completed.

- Suitable training on the 'good housekeeping' aspects referred to in **CHAPTER 8: MANAGING FIRE SAFETY (AWARENESS, PREVENTION AND FIRE SAFETY POLICY)** should also be given.

- The location and use of all fire-fighting equipment provided throughout the premises, including instruction on any fixed fire suppression systems that may be installed.

Further information on a suggested training syllabus for security staff can be found in **CHAPTER 9: MANAGING FIRE SAFETY (FIRE SAFETY PROCEDURES, TRAINING AND FIRE MANUAL)**.

Insurance

11.75 In order to ensure the survival of a business in the event of fire, property protection must be considered and a business risk assessment carried out. The involvement of a company's insurer is essential as insurers have considerable experience in this field. The business risk assessment follows a similar pattern to that used in a fire risk assessment, but it assesses the importance of all aspects of the business in order to determine how vulnerable the business is to fire.

For example, damage to an area where essential records or documents are stored is likely to have a serious effect on the business. Essential equipment (for example, plant or stock which if destroyed or damaged by fire might be difficult to replace or have a serious effect on production) would require special consideration. This will often mean the provision of a higher degree of fire protection measures for special areas or for particular equipment. In fact, it is not unknown for insurers to refuse insurance cover unless certain protection measures are implemented.

A business management team, in consultation with the company insurers, should determine these risks. A report should be produced by each departmental head outlining areas, records and other essential documents which may require special consideration and can be considered vital to the business's future well being.

More detailed information on the subject of risk management can be found at **8.19** above.

12 Managing fire safety (maintenance, servicing routines and record keeping)

In this chapter:	
Introduction	12.1
New systems	12.2
Fire alarm and automatic fire detection systems	12.3
Fire alarms and automatic fire detection system: servicing routines and record entries	12.11
Fire-fighting equipment (portable fire extinguishers)	12.12
Escape or emergency lighting	12.14
Stand-by generator	12.20
Automatic sprinkler systems	12.24
Pressurisation systems (smoke control using pressure differentials)	12.31
Fire doors, fire shutters and dampers	12.37
Smoke control systems	12.39
Conclusion	12.42

Introduction

12.1 The keeping of records is essential and is a specific requirement in a fire certificate issued under the provisions of the *Fire Precautions Act 1971*. To comply with the provisions of the fire regulations, ie *Fire Precautions (Workplace) Regulations 1997* ('1997 Regulations') (*SI 1997 No 1840*) that came into force on 1 December 1997 and *Fire Precautions (Workplace) (Amendment) Regulations 1999 (SI 1999 No 1877)*, which came in to effect on 1 December 1999, the maintenance records of certain items will need to be recorded on a regular basis.

These documents indicate, in detail, the wide range of tests and inspection procedures that need to be adopted. Following any major incident, records are invariably examined in detail to highlight any discrepancies in the maintenance regime. Some insurance companies are also insisting upon record keeping, both to minimise the likelihood of failure of fire safety systems and their liability.

The servicing and logbook are an amalgam of the recommendations contained in British Standards 'codes of practice' together with the requirements that are imposed by many fire authorities.

New systems

12.2 It is essential that whenever a new system is installed, or maintenance is carried out on any fire safety system, that the appropriate certification documents are obtained and kept with the relevant section of the logbook.

A similar quality of servicing will be required for other systems such as fire-fighting equipment (see **12.12** below), escape or emergency lighting (see **12.14** below), standby generators (see **12.19** below) and automatic sprinkler systems (see **12.24** below).

Fire alarm and automatic fire detection systems

12.3 The servicing programme should be recorded in a logbook, maintained and supervised by a responsible executive who should ensure that every entry is properly recorded. Fire alarms (real or false), faults, pre-alarm warning, tests, temporary disconnections and dates of the installing or servicing engineer's visits should all be included. A brief note of work carried out, and/or outstanding, should be made at the time of the visit.

To ensure greater reliability, correct servicing is essential. Normally an agreement will be made with a manufacturer, supplier or other competent contractor for regular servicing. The agreement should specify the method of liaison to provide access to the premises. The name and telephone number of the servicing organisation should be prominently displayed at the control and indicating equipment, and emergency telephone numbers entered in the front of the logbook.

For premises in continuous use, eg residential apartments and hotels, the agreement should preferably include a requirement that an engineer will be on call at all times, both during and outside normal working hours, and that telephone requests for emergency service should be executed promptly.

In any case, agreement should be made that repair services will be available within 24 hours.

12.4 *Managing fire safety (maintenance, servicing routines and record keeping)*

Where it is not possible to obtain service from engineers on call at all times or, because of special circumstances, no service contract has been arranged, then the responsible person should ensure that at least one person is employed who has had suitable experience of electrical equipment and who has had special training with the manufacturer, supplier or installer to deal with simple servicing. Employees should be instructed not to attempt to exceed the scope of their training.

Prevention of false alarms during routine testing

12.4 It is important to ensure that operation during testing does not result in a false alarm of fire.

Where the fire alarm system is connected to a 999 automatic dialling unit, then transmission should be prevented (for instance by disconnection) before the routine test is carried out, since under normal conditions 999 test calls are not permitted. In certain equipment using automatic dialling, it is possible to prevent transmission of a signal by lifting a telephone receiver.

Use of this function to inhibit transmission is to be avoided, but where used the inhibited state should be indicated by the use of a notice on the control equipment that 're-connection shall be made immediately the test has been completed'.

When transmission of signals to a remote staffed centre is prevented during a test, a visual indication of this state should be given at the control equipment. If a link to a remote staffed centre is to be used during the test, then it is essential to notify the centre before undertaking the test, unless a recognised test procedure is regularly carried out at an agreed time.

There are specific tests for daily, weekly, quarterly, annual and five-yearly inspections.

Daily attention by the user

12.5 A check should be made every day to ascertain the following.

(a) That either the panel indicates normal operation, or if not, that any fault indicated is recorded in the logbook and the servicing engineer is notified.

Note: in program-controlled systems, failure to correctly execute software is indicated either on an event counter, or on an automatic reset indicator.

(b) That any fault warning recorded the previous day has received attention.

If any connection to the public fire brigade or other remotely staffed centre is not continuously monitored, then it should be tested daily in accordance with the supplier's instructions.

Note: on one day each week the daily test will be incorporated into the weekly test.

Weekly attention by the user

12.6 At least one detector, call point or end of line switch on one circuit should be operated to test the ability of the control and indicating equipment to receive a signal and to sound the alarm and operate any other warning devices. When there are more than 13 zones then more than one zone may need to be tested in any one week so that the interval between tests on one circuit does not exceed 13 weeks. It is preferable that each time a particular circuit is tested a different trigger device is used.

An entry should be made in the logbook quoting the particular trigger device that has been used to initiate the test.

When the operation of the alarm sounders has been prevented by disconnection then a further test should be carried out to prove the final reinstatement of the sounders, and, if permissible, of the alarm transmission circuits.

If the batteries are open or accessible, then a visual examination of the battery and its connections should be made to ensure that they are in good condition. Action should be taken to remedy any defect, including low electrolyte level.

The printer (if used) should be checked to ensure that its reserves of paper, ink or ribbon are adequate for at least two weeks' normal usage.

Should the actuation of the fire alarm activate other active systems ie electro-magnetic door holders, pressurisation of staircases, ventilation systems, pre-action sprinkler systems, or shut-down procedures of plant or machinery, the activation of these systems should be noted in the logbook. Action should be taken to remedy any defect.

Monthly attention by the user

12.7 Where an automatically started emergency generator is used as part of the standby supply then it should be started up once a month by a simulation of a failure of the normal power supply, and allowed to energise the fire alarm supply for a continuous period of at least one hour. The fire alarm system should be monitored to identify any malfunction caused by the use of the generator. At the end of the test period the normal supply should be restored and the charging arrangements for the starting battery checked for proper functioning. The fuel tanks should be left filled and the oil and coolant topped up as necessary.

12.8 *Managing fire safety (maintenance, servicing routines and record keeping)*

Note: frequent starting of the generator followed by a few minutes 'on-load' is not recommended. It is important that when the engine is running, the generator is loaded to at least 50 per cent of the engine's capacity to prevent sooting up with the obvious resultant loss of performance.

Quarterly inspection and test

12.8 The responsible person should ensure that every three months a competent person carries out the following checks.

(a) Entries in the logbook should be checked and any necessary action taken.

(b) Batteries and their connections should be examined and tested as specified by the supplier to ensure that they are in good serviceable condition and not likely to fail before the next quarterly inspection.

(c) Where applicable, secondary batteries should be examined to ensure that the specific gravity of electrolyte in each cell is correct. Any necessary remedial action should be taken.

(d) Primary batteries, including reserves, should be tested to verify that they are satisfactory for a further period of use by taking measurements that are indicative of the conditions of each cell, eg its voltage on a known and very high rate of discharge.

Note: the test conditions and the significance of the readings will depend on the type of cell and the use to which it is being put. These should be clearly specified by the supplier or commissioning company and applied with care. Primary batteries should in any case be replaced within the period of shelf-life stipulated by the battery manufacturer.

(e) Alarm functions of the control and indicating equipment should be checked by the operation of a detector or call point in each zone as described earlier. The operation of the alarm sounders and any link to a remote staffed centre, other than a 999 auto-dialler, should be tested. All ancillary functions of the control panel should also be tested and, where practicable, all fault indicators and their circuits should be checked, preferably by simulation of fault conditions. Control and indicating equipment should be visually inspected for signs of moisture ingress and other symptoms of deterioration.

(f) A visual inspection should be made to check whether structural or occupancy changes have affected the requirements for the siting of manual call points, detectors and sounders. The visual inspection should also confirm that a clear space of at least 750 millimetres is preserved in all directions below every detector and that all manual call points remain unobstructed and conspicuous.

(g) All further checks and tests specified by the installer, supplier or manufacturer should be carried out, and the results entered or attached to the premises logbook.

Note: the recommendations in items (b), (c) and (d) above need not be applied to batteries which power individual items of equipment (such as detectors or sounders) and which have provision for monitoring.

Any defect should be recorded in the logbook and reported to the responsible person, and action should be taken to correct it. On completion of the work, a certificate of testing should be given to the responsible person, and attached to the premises logbook.

Annual inspection and test

12.9 A responsible person should ensure that the following check and test sequence is carried out every year by a competent individual:

(a) inspection and test routines detailed for quarterly inspection (see **12.8** above);

(b) each detector should be checked for correct operation in accordance with the manufacturer's recommendations; and

(c) a visual inspection should be made to confirm that all cable fittings and equipment are secure, undamaged and adequately protected.

Any defect should be recorded in the logbook and reported to the responsible person, and action should be taken to correct it. On work completion, a certificate of testing should be given to the responsible person, and recorded in and attached to the premises logbook.

Wiring check

12.10 The responsible person should ensure that every five years (or more frequently if the premises electrical system is tested at shorter intervals) the installation should be tested in accordance with the testing and inspection requirements of the Institution of Electrical Engineers (IEE) Wiring Regulations.

Any defect should be recorded in the logbook and reported to the responsible person, and action should be taken to correct it. On completion of the work, a certificate of testing should be given to the responsible person, and attached to the premises logbook.

12.11 *Managing fire safety (maintenance, servicing routines and record keeping)*

Fire alarms and automatic fire detection system: servicing routines and record entries

12.11

Date	Actuation point	Associated systems	Result	Defect/ action	Signature

Remarks and comments	
REFERENCE DATA	
Responsible person:	Name:
Address:	
System installed by:	
System commissioned by:	
Maintained under contract:	Until:
Telephone number for immediate service:	
Details of test procedures are on next page:	

Fire-fighting equipment (portable fire extinguishers)

12.12 To give greater assurance of reliability, correct servicing is essential. Normally an agreement should be made with the supplier, or other competent contractor for regular servicing. This agreement should also include the recharging of extinguishers following discharge.

All tests should be recorded in the logbook together with any test certificate issued by the engineer.

A competent person should rectify defects in equipment as soon as possible.

A competent engineer should examine fire extinguishers annually and any faults immediately rectified.

Fire-fighting equipment (portable)

Log for annual inspection of portable fire-fighting equipment

12.13 Also to be recorded is any discharge of equipment or faults other than at annual inspections.

Date	Position of Equipment	Result	Defect	Action Taken	Signature

Remarks and comments:	
REFERENCE DATA:	
Responsible person:	Name:
Address:	
System installed by:	
System commissioned by:	
Maintained under contract:	Until:
Telephone number for number for immediate service:	

12.14 *Managing fire safety (maintenance, servicing routines and record keeping)*

Escape or emergency lighting

12.14 To give greater assurance of reliability, correct servicing is essential. Normally an agreement should be made with the supplier or other competent contractor for regular servicing.

Frequency of inspection

12.15 Escape lighting should be inspected daily, monthly, six-monthly, three-yearly and, after the first three years' test, on an annual basis.

Daily attention by the user

12.16 An inspection should be made every day to ascertain that:

(a) any fault recorded in the logbook has been given urgent attention and the action noted;

(b) every lamp in a maintained system is lit;

(c) the main control or indicating panel of each central battery system or engine driven generator plant indicates normal operation; and

(d) any fault found is recorded in the logbook and the action taken noted.

Monthly attention by the user

12.17 An inspection should be made at monthly intervals to a systematic schedule.

Tests should be carried out as follows.

(a) Each self contained luminaire and internally illuminated exit sign should be energised from its battery by simulation of a failure of the supply to normal lighting. This should be for a period sufficient only to ensure that each lamp is illuminated. (This may be done by actuation of a test switch or removal of the local lighting circuit fuse.)

The period of simulated failure should not exceed one quarter of the rated duration of the luminaire or sign. During this period all luminaires and/or signs should be examined visually to ensure that they are functioning correctly.

At the end of this test period the supply to the normal lighting should be restored and any indicator lamp or device checked to ensure it is showing that the normal supply has been restored.

(b) Each central battery system should be energised from its battery by simulation of a failure of the supply to the normal lighting for a period sufficient only to ensure that each lamp is illuminated.

The period of simulated failure should not exceed one quarter of the rated duration of the battery. During this period all luminaires and/or signs should be visually examined to ensure that they are functioning correctly. If it is not possible to examine visually all luminaires and/or signs in this period, further tests should be made after the battery has been fully recharged.

At the end of each test period the supply to the normal lighting should be restored and any indicator lamp or device checked to ensure it is showing that the normal supply has been restored. The charging arrangements should be checked for proper functioning.

(c) Each engine-driven generating plant should be started up by a simulation of a failure of the supply to the normal lighting and allowed to energise the emergency lighting system for a continuous period of at least one hour.

During this period of time all luminaires and/or signs should be visually examined to ensure that they are functioning correctly.

At the end of the test period the system should be restored to normal operation and the charging arrangements for the 'back-up' and the engine-starting battery checked for proper functioning. The fuel tanks should be left filled and the oil and coolant levels topped up as necessary.

Frequent starting of the plant followed by a few minutes on load is not recommended.

(d) The engine of each engine-driven generating plant with 'back-up' batteries should be prevented from starting.

The escape lighting should then be energised solely from the 'back-up' battery by simulation of a failure of the supply to the normal lighting. This should be for a period sufficient only to ensure that the change-over from normal supply to battery is functioning properly.

After this check that the starting system of the engine is returned to normal operation and the engine allowed to start up in the normal way to energise the emergency lighting system. Allow to run for a continuous period of at least one hour. During these periods all luminaires and/or signs should be visually examined to ensure that they are functioning correctly.

At the end of the test period the system should be restored to normal operation and the charging arrangements for the back-up and engine starting batteries checked for proper functioning. Fuel tanks should be left filled and oil and coolant levels topped up as necessary.

Test(s) for the appropriate system should be recorded in the logbook and any defects noted and appropriate action taken to rectify the fault.

12.18 Managing fire safety (maintenance, servicing routines and record keeping)

Subsequent tests

12.18 Six-monthly tests, three-yearly tests and subsequent annual tests should be carried out by a competent engineer.

Any defect should be recorded in the logbook and reported to the responsible person, and action should be taken to correct it. On completion of the work, a certificate of testing should be given to the responsible person, and attached to the premises logbook.

Log for monthly, biennial, annual and triennial inspections (escape or emergency) lighting

12.19

Date	Position of unit	Duration	Result	Defect	Action taken	Signature

Remarks and comments	
REFERENCE DATA	
Responsible person:	Name:
Address:	
System installed by:	
System commissioned by:	
Maintained by:	Until:
Telephone number for immediate service:	

Stand-by generator

12.20 To give greater assurance of reliability, when this type of equipment is installed, correct servicing is essential.

Frequency of inspection

12.21 Normally an agreement should be made with the supplier or other competent contractor for regular servicing. In some cases the stand-by generator should be tested weekly, however, the generally accepted period is monthly.

(a) Each engine-driven generating plant should be started up by a simulation of a failure of the supply to the normal lighting and allowed to energise the emergency lighting system for a continuous period of at least one hour.

During this period all equipment, eg pressurisation fans, should be checked to ascertain that they are operating satisfactorily and the ventilation system has operated. As mentioned at **12.14** above the luminaires and/or signs should be visually examined to ensure that they are functioning correctly.

At the end of the test period the system should be restored to normal operation and the charging arrangements for the back-up and the engine-starting battery checked for proper functioning. The fuel tanks should be left filled and the oil and coolant levels topped up as necessary.

Frequent starting of the plant followed by a few minutes on load is not recommended.

(b) The engine of each engine-driven generating plant with back-up batteries should be prevented from starting.

The escape lighting should then be energised solely from the back-up battery by simulation of a failure of the supply to the normal lighting for a period sufficient only to ensure that the change-over from normal supply to battery is functioning properly.

After this check that the starting system of the engine should be returned to normal operation and the engine allowed to start up in the normal way. This is to energise the emergency lighting system for a continuous period of at least one hour. During this period all luminaires and/or signs should be visually examined to ensure that they are functioning correctly.

At the end of the test period the system should be restored to normal operation and the charging arrangements for the back-up and engine starting batteries checked for proper functioning. Fuel tanks should be left filled and the oil and coolant levels topped up as necessary.

12.22 Managing fire safety (maintenance, servicing routines and record keeping)

Test(s) for the appropriate system eg stand-by generator, pressurisation system, fire alarm and escape lighting etc should be recorded in the appropriate sections of the logbook(s) and any defects noted and appropriate action taken to rectify the fault.

Subsequent tests

12.22 The six-monthly test, three-yearly test and subsequent annual tests should be carried out by a competent engineer.

Any defect should be recorded in the logbook and reported to the responsible person, and action should be taken to correct it. On work completion, a certificate of testing should be given to the responsible person, and attached to the premises logbook.

Log for monthly, biennial, annual and triennial inspections (stand-by generator)

12.23

Date	Position of unit	Duration	Result	Defect	Action taken	Signature
Remarks and comments						
REFERENCE DATA						
Responsible person:				Name:		
Address:						
System installed by:						

System commissioned by:	
Maintained by:	Until:
Telephone number for immediate service:	

Automatic sprinkler systems

12.24 These are normally the most expensive items of active fire protection that are installed in buildings. Where an automatic sprinkler system(s) is installed it should be commissioned in accordance with the provisions outlined in British Standard 5306 'Fire extinguishing installations and equipment on premises' Part 2: 1990 'Specification for sprinkler systems'.

Commissioning and acceptance tests

12.25 In carrying out the commissioning and acceptance tests the following results or better should be obtained.

Installation pipework

12.26 All installation pipework shall be pressure tested as follows:

(a) Dry pipework. Pneumatically, to not less than 2.5 bar for not less than 24 hours;

(b) Wet pipework. Hydraulically, to not less than 15 bar, or one and a half times the working pressure, whichever is the greater for not less than one hour.

Any faults disclosed, such as permanent distortion, rupture or leakage, shall be corrected and the test repeated.

Initial testing

12.27 Initial testing should be conducted to regular routine procedures. The automatic sprinkler system should be tested by the company (or an agreed representative) who commissioned the installation. The tests that are embodied in the daily, weekly and quarterly routine tests and any faults corrected.

In this connection the installer should provide the user with an inspection and checking programme for the system. This programme should include instructions on the actions to be taken in respect of faults, the operation of the system and, in particular, the procedure for emergency manual starting of any pumps, and details of the routines. These programmes should, where applicable, include the following items during the periodic routine inspections, tests and checks.

Weekly routine

12.28 General. Each part of the weekly routine should be carried out at intervals of not more than seven days.

Checks. The following should be checked and recorded.

(a) All water and air pressure gauge readings on installations.

Water motor alarm test. Each water motor alarm should be tested for not less than 30 seconds.

Note: this verifies that the alarm will not ring intermittently.

Automatic pump starting test. Tests on automatic pumps should include instructions to:

(a) check the fuel and engine lubricating oil levels in diesel engines;

(b) reduce water pressure on the starting device, thus simulating the condition of automatic starting;

(c) when the pump starts record the starting (cut-in) pressure, and check that this is correct; and

(d) on diesel pumps, check the oil pressure where gauges are fitted, and the flow of cooling water through open circuit cooling systems.

Diesel engine restarting test. Immediately after the pump start test, diesel engines should be tested as follows:

(a) run the engine for 30 minutes, or for the time recommended by the manufacturer, whichever is the longer; shut down the engine and immediately use the manual start test button and check that the engine restarts; and

(b) check the water level in the primary circuit of closed circuit cooling systems.

Note: oil pressure (where gauges are fitted), engine temperatures and coolant flow should be monitored throughout the test.

Oil hoses should be checked and a general check made for leakage of fuel, coolant or exhaust fumes.

Lead acid plante batteries. The electrolyte level and density of all lead acid plante cells (including diesel engine starter batteries and those for control panel power supplies) should be checked. If the density is low the battery charger should be checked and, if this is working normally, the battery or batteries affected should be replaced.

Acceptance tests

12.29 The authorities should be invited to witness the commissioning tests (see **12.25** above) and to inspect the system.

Completion certificate and documents

12.29A General. The installer of the system, or their supervising supplier, shall provide to the user the following:

(a) a completion certificate stating that the system complies with all of the appropriate requirements contained in British Standard 5306: Part 2: 1990 'Specification for sprinkler systems', and giving details of any departure from the appropriate recommendations contained in this standard;

note: this certificate should also contain appropriate certificates that components used in the system are suitable for sprinkler service;

(b) a copy of the independent laboratory test report where pipe assemblies do not comply with the recommendations of the standard; and

(c) a complete set of operating instructions and 'as-installed' drawings including identification of all valves and instruments used for testing and operation and a user's programme for inspection and checking.

Following tests and inspections any defect should be recorded in the logbook and reported to the responsible person, and immediate action should be taken to correct it. On completion of work, a certificate of testing should be given to the responsible person, and attached to the premises logbook.

Automatic sprinkler system-log for routine examinations, inspections and tests

12.30

Date	Duration	Result	Defect	Action taken	Signature

12.31 *Managing fire safety (maintenance, servicing routines and record keeping)*

Date	Duration	Result	Defect	Action taken	Signature

Remarks and comments	
REFERENCE DATA	
Responsible person:	Name:
Address:	
System installed by:	
System commissioned by:	
Maintained under contract by:	Until:
Telephone Number:	

Pressurisation systems (smoke control using pressure differentials)

12.31 These are probably the most important of all the 'active' fire protection measures installed within the premises as they normally protect escape routes from the upper and below ground levels. Therefore, it is essential that the system, as installed, conforms fully to British Standard 5588: Part 4: 1998 'Code of practice for smoke control using pressure differentials'. In addition, there has to be a full and continuous maintenance programme to ensure that it is always available for use.

There is a range of different systems. There are currently five different classifications outlined in the code of practice.

It is essential that the correct detail shown for the particular class of system is followed and when the system is commissioned and accepted, together with all routine tests and examinations, are fully and correctly indicated in the logbook.

The advice that follows is in general terms, as there are variations between the design recommendations for the different classifications of system.

Acceptance test

12.32   General. The only satisfactory way of establishing that a pressurisation installation is operating correctly and installed according to the design concept is to make physical measurements of the pressure differentials across closed doors and of the air velocities through open doors.

A test using cold smoke will demonstrate only the air movement in the building and short of an actual fire test a hot smoke test is almost impossible. However, cold smoke tests can sometimes reveal unwanted smoke flow paths caused by faulty construction.

The design criteria for a pressurisation system contains an allowance for adverse weather conditions and because it cannot be ensured that such weather conditions will occur on the selected day, even a fire test cannot be regarded as completely satisfactory as an acceptance test.

The acceptance test should consist of the following information.

Operating the pressure differential system and allow the fans to operate for at least ten minutes to establish a steady air pressure. Switch off the fans and leave all the other components of the system in their operational mode. Fully open all doors and allow them to close and then take:

(a) a measurement in all the pressurised spaces of the pressure differential between each space and the adjacent unpressurised space, all doors being closed (this may require the summation of pressure differences measured across sets of doors); and

(b) a measurement of the air velocity out of a representative selection of open doors that, when closed, separate the pressurised space and the accommodation space of the building.

Within 15 minutes switch the fans on and repeat the pressure measurements. The change in measurement between the first and second pressure readings should be compared with the performance requirements specified for the design pressure difference.

This test should only be carried out when the building is completed, the air conditioning and pressurisation systems balanced, and the whole system in working order, with every component functioning satisfactorily and directed by the initiating system into its correct emergency mode.

Where mechanical systems are used for normal ventilation or air conditioning it is normally advisable for the engineer responsible for the system to be present during the test(s).

12.32 *Managing fire safety (maintenance, servicing routines and record keeping)*

Measurement of pressure differentials. The measurement of pressure differentials between the various pressurised spaces and the adjoining unpressurised spaces should be carried out using an adjustable liquid manometer or other sensitive and properly calibrated device.

A convenient place to measure the pressure differentials will normally be across a closed door; small probe tubes are led to each side of the door, one tube passing through a door crack or under the door.

The two probes are then connected to the manometer by flexible tubes. It is important that the tube inserted through a door crack should pass through the crack and far enough into the space beyond for the open end to lie in a region of still air. In addition, it is suggested that this tube should contain an L-bend (at least 50 millimetres long) so that after insertion through the crack the tube can be rotated at right angles to the crack. This procedure will bring the open end into a region of still air.

It is important that the insertion of the probe tube in the door crack does not modify the leakage characteristic of the door, for instance by holding the door face away from the frame rebate. The position for the measuring probe should be chosen accordingly.

Attainment of correct pressure differential level. If there is any serious divergence from the design pressurisation level in any of the pressurised spaces the reason for this should be established.

There are three main reasons for failure to achieve the design pressurisation level; these are as follows:

(1) the rate of input of fresh air to the pressurised space is too low, and remains low even when substantial leakage paths are opened from this space to the open air for example by opening doors and windows;

(2) the leakage areas out of the pressurised spaces are greater than those assumed in the design calculation; or

(3) the leakage areas out of the rest of the building are insufficient meaning the rate of input of fresh air to the pressurised space will be lower than the design value, but will increase if substantial leakage paths are opened, for example by opening doors and windows.

Measurement of the air supplies to the pressurised space. The measuring procedure to be used in estimating the airflow at any point in a ventilation duct system is designed to determine the volume rate of airflow into any pressurised space and the following steps are required.

(a) Measure the total fan air supply volume rate of flow.

(b) Subtract the system leakage.

(c) Proportion the net airflow to the terminals from the measurements obtained during the regulating of the system.

The measurement of air supply to the pressurised space should first be made with the doors leading to the pressurised space closed and with normal leakage from the accommodation. If the measured air supply is less than the design value, it should be measured again with substantial leakage paths opened between the pressurised space and the outside.

The leakage paths for the pressurising air. If the measurements made show that the design input rate has been satisfactorily achieved in the installation, the reason for a low pressurisation level should be sought in the leakage areas out of either the pressurised spaces or the unpressurised space.

The cracks around doors and windows should be examined, with special attention to the gap at the bottom of all the doors. When any of the door or window gaps are found to be unacceptably large they should be reduced in size.

If all the doors and windows are found to be normally close-fitting, the enclosure of the pressurised space should be examined for other leakage paths that were not identified in the original design calculations. Extra leakage paths found should be sealed or the air volume input rate should be increased to allow the correct value of the pressurisation level to be achieved with the additional leakage paths effective.

Finally, the air leakage from the unpressurised spaces should be examined to ensure that this is in accordance with the values set out in the system design. If this is inadequate, the leakage should be increased to conform to the appropriate values.

Where an excess leakage from a pressurised space is corrected by increasing the air input supply an increase in the leakage from the unpressurised space(s) may also be required.

The tolerance permissible in pressurisation level. In theory there should not be any significant variation when the pressure differentials are measured. However, most enforcing authorities accept that the measured pressurisation level should not be lower than 80 % of the design value or greater than 60 Pascal's.

Measurement of airflow through an open door. This measurement should be made with a rotating vane anemometer or other suitable instrument, which has been properly calibrated.

The average velocity through an open door should be found by combining measurements at a sufficient number of points over the opening to ensure that vertical and horizontal asymmetry in the flow does not cause substantial inaccuracy. For high accuracy at least ten measurements at uniformly

12.33 *Managing fire safety (maintenance, servicing routines and record keeping)*

distributed positions in the door opening should be taken and an average value obtained for the airflow: high accuracy will usually require stable wind conditions and an empty building.

Following tests any defect should be recorded in the logbook and reported to the responsible person, and action should be taken to correct it. On completion of the work, a certificate of testing should be given to the responsible person, and attached to the premises logbook.

Information to be made available

12.33 To the approving authority. The approving authority should be provided with full details of the installation. These should include:

(a) full calculations showing the design criteria used;

(b) full specification details of the equipment used;

(c) complete plans showing position and protection of the fan and associated electrical control equipment, and the location of fresh air inlets;

(d) constructional details of the duct-work and duct terminals used for the pressurisation system;

(e) any other relevant constructional information required by the authority;

(f) full operational details describing in words and by diagram the exact sequence of actions that will occur in the pressurisation system and in the normal ventilating system when a fire occurs in the building;

(g) a complete maintenance schedule indicating the maintenance check needed for each item of the equipment and the frequency of this check; and

(h) on completion, the results of the tests carried out on the pressurisation system.

Information to be made available

12.34 To the building occupier/owner. The occupier/owner of the building should be provided with a clear description of the purpose and operation of the installation. This should include the following.

(a) A clear description of the purpose of the installation.

(b) A concise statement in words assisted by diagrams of the operation of the installation giving a clear indication of the sequence of events that will follow a fire alarm.

(c) A description of the function of each individual item of the installation with an indication of where in the building each part is situated.

(d) A complete maintenance schedule indicating the maintenance check needed for each item of the equipment and the frequency of this check.

(e) A checklist in the maintenance schedule of the actions necessary for maintenance, together with a logbook that will form a record of the maintenance carried out and in which any faults found are recorded. Any corrective actions, which may be required should also be recorded in this logbook.

(f) A set of drawings for retention on the site.

(g) A warning that alterations to partitioning or floor coverings under doors may affect the operation of the pressurisation system.

(h) A recommendation to inform occupants that a pressurisation system is installed and that, in the event of a fire, doors may be slightly harder or easier to open, and that there may be noise from the fans.

Frequency of tests

12.35 There are wide a range of tests that are required when a smoke control system using pressure differentials is installed in a building. These range from weekly, monthly, three-monthly, six-monthly and annually involving both maintenance and functional testing. The precise nature of the testing regime will be indicated in the information deposited with the approving authorities (see (g) at **12.33** above) and the owner/occupier (see (d) and (e) at **12.34** above) of a building.

Log for routine tests (pressure differential systems)

12.36

Date	Duration	Result	Defect	Action taken	Signature

Date	Duration	Result	Defect	Action taken	Signature

Remarks and comments	
REFERENCE DATA	
Responsible person:	Name:
Address:	
System installed by:	
System commissioned by:	
Maintained under contract by:	Until:
Telephone number for immediate service:	

Fire doors, fire shutters and dampers

12.37 Fire doors are provided throughout offices to staircases, lobbies and to a number of the escape routes from the premises. Fire doors, fire shutters and fire dampers are also provided to service risers, in compartment walls and floors and air conditioning and ventilation ductwork etc.

Fire doors and the associated protective equipment are considered to be one of the most important links in the range of fire protection measures provided in premises. Care in their selection, installation and maintenance cannot be over emphasised.

Failure of fire doors under fire conditions usually occurs either at the gap between the door and frame or where door fittings are provided eg hinges and locks.

The ability of a fire door to fulfil its design function depends upon the door being fully closed at the time when a fire occurs.

All fire doors installed in common parts of the premises should be examined daily to ensure that they are still self-closing. These doors should be examined weekly to ensure that they are closing fully in their frames.

Self-closing doors should be inspected at monthly intervals to ensure that they are capable of closing the door into its frame without undue delay. More frequent inspections should be introduced if there is a history of mechanical damage to doors that are in constant use. All fire doors, shutters and, as far as practicable, fire dampers, should be examined annually. Self-closing doors should be inspected and overhauled as necessary and the person supervising

Managing fire safety (maintenance, servicing routines and record keeping) **12.38**

the examination should issue a certificate stating that the equipment, particularly fire doors, is still capable of performing its design function.

A substantial number of fire doors are held open by automatic release mechanisms. These automatic release mechanisms are invariably linked to the fire alarm and any automatic fire detection system installed within a building. The majority of these systems conform to the provisions of British Standard 5839 'Fire detection and alarm systems for buildings; Part 3: Specifications' for automatic release mechanisms for certain fire protection equipment. The automatic releases should be tested and examined in accordance with the testing regime outlined in the document.

Log for routine examinations, inspections and tests

12.38

Date	Location of fire door, fire shutter or damper	Defect	Action taken	Signature

Remarks and comments		
REFERENCE DATA		
Responsible person:	Name:	
Address:		
System installed by:		
System commissioned by:		
Maintained under contract by:	Until:	
Telephone Number:		

Smoke control systems

12.39 Many buildings are now provided with sophisticated smoke control systems. These systems are used in a wide range of premises ranging from large shopping complexes, airport terminals to basement car parks etc. In some instances, eg basement car parks, an existing ventilation system is used for the removal of petrol vapours and fumes. In this example should a fire occur, the system increases the rate of extraction from six air changes per hour to the equivalent of ten air changes per hour.

The extensive range of systems means that in most cases the smoke control system has to be specifically designed for the premises being protected. The various acceptance, functional and routine tests, together with the maintenance procedures and examinations that are suggested are, 'without prejudice', to any additional requirements which may be imposed by the enforcing authorities.

Acceptance test

12.40 General. The enforcing authorities have to be satisfied that the system can achieve these rates of extraction for which it was designed when it operates in the 'fire mode'. The only satisfactory way of ensuring that the system is installed correctly is to take physical measurements of the rate of extraction.

A test using cold smoke will only demonstrate the air movement in the building and short of an actual fire, a hot smoke test is virtually impossible. Nevertheless, a smoke test of this type can sometimes reveal unwanted smoke paths caused by faulty construction.

Editorial Note. There are now some companies who specialise in hot smoke tests. These tests do not, apparently, affect the décor within the building.

Routine testing of the smoke control system

12.41 Weekly test. Each week the smoke system should be actuated in the fire mode. While the system is operating, checks should be made to ensure that the fans are running satisfactorily.

During this test a general examination should be made of the fans and associated drive mechanism with particular attention being given to the belt drive, motors, extraction system and actuation and control system associated with, eg the fire alarm system, automatic fire detection system or automatic sprinkler system.

Monthly test. Every month, in addition to the weekly test, the emergency power supply and stand-by equipment should be tested as follows.

(a) A failure of the primary power supply should be simulated and a check made that the system has automatically changed over to the secondary power supply. When a secondary power supply is provided by a diesel generator it should, on operation, energise the system for a minimum of one hour.

(b) A zero airflow condition should be simulated and a check made that the stand-by fans are operating.

Three-monthly. Every three months the motors, belt drives and fans should be subjected to a full maintenance programme ie this over and above the functional testing, which they receive weekly.

Six-monthly. At six-monthly intervals the emergency power system, dampers, grilles and the actuation and control system should be subjected to a full maintenance programme.

Annual test. Every twelve months the ventilation system including ductwork should be maintained and the maintenance engineer should provide a certificate indicating that the smoke control system is satisfactory and conforms to its design and performance criteria.

Following maintenance and tests, any defect should be recorded in the logbook and reported to the responsible person with action being taken to correct it. On completion of work, a certificate of testing should be given to the responsible person and attached to the premises logbook.

Conclusion

12.42 If the servicing and maintenance routines mentioned in this chapter are correctly implemented, these documents will virtually serve as checklists, in smaller premises, for relevant items of fire safety equipment. They will also complement the relevant 'checklists' used for fire risk assessments in larger premises.

13 Specialist information

In this chapter:	
Introduction	13.1
Historic buildings	13.2
Means of escape	13.9
Risk management	13.10
Structural alterations and extensions	13.14
Legal responsibility	13.16
Building control	13.17
Planning control	13.18
Dealing with contractors	13.19
Code of practice	13.20
Advising contractors of your fire safety policy	13.25
Other potential hazards with contractors	13.31
Construction (Design and Management) Regulations 1994 (CDM Regulations)	13.32
To what do the Regulations apply?	13.34
Health and safety plan	13.39
Preparation of health and safety plan by the planning supervisor	13.41
Where specific guidance be found	13.44
Fire engineering	13.69
Useful sources of information	13.72

Introduction

13.1 With many existing buildings it is necessary to remember that there are a significant number that are classified as being of special architectural or historic interest. To overcome the problems that this may cause, in particular cases, the authors of the current edition of *Approved Document B* (2000 Edition) have accepted that adherence to the guidance in this document could be unduly restrictive.

Guidance on the approach that may be relevant when dealing with these types of buildings is considered in this chapter.

The question of structural alterations and extensions to existing buildings is also considered, as well as guidance regarding a wide range of premises including information where specific advice can be obtained.

A number of new buildings are now being designed using a fire engineering approach. This is an alternative to the normal prescriptive approach and some of the pros and cons regarding this way of achieving an appropriate level of fire safety will also be outlined.

Historic buildings

13.2 On average each year, at least one important historic building of national importance is either totally destroyed or seriously damaged by fire. Like any old building, historic buildings are particularly vulnerable to fire, due to their age and the methods of construction used.

Throughout history timber has been used extensively in building construction which, due to its great age, will have become 'tinder dry' and very vulnerable to fire. Throughout the lifespan of an old building numerous alterations and extensions will have taken place. These alterations frequently leave concealed spaces through which fire can travel undetected. These concealed spaces are known to contribute significantly to the rapid development and spread of fire.

Throughout the UK many old buildings, which are not particularly important historically but are nevertheless fine buildings have been converted and adapted for business and commercial use. Common examples of this are where large country houses have been converted to hotels, conference centres and schools. It is a reasonable assumption that many buildings would have become dilapidated and beyond repair had it not been for the timely intervention of a commercial venture.

However, the occupation of an old building by a business presents additional problems, which are not normally encountered in modern buildings. This is particularly relevant when it comes to providing structural fire precautions, as very often the building is 'listed' by the local planning authority.

What are listed buildings

13.3 'Listed' buildings are generally older buildings that are protected for their architectural or historical value under the *Planning (Listed Buildings and Conservation Areas) Act 1990*.

13.4 *Specialist information*

Although a building has been listed, that does not mean it must remain entirely unaltered. There are strict controls with regard to demolition, either in whole or in part and with regard to the removal of features of architectural importance. Any alteration must preserve the character of the building and special consent is needed for alterations.

Buildings are now classified by grades, to show their relative importance. Where an employer is the occupier or owner of a listed building, the local planning authority will have made the company aware of the fact.

Listed building consent

13.4 Anyone wishing to demolish a listed building or to alter one in any way, which affects its interior or exterior character must first obtain listed building consent. The procedures are similar to those for obtaining planning permission. Application forms are obtained from the local authority, ie planning department and must be submitted with drawings, as would apply for normal planning permission.

Grants

13.5 An employer should also be aware that there might be some limited money available from the local authority for grants for restoration work to residential listed buildings and properties in conservation areas.

Grants for listed buildings may also be available from English Heritage who can be contacted on tel: 020 7973 3000.

Special problems

13.6 Conventional fire protection in buildings is usually achieved by means of structural fire precaution measures which are designed to restrict the spread of fire and smoke by the use of structural building elements such as fire resisting walls, doors and floors. These measures may well be necessary to comply with *Building Regulations 1991 (SI 1991 No 2768)* or fire authority requirements.

Upgrading structural elements

13.7 However, in historic or 'listed' buildings the measures required to achieve the necessary period of fire resistance often cannot be used or achieved. This is because upgrading the elements of structure very often conflicts with the architectural nature of the building.

These problems inevitably lead to a potential source of conflict between the need to provide adequate fire safety and the opposing need to preserve the architectural and historic character of the building.

Proprietary systems

13.8 However, there are a number of ways of upgrading structural elements by various special proprietary-coating systems. For example, there is a method of upgrading an architecturally attractive door to achieve 30 minutes' fire resistance by means of a layered intumescent paint system. For solutions to these types of problems the best advice is to seek advice from the local building control officer or local fire safety officer. In addition, it is always worth approaching the local planning officers who may be able to provide solutions to any problems, which they may have encountered in similar situations in other buildings.

Means of escape

13.9 The provision of adequate means of escape in case of fire can be a particular problem. It may be that an additional staircase is required to provide an escape route from the upper floors and the only way of achieving this is by constructing an external staircase. Sometimes it is possible to satisfy all parties by the provision of an architecturally acceptable staircase, which will blend in with the existing building. The use of a 'spiral staircase' can occasionally achieve this objective, provided that the staircase is only required for relatively small numbers of people.

This type of staircase has the added benefit of being fairly unobtrusive and takes up less space than a conventional steel 'dog-leg' staircase.

Another area of conflict can be found when installing fire defence equipment and signage in an architecturally sensitive building. The installation of fire alarm systems, emergency lighting, fire-fighting equipment and exit notices can have a decidedly detrimental effect on the interior appearance of an historic building.

Risk management

13.10 This is an area where the use of 'risk' management methods may be appropriate, in order to provide a mutually acceptable solution.

To illustrate the 'risk' management concept, a simple example might be that the provision of a sophisticated addressable fire detection system will be acceptable to the enforcing authority as a 'compensatory measure' for more conventional structural methods. By doing this it could be argued that the provision of such a system that gives very early detection of a fire, would

13.11 Specialist information

ensure both the speedy evacuation of the occupants, an immediate call to the fire service and, consequently, the early attendance of the fire service.

As risk management is potentially a very complex area it would probably be appropriate, if an employer were to contemplate this course of action, to employ a specialist fire safety consultant to negotiate with the enforcing authority on their behalf.

It is also advisable to discuss the problems with the relevant enforcing authority, as by discussion it may be possible to reach a mutually acceptable solution.

How can we help ourselves?

13.11 If a business occupies an older building, where a fire could develop and spread very rapidly, there are many things that can be done 'before the fire' with a view to minimising the potential effect on the business.

This is particularly relevant if your premises are located, as many are, in rural areas where the water supplies for fire-fighting may be limited or non-existent and situated a long way from the nearest fire station.

Making improvements

Checklist

13.12

> The list below should assist by identifying areas in which improvements can be made, very often at relatively little cost.
>
> - Ensure that the business name is indicated clearly and conspicuously on signs at each entrance to the site. This will greatly assist the fire and other emergency services in finding the premises in an emergency and ensure that there is no delay in receiving assistance.
>
> - Ensure that the buildings are provided with good vehicular access. Remember that fire engines are very large, wide heavy vehicles, which carry their own water supplies. Therefore, the closer they can get to a building on fire, the quicker the fire service can get to work to put the fire out. An employer should liase with the local fire service, as they will advise on widths of roadways, vehicle entrances and the weight requirements that are relevant for their vehicles.
>
> - An employer should identify any nearby open water supplies that could potentially be used for fire-fighting. As buildings located in

Specialist information **13.14**

> rural areas are generally only provided with a low volume mains water supply for domestic usage, the fire service will need to locate a nearby water supply of sufficient capacity to be able to fight a developing fire. Examples of potential supplies are rivers, large streams, ponds, lakes, and swimming pools. If none of these are available nearby, you may wish to consider providing a static water storage tank for fire-fighting purposes. Again, if you are considering this option, it is worth discussing the matter with your local fire service.
>
> - Consider setting up an 'in-house' 'fire team' from amongst employees.
>
> Where premises are located a long way from a fire station, a properly trained fire team could be quite an effective way of extinguishing or controlling a fire until the fire service arrive. The fire service should be consulted before the setting up of fire teams as they will be prepared to give advice on training, or they may even provide the training. However, for this service, they will probably charge a fee.
>
> The local fire service should be invited to visit the premises for familiarisation visits. Most rural locations are served by 'retained' firefighters, who have normal jobs and respond in an emergency to the fire station, when they are alerted on their pagers. The local fire officer will welcome an invitation for his crew to exercise their equipment and systems using the premises. These visits will have the added benefit of highlighting any access or water supply problems, which an employer will be able to remedy, before an emergency occurs.

Further reading

13.13

- *Heritage under fire: A guide to the protection of historic buildings* (Fire Protection Association ISBN 0 90 216790 1).

Structural alterations and extensions

13.14 There are occasions when many businesses when, due to expansion, alterations, improvements or refurbishment, carry out structural building work on the fabric of the buildings they occupy. Owners and occupiers of buildings do not appreciate that this is a time when accidental fires frequently occur.

Statistically, there are about eleven construction fires every day in the UK. These fires not only kill and injure people, but they can also be financially disastrous.

Due to the nature of building alterations and extensions, when walls and doors are removed as part of these processes, some or all of the structural fire safety measures provided to protect a building and occupants could be compromised. As a result, fires can grow and spread extremely rapidly. This highlights the need for both the contractor and the building occupier to be extremely vigilant to the possibility of a fire occurring during construction work. The risk of accidental fires is increased by the fact that many of the processes carried out by contractors involve the use of a naked flame or other hot work systems.

The Regulations

13.15 For the contractors involved, there are strict Regulations with which they must comply in order to ensure a safe working environment. The *Construction (Health, Safety and Welfare) Regulations 1996 (SI 1996 No 1592)* contain measures to prevent fires from happening and to make sure that all people on construction sites (including visitors) are protected if they do occur. The *Construction (Design and Management) Regulations 1994 (CDM) (SI 1994 No 3140)* also require that those who design, plan and carry out projects take construction fire safety into account.

Legal responsibility

13.16 An employer should also be aware that he has a legal responsibility to protect a contractor's employees under the *Health and Safety at Work etc Act 1974*. HSWA 1974, s 3(1) states that:

> 'It shall be the duty of every employer to conduct his undertaking in such a way as to ensure, so far as is reasonably practicable, that persons not in his employment who may be affected thereby are not thereby exposed to risks to their health or safety.'

Employers therefore have a clear duty to contractors and their employees, as well as to visitors, customers, members of the emergency services, neighbours, passers-by and the public at large.

This chapter sets out some basic advice for the owners and occupiers of buildings that are undergoing structural building alterations. It also provides guidance on how to address the problems associated with construction work.

Building control

13.17 The building control system exists to ensure that buildings are properly designed and constructed in order to ensure the health, safety, welfare and convenience of people using them.

Building control generally operates as a two-stage process, referred to as the full plans application. The first stage, known as the plan stage, requires the applicant to submit detailed plans for approval. These plans are checked by a building control officer to ensure that all necessary information is shown, and that it complies fully with the *Building Regulations 1991 (SI 1991 No 2768)*.

Wherever possible, applicants are given the opportunity to make any required amendments, before 'either' an approval, conditional approval or rejection is given.

The second stage, the inspection stage, starts when work commences on-site after which a series of site visits are made to check that the work proceeds in accordance with the plan, and therefore complies with *SI 1991 No 2768*.

Planning control

13.18 Planning in England and Wales is principally based on the provisions of the *Town and Country Planning Act 1990* as amended by the *Planning and Compensation Act 1991*. The main purpose of the system is to regulate the use and development of land and buildings in the public interest.

Planning permission is required for any 'development'. Development is defined in the *Town and Country Planning Act 1990, section 55* as:

> 'the carrying out of building, engineering, mining or other operations in, on, over or under land or the making of any material change in the of any buildings or other land.'

Planning permission is normally not required for internal alterations unless the building is listed as being of architectural or historic interest or unless the work involves sub division of residential accommodation or a change of use is involved. However, permission may still be required under the *Building Regulations 1991 (SI 1991 No 2768)*.

More detailed information on the *Building* and *Planning Regulations* can be found in the **CHAPTER 10: MANAGING FIRE SAFETY (PREVENTION OF FIRE AND LIAISON WITH AUTHORITIES)**.

Dealing with contractors

13.19 Construction work covers the activities of all types of construction sites, from the smallest internal jobs to large-scale complex projects including:

- general building and construction work;
- refurbishment work;
- maintenance and repair work;

13.20   *Specialist information*

- engineering construction work;
- civil engineering work; and
- demolition and dismantling work.

It is important that the owner or occupier of premises on which work is to be carried out ensures that any fire safety issues are addressed with the contractor *before* work commences on site. This may mean that fire safety clauses need to be inserted into the contract between the two parties involved.

Code of practice

13.20   As guidance on the inclusion of fire safety clauses in a contract, the *Joint Code of Practice on the Protection from Fire of Construction Sites and Buildings Undergoing Renovation,* is recommended. It is published by The Fire Protection Association, Bastille Court, 2 Paris Gardens, London SE1 8ND (tel: 020 7902 5300; website: www.thefpa.co.uk).

This guidance proposes that the main contractor should appoint a 'site fire safety co-ordinator' to be in charge of assessing the degree of fire risk and formulating and regularly updating the 'site fire safety plan'.

Fire safety plan

13.21   The site fire safety plan should detail:

- organisation of and responsibility for fire safety;
- general site precautions, fire detection and warning alarms;
- requirements for a 'hot work permit' system;
- site accommodation;
- fire escape and communications system, including evacuation plans and procedures for calling the fire brigade;
- fire brigade access, facilities and co-ordination;
- fire drills and training;
- effective security measures to minimise the risk of arson; and
- storage of materials and waste control system.

Role of site fire safety co-ordinator

13.22   The site fire safety co-ordinator must:

- ensure that all procedures, precautionary measures and safety standards, as specified in the site fire safety plan, are clearly understood and complied with by all those on the project site;
- ensure establishment of 'hot work permit' systems;
- carry out weekly checks of all fire-fighting equipment and test all alarm and detection devices;
- conduct weekly inspections of the escape routes, fire brigade access, fire-fighting facilities and work areas;
- liase with the local fire brigade for site inspections;
- liase with security personnel;
- keep a written record of all checks, inspections, tests and fire drill procedures;
- monitor arrangements and procedures for calling the fire brigade;
- during the alarm, oversee safe evacuation of site during the alarm, ensuring that all staff and visitors report to the assembly points; and
- promote a safe working environment.

Emergency procedures

13.23 The following emergency procedures should be implemented, where necessary:

- establish a means of warning of fire, eg hand bells, whistles etc;
- display written emergency procedures in prominent locations and give copies to all employees;
- maintain clear access to site and buildings;
- alert security personnel to unlock gates and doors in the event of an alarm; and
- install clear signs in prominent positions, indicating locations of fire access routes, escape routes and the positions of any dry riser inlets and fire extinguishers.

Designing out fire

13.24 Construction works should be designed and sequenced to accommodate:

- permanent fire escape stairs, including compartment walls;
- fire compartments in buildings under construction, including the installation of fire doors;

- fire protective materials to structural steelwork;
- planned fire-fighting shafts duly commissioned and maintained;
- lightning conductors;
- automatic fire detection systems;
- automatic sprinkler and other fixed fire-fighting installations; and
- adequate water supplies including hydrants, which are suitably marked and kept clear of obstruction.

The early use of portable fire extinguishers can represent the difference between a major incident and a fire being kept under control. To assist in this procedure, construction site personnel must be trained in the use of portable fire-fighting equipment and an adequate number of suitable types of portable extinguishers should be available. These fire appliances should be located in conspicuous positions near exits on each floor. When fire extinguishers are situated in the open air, they should be fixed on brackets approximately 500 millimetres above ground level at a suitably marked 'fire point' and protected from both work activities and adverse weather conditions.

All mechanically propelled site plant should carry an appropriate fire extinguisher, and extinguishers, hydrants and fire protection equipment should be maintained and regularly inspected by the site fire safety co-ordinator.

There must be an agreed procedure to ensure ongoing liaison, for the duration of the project, between the individual with the responsibility for fire safety within the company and the contractor's site fire safety co-ordinator.

Advising contractors of your fire safety policy

13.25 Whenever construction work is carried out within premises, particularly if premises are in normal use during the construction, it is important that the fire safety policy is compatible with any policy used by the contractors. This will ensure that if a fire occurs in any part of the premises, including those within the area of responsibility of the contractor, both sets of employees are made immediately aware of the incident so that a total evacuation can take place.

'Hot work' permits

13.26 It is essential that employees and contractors follow clear rules for 'hot work' such as cutting, welding, brazing, grinding, sawing, soldering, thawing frozen pipe, applying roof covering or sealing plastic shrink-wrap etc.

A formal permit-to-work systems system should be established, for employees and contractor's employees. The system that is established must ensure that no hot work is commenced until; a competent person within the company gives 'written authorisation'.

Formulating a 'hot work' permit system

13.27 The following guidelines are suggested as the basis for the formulation of a 'hot work' permit system.

Purpose of the system

13.28 To ensure that safety rules are followed in the prevention of fires and false alarms during torch, welding or other hot works.

Policy

13.29 All the necessary safety precautions set out below must be exercised during and after grinding, cutting, welding, brazing, or other open flame torch operations. If there is a practical way to do the job without 'hot work,' that process should be used in preference.

Checklist

13.30

Contractors, employees and others who perform welding or other 'hot work' operations must follow the following fire safety requirements.

- It is important to make sure extinguishers are at hand and that sparks or heat cannot set fire to surrounding materials. After the work has finished (usually an hour later) check the work site to make sure that there is no smouldering.

- When practical, objects to be cut, soldered, brazed, welded, or heated should be taken to a safe location to be worked on, preferably outside of the building. All combustible material should be removed from the area before the hot work starts.

- If it is impractical to move the object and if it is not possible to remove all fire hazards, positive means should be taken to confine the heat, sparks, and slag.

- A suitable fire extinguisher should be immediately available in the work area and maintained ready for use.

13.31 *Specialist information*

> - A fire blanket should be used to protect adjacent combustible materials or equipment from slag, sparks, and heat.
>
> - If the welding or open flame operation is such that normal fire prevention precautions are insufficient, such as welding on roofs of combustible material or wooden structures, another person should be assigned to guard against fire while the welding is being performed. After the welding or open flame operation is complete, sufficient time should be spent to observe possible flare-ups or other fire situation possibilities.
>
> - If the heat from welding or cutting on walls, floors, or ceilings could penetrate to adjacent areas, the same precautions must be taken on the opposite side from where the hot work is taking place.
>
> - If conditions exist where an automatic fire alarm detection system might be activated, the fire alarm detector head for that area should be bypassed, isolated or the detector head removed. It is important to remember to *reinstate* the system immediately following completion of the work.

Other potential hazards with contractors

13.31 In addition to the hazards detailed earlier with regard to 'hot work', there are other potential fire hazards associated with construction work, much of which is related to the extensive use of liquefied petroleum gas ('LPG') in the industry.

The risks can be prevented or minimised by:

- Making sure that liquefied petroleum gas (LPG) cylinders and other flammable materials are properly stored. LPG should be stored outside buildings in well-ventilated and secure areas.

- Storing flammable materials such as solvents and adhesives in lockable steel containers.

- Turning off liquefied petroleum gas (LPG) supplies at the cylinder when not in use. This is particularly important out of hours. Serious explosions have occurred after site huts have gradually filled with gas because a liquefied petroleum gas (LPG) heater has not been turned off. Site huts should be provided with adequate low level ventilation and LPG cylinders should not be kept inside them if it is not necessary.

- Making sure that liquefied petroleum gas (LPG) equipment and fittings are properly maintained. Damaged hoses and fittings or makeshift connections are extremely dangerous because they can easily lead to leaks in a tough construction environment.

- If there is any suspicion that liquefied petroleum gas (LPG) is leaking *stop* using it and check. Leaks can be identified by listening for a hiss, detecting a smell or by using soapy water to identify them but *never* by attempting to use a naked flame. Only light up when you are certain that there are no leaks and that any vapour which has leaked has dispersed.

- Tar boilers should not be left unattended.

- Keeping the site tidy by making sure rubbish is cleared away promptly and regularly.

- Avoiding the unnecessary stockpiling of combustible materials such as polystyrene and storing what is necessary away from any ignition sources. Limit what is taken onto site from the store to what is needed for a day's work.

- Consider the need for special precautions in areas where flammable atmospheres may develop, such as the use of volatile solvents or adhesives in enclosed areas.

- Avoid burning waste materials on site wherever possible. *Never* use petrol or similar accelerants to start or encourage fires.

- Make sure everyone abides by site rules on smoking.

- Plant used by contractors on construction sites also constitutes a potential danger. It is therefore important that all internal combustion engines of powered equipment be positioned in the open air or in a well-ventilated non-combustible enclosure. They should be separated from working areas and sited so that exhaust pipes are kept clear of combustible materials. Moreover, fuel tanks should not be filled whilst engines are running and compressors should be housed away from other plant in separate enclosures.

Construction (Design and Management) Regulations 1994 (CDM Regulations)

13.32 In recent years a growing number of larger companies, local authorities and other public bodies have had increasingly formalised procedures for checking the health and safety standards of contractors wishing to work for them. This process has been accelerated by the introduction of the *Construction (Design and Management) Regulations 1994 (CDM Regulations) (SI 1994 No 1340)* which require clients to satisfy themselves that potential principal contractors are capable of dealing with the health and safety issues associated with projects relating to their premises. SI 1994 No 1340 also places responsibilities on principal contractors in respect of sub-contractors. Consequently contractors are frequently required to provide details of their health and safety policies and generic risk assessments together with risk assessments and method statements for specific projects or activities. Many clients also take a hands-on approach in policing the work of contractors on their premises.

Objectives of the CDM Regulations

13.33 The *Construction (Design and Management) Regulations 1994 (CDM Regulations)* (*SI 1994 No 1340*) introduce a control framework that requires the effective management of all stages of a construction project from conception and design through to the commissioning of work, its planning and execution. They make particular reference to construction activities that are likely to pose significant risks to workers.

The overall aim of *SI 1994 No 1340* is to raise construction safety standards by improving co-ordination between various parties involved at both the preparation and the execution stages.

Clients are required to appoint a 'planning supervisor' and a 'principal contractor'. Both have duties in respect of 'the health and safety plan', which is the key to the operation of *SI 1994 No 1340*. Planning supervisors deal mainly with the designers and principal contractors deal mainly with sub-contractors.

The health and safety plan is initiated by the planning supervisor and forms part of the tendering documentation. Before actual construction work begins the plan should be scrutinised by the principal contractor in order to ensure that it is relevant to contractors and the site activities. Both planning supervisors and principal contractors should be appointed as early as possible during the planning stage of the project.

It is incumbent on clients to specify that contractors comply with the health and safety plan. In turn, contractors should ensure that the price of compliance is included within their tenders. Tenders should include identifiable sums for the management of health and safety and to deal with specific hazards. In this way, contracts should only be awarded to contractors or tenderers who are prepared to comply with health and safety requirements and standards.

To what do the Regulations apply?

13.34 The *Construction (Design and Management) Regulations 1994 (CDM Regulations)* (*SI 1994 No 1340*) apply to construction sites only when the 'construction work', is required to be notified to the Health and Safety Executive and where five or more people are likely to be employed at any one time. *SI 1994 No 1340)* implements (with minor exceptions) European Union Directive *92/57/EEC* on the minimum safety and health requirements at temporary or mobile construction sites. An Approved Code of

Practice, *L54: Managing construction for health and safety* (ISBN 071760792 5) has been issued to support these Regulations.

Generally, *SI 1994 No 1340* applies to construction work carried out on a construction site which is notifiable to the Health and Safety Executive (HSE) that is, a construction project which:

- is scheduled to last for more than 30 days;
- will involve more than 500 days of work;
- includes any demolition or dismantling of a structure regardless of the size or duration of the work; or
- involves five or more workers carrying out construction work at any one time.

SI 1994 No 1340 always applies when construction design work is involved.

Exclusions

13.35 The *Construction (Design and Management) Regulations* 1994 (*CDM Regulations*) (*SI 1994 No 1340*) in the main do *not* apply to construction work:

- where a client has reasonable grounds for believing that:
  (a) a project is not notifiable;
  (b) no more than four people are working at any one time (except for demolition and dismantling);
- of a minor nature where the local authority is the enforcing authority;
- carried out for a domestic client, unless (as a result of agreement or arrangement with the developer):
  (a) land is transferred to the client;
  (b) the developer undertakes to build on the land; and
  (c) after construction, the land will incorporate premises to be occupied by the client.

Planning supervisors role

13.36 Under the *Construction (Design and Management) Regulations 1994 (CDM Regulations) (SI 1994 No 1340)*, *Regulation 14* the planning supervisors must:

13.37 *Specialist information*

- ensure that the project is notified to the Health and Safety Executive (HSE) unless he has reasonable grounds for believing that the project is not notifiable (see **13.33** and **13.34** above);
- ensure that the design of a project includes:
  (a) reference to health and safety management;
  (b) adequate information regarding structure and materials;
- ensure co-operation between designers, for the purposes of compliance with their duties as designers;
- be able to give adequate advice to:
  (a) any client or contractor regarding competence of the designer(s) and their allocation of resources;
  (b) any client regarding competence of a contractor and a contractor's allocation of resources as well as the health and safety plan;
- ensure preparation of a health and safety file in respect of each structure comprised in the project, containing:
  (a) information concerning aspects of the project, structure or materials that may affect health and safety;
  (b) any other information that it is reasonably foreseeable will be necessary to ensure the health and safety of persons involved in construction, cleaning and maintenance or demolition work;
- review, amend or add to the health and safety file as necessary;
- ensure that, on completion of the construction work on each structure comprised in the project, the health and safety file is delivered to the client; and
- ensure that the health and safety plan is made available to every contractor before arrangements are made for them to manage or carry out construction work.

Principal contractors' role

13.37 The principal contractor's role is set out within *Regulation 14* of the *Construction (Design and Management) Regulations 1994 (CDM Regulations) (SI 1994 No 1340)*, which states the 'Requirements on and powers of principal contractor'.

The principal contractors must ensure the following.

- Take reasonable steps to ensure co-operation between all contractors, for the purposes of compliance with statutory requirements and/or prohibitions.

- So far as is reasonably practicable, ensure that every contractor and every employee complies with the health and safety plan.

- Take reasonable steps to ensure that only authorised persons are allowed where construction work is carried on.

- Ensure that notification particulars are displayed prominently so that they can be read by construction personnel.

- Provide the planning supervisor promptly with any information that:

  (a) is in his possession, or which he could ascertain by making reasonable enquiries of a contractor;

  (b) it is reasonable to believe the planning supervisor would include in the health and safety file; and

  (c) is not in the possession of the planning supervisor.

- For the purposes of enabling the principal contractor to comply with their duties they can:

  (a) give any reasonable directions to any contractor; and

  (b) include in the health and safety plan rules for the management of construction work reasonably required for health and safety management, with such rules being in writing and brought to the attention of persons who may be affected by them.

- So far as is reasonably practicable, ensure that every contractor is provided with comprehensible information on the health and safety risks to that contractor or to employees or other persons under that contractor's control.

- So far as is reasonably practicable, ensure that every contractor who is an employer provides their employees who are engaged in construction work with:

  (a) information relating to:

  (i) health and safety risks;

  (ii) preventative and protective measures;

  (iii) procedures to be followed in imminent danger and in danger areas;

  (iv) the identity of persons appointed to implement those procedures, and

  (b) health and safety training, both on recruitment and/or on exposure to new/increased risks, with training being repeated periodically where appropriate.

- Ensure that:

(a) employees/self-employed personnel are able to discuss and offer advice on matters which it can be reasonably foreseen will affect their health or safety; and

(b) there are arrangements for the co-ordination of employees' views (or their representatives' views).

Contractors' role

13.38 The *Construction (Design and Management) Regulations 1994 (CDM Regulations), Regulation 19 (SI 1994 No 1340)* sets out the 'Requirements and prohibitions on contractors'.

The contractors must:

- co-operate with the principal contractor, so that both can comply with their statutory duties;

- so far as is reasonably practicable, provide the principal contractor promptly with any information (including risk assessments for the purposes of the *Management of Health and Safety at Work Regulations 1999 (SI 1999 No 3242)*, which might affect the health and safety of construction workers, or those who might be affected by construction work, or which might justify a review of the health and safety plan;

- comply with any directions of the principal contractor given for the purpose of enabling the principal contractor to comply with his duties;

- comply with rules applicable to him in the health and safety plan;

- provide the principal contractor promptly with information relating to deaths, injuries, conditions and dangerous occurrences which the contractor is required to notify under the *Reporting of Injuries, Diseases and Dangerous Occurrences Regulations 1995 (SI 1995 No 3163)*;

- provide the principal contractor promptly with any information which:

    (a) is in the possession of the contractor, or which he could ascertain by reasonable enquiries of persons under their control;

    (b) it is reasonable to suppose the principal contractor would provide to the planning supervisor, for the purposes of inclusion in the health and safety file;

    (c) is not in the possession of the planning supervisor or the principal contractor;

- not allow any employee to work on construction work unless provided with:

    (a) the name of the planning supervisor;

    (b) the name of the principal contractor; and

(c) the contents of the health and safety plan (see **13.39** below) relating to work being carried out by the employee.

Self-employed personnel must also be provided with this information.

Health and safety plan

13.39 Before construction work starts, a duty is placed on the 'client' to ensure that the appointed planning supervisor prepares a document known as the 'health and safety plan'. The health and safety plan serves two purposes:

(a) during the pre-construction phase it brings together the health and safety information obtained from the client and designers; and

(b) during the construction phase it includes the principal contractor's health and safety policy and risk assessments.

The health and safety plan will continue to evolve and provide a focus for the co-ordination of health and safety matters as the construction work progresses.

Checklist

13.40

> The 'health and safety plan' forms the basis of the health and safety management structure, being part of the tender documents. The following should be applied to the plan.
>
> - Indicate, in general terms, the approach to health and safety to be adopted by everyone in accordance with the *Management of Health and Safety at Work Regulations 1999 (SI 1999 No 3242)*.
>
> - Identify the main health and safety hazards likely to occur to employees, self-employed operatives and the general public. These matters, will have been specified earlier by the client or designers. An example of this might be any work in compressed air.
>
> - Specify precautions to be taken.
>
> - Require work to be done to recognised technical standards and in accordance with published guidance, which should be specified within the plan.
>
> Before the commencement of work, the planning supervisor should acquaint both principal contractor and contractors with the health and safety plan, so that they can agree to it or to modifications within it, and then draw up tenders.

Preparation of health and safety plan by the planning supervisor

13.41   The planning supervisors should:

- ensure the preparation of the health and safety plan;
- ensure that the plan forms part of the tender documentation;
- investigate significant differences in tender documents relating to the plan;
- assess the adequacy of sums specified in tender's vis-à-vis the plan;
- advise the client on the adequacy of tenders; and
- review the plan if the basis of the original advice changes.

The principal contractor

13.42   The principal contractor should ensure that the health and safety plan:

- translates into an intelligible working document for all those involved in the construction phase;
- incorporates arrangements submitted by individual contractors for the overall management of health and safety;
- includes arrangements enabling them to comply with the duties placed upon them;
- specifies detailed arrangements for monitoring compliance with health and safety law;
- includes arrangements for assessing competence of subcontractors; and
- can be modified or updated in the light of the experience and information provided by contractors.

Health and safety file

13.43   The health and safety file is a permanent record containing information about the particulars and arrangements relating to the design, methods and materials, maintenance and other information relating to each structure within the project. In practice, the 'file' amounts to a manual to alert those who will be responsible for the structure after construction to safety matters which must be managed after hand over. The manual should contain appropriate information regarding maintenance, repair, renovation and demolition.

Specialist information **13.46**

Where specific guidance be found

13.44 This chapter contains a miscellany of fire safety information on subjects where, specific guidance is somewhat limited. The intention is to provide the reader with basic helpful guidance and a direction in which to look for further specific help. It lists particular types of premises, businesses or building types, which are not specifically covered elsewhere in this handbook.

Where no further explanation is considered necessary, relevant legislation and guidance publications are merely listed under the various headings. However, where it is considered appropriate, a brief explanation of the fire safety requirements for the particular type of premises is provided together with any legislation and guidance publications.

It should be noted that if any of the business premises listed below employ at least one person, the *Fire Precautions (Workplace) Regulations 1997 (SI 1997 No 1840) will* apply.

For further information, please see **CHAPTER 3: LEGISLATION AND REGULATIONS (THE FIRE PRECAUTIONS (WORKPLACE) REGULATIONS 1997 AND 1999)** and **CHAPTER 3: HOW TO CARRY OUT A FIRE RISK ASSESSMENT**.

Additionally, depending on the type of business and the number of employees, a fire certificate may be required under the *Fire Precautions Act 1971*. For further information, please see **CHAPTER 2: FIRE PRECAUTIONS ACT 1971**.

Air supported structures

13.45

- *Approved Document B: The Building Regulations 1991*

The *Approved Document B* (2000 edition) document, includes new guidance on the use of air supported structures, structures covered with flexible membranes and Polytetraflouroethylene (PTFE) based materials.

Atrium buildings

13.46

- London District Surveyors' Association – *Guide No. 2 – Atrium Buildings*.
- Fire Protection Association *Code of Practice for the Construction of Buildings Appendix 10 – Atrium Buildings*.
- Building Research Establishment Code: *BR 258 – Design approaches for smoke control in atrium buildings*.

13.47 *Specialist information*

- British Standard: *BS 5588 – Fire precautions in the design and construction of buildings.* Part 7: 1997 Code of practice for atrium buildings.

Caravan sites

13.47

- Health and Safety Executive (HSE) Chemical Safety booklet.
  - CHIS 5 – *Small-scale use of LPG in cylinders 1999.*
- Health and Safety Executive (HSE) Chemical Safety booklet.
  - CHIS 4 – *Use of LPG in small bulk tanks 1999.*

Children's homes

13.48 These local authority controlled establishments must carry out fire drills and practices and, in addition, consult with the fire authority.

- *Children's Homes Regulations 1991 (SI 1991 No 1506).*

Cinemas

13.49 Safety in cinemas is controlled by:

- *Cinematograph (Safety) Regulations 1955 (SI 1955 No 1129)*, as subsequently amended.

The regulations state that cinemas must be provided with:

[Regulation 2]

(a) adequate, clearly marked exits, so placed as to afford safe means of exit;
(b) doors, which are easily and fully openable outwards;
(c) passages and stairways kept free from obstruction;

[Regulation 5]

(d) suitable and properly maintained fire appliances;
(e) proper instruction of licensee and staff on fire precautions;
(f) treatment of curtains so that they will not readily catch fire;
(g) use of non-flammable substances for cleaning film or projectors;

[Regulation 6]

(h)   prohibition on smoking in certain parts of the premises; and

[Regulation 24]

(j)   appropriate siting of heating appliances.

Computer suites

13.50

- British Standard: *BS 6266:1982: Code of practice for fire protection for electronic data processing installations.*
- British Standard: *BS 5306: Fire extinguishing and equipment on premises*:
  - Part 0:1986: Guide for the Selection of Installed Systems and Other Fire Equipment;
  - Part 4:1986 – Requirements for Carbon Dioxide Systems;
  - Part 5.1:1992 – Specification for Halon 1301 Total Flooding Systems; and
  - Part 5.2:1984 – Halon 1211 Total Flooding Systems.

Community homes

13.51   Similar to children's homes, these local authority controlled establishments must carry out fire drills and practices and, in addition, consult with the fire authority.

- *Children's Homes Regulations 1991 (SI 1991 No 1506)*.

Construction sites

13.52

- *Construction (Design and Management) Regulations 1994 (CDM Regulations) (SI 1994 No 3140)* as amended.
- *Construction (Health, Safety and Welfare) Regulations 1996 (CHSW) (SI 1996 No 1592)*.
- *Highly Flammable Liquids and Liquefied Petroleum Gases Regulations 1972 (SI 1972 No 917)* as amended.
- *Electricity at Work Regulations 1989 (SI 1989 No 635)* as amended.
- *Health and Safety (Safety Signs and Signals) Regulations 1996 (SI 1996 No 341)*.
- *Work in Compressed Air Regulations 1996 (SI 1996 No 1656)*.

13.53 *Specialist information*

- *Confined Spaces Regulations 1997 (SI 1997 No 1713).*
- *Petroleum (Consolidation) Act 1928.*
- Health and Safety Executive (HSE) Health and Safety: Guidance Booklet:
  - *HS(G) 168 – Fire safety in construction: guidance for clients, designers and those managing and carrying out construction work involving significant fire risks 1997.*
- Health and Safety Executive (HSE) Health and Safety: Guidance Booklet – L 54:
  - *Managing construction for health and safety. The Construction (Design and Management) Regulations 1994 – Approved Code of Practice 1995.*
- Health and Safety Executive (HSE) Chemical Safety Booklet:
  - *CHIS 4 – Use of LPG in small bulk tanks 1999.*
- Health and Safety Executive (HSE) Construction Safety booklet:
  - *CIS51 – Construction fire safety 1997.*
- Health and Safety Executive (HSE) Construction Safety booklet:
  - *CIS17 – Construction health and safety checklist. Revised 1996.*

Factories

13.53 *Fire Precautions (Factories, Offices, Shops and Railway Premises) Order 1989 (SI 1989 No 76).*

Fire Precautions Act 1971. Guide to fire precautions in existing places of work that require a fire certificate. Factories, offices, shops and railway premises.

Gaming houses (casinos, bingo halls etc)

13.54 The issue, or retention on renewal, of a licence to operate a gaming facility is subject to the applicant having complied with all fire safety requirements under the:

- *Gaming Act 1968.*

Garages/filling stations/workshops

13.55

- Health and Safety Executive (HSE) Health and Safety: Guidance Booklet:
  - *HS (G) 67 – Health and safety in motor vehicle repair 1997.*

- Health and Safety Executive (HSE) Health and Safety: Guidance Booklet:
  - HS (G) 139 – *The safe use of compressed gases in welding, flame cutting and allied processes 1997*.
- Health and Safety Executive (HSE) Health and Safety: Guidance Booklet;
  - HS (G) 146 – *Dispensing petrol: assessing and controlling the risk of fire and explosion at sites where petrol is stored and dispensed as a fuel 1996*.
- Health and Safety Executive (HSE) Health and Safety: Guidance Booklet:
  - HS (G) 41 – *Petrol filling stations: construction and operation 1990*.
- Health and Safety Executive (HSE) Engineering Safety Booklet:
  - EIS1 – *Hot work on vehicle wheels 1997*.
- Health and Safety Executive (HSE) Chemical Safety booklet:
  - CHIS 4 – *Use of LPG in small bulk tanks 1999*.
- Health and Safety Executive (HSE) Chemical Safety booklet:
  - COP 6 – *Plastic containers with nominal capacities up to 5 litres for petroleum spirit: requirements for testing and marking or labelling (in support of SI 1982 No 830) – Approved Code of Practice 1982*.

Hospitals

13.56 FIRECODE (and FIRECODE Scotland).

A suite of documents aimed at healthcare premises. Available from HMSO, ISBN various – contact NHS Estates (tel: 0113 254 7000) for details.

Hotels, guest houses, boarding houses and bed and breakfast accommodation

13.57

- *Fire Precautions (Hotels and Boarding Houses) Order 1972 (SI 1972 No 238)*.

  Fire Precautions Act 1971. Guide to Fire Precautions in Premises used as Hotels and Boarding Houses which require a Fire Certificate.

Available from HMSO.

13.58 *Specialist information*

Fire Precautions Act 1971. Fire Safety Management in Hotels and Boarding Houses.

Available from HMSO.

Railway premises

13.58

- *Fire Precautions (Factories, Offices, Shops and Railway Premises) Order 1989 (SI 1989 No 76).*

- *Fire Precautions (Sub-surface Railway Stations) Regulations 1989 (SI 1989 No 1401) as amended.*

  Fire Precautions Act 1971. Guide to fire precautions in existing places of work that require a fire certificate. Factories, offices, shops and railway premises.

Available from HMSO.

Residential and nursing homes

13.59 Similar requirements to children's and community homes, in that these local authority controlled establishments must carry out fire drills and practices and, in addition, consult with the fire authority. In addition, satisfactory arrangements must be made for the evacuation of patients and staff in the event of fire.

- Residential homes: *Residential Care Homes Regulations 1984 (SI 1984 No 1345).*

- Nursing homes: *Nursing Homes and Mental Nursing Homes Regulations 1984 (SI 1984 No 1578).*

- HSE Health and Safety: Guidance Booklet.
  - *HS (G) 104 – Health and safety in residential care homes 1993.*

Schools

13.60 In the case of local authority controlled schools, including special schools, the 'health and safety of their occupants, and in particular, their safe escape in the event of fire, must be reasonably assured', with particular reference to the design, construction, limitation of surface flame spread and fire resistance of structure and materials therein.

- *New Constructional Standards for School Buildings – Department for Education and Employment.*

- *Standards for School Premises Regulations 1972 (SI 1972 No 2051).*

- *Fire and the Design of Educational Buildings: Building Bulletin 7.*★

★ Although it is still available this document has been largely superseded by *Approved Document B* (2000 Edition).

Available from HMSO.

Shopping centres

13.61

- British Standard: *BS 5588 – Fire precautions in the design and construction of buildings*; Part 10 – Shopping Complexes.
- Building Research Establishment Codes.
  - BR 186 – *Design Principles for Smoke Ventilation in Enclosed Shopping Centres.*
- British Standard: *BS 5588: Fire precautions in the design and construction of buildings.*
  - Part 11:1997 Code of practice for shops, offices, industrial, storage and other similar buildings.
- London District Surveyors' Association.
  - *Guide No. 3 – Phased Evacuation.*

Shops

13.62

- *Fire Precautions (Factories, Offices, Shops and Railway Premises) Order 1989* (SI 1989 No 76).

  Fire Precautions Act 1971. Guide to fire precautions in existing places of work that require a fire certificate. Factories, offices, shops and railway premises.

Available from HMSO.

- British Standard: *BS 5588: Fire precautions in the design and construction of buildings.*
  - Part 11:1997 Code of practice for shops, offices, industrial, storage and other similar buildings.
- London District Surveyors' Association.
  - *Guide No. 3 – Phased Evacuation.*
- *Explosives Act 1875* (Storage of fireworks).

Sports facilities

13.63

- Fire Safety and Safety of Places of Sport Act 1987.
  Guide to Safety at Sports Grounds.

Available from HMSO.

Theatres

13.64 Under *Theatres Act 1968, s 12(1)* premises used for the public performance of a play must be licensed. The conditions for obtaining or having a licence renewed, include compliance with rules relating to the safety of persons in the theatre.

The rules cover:

- staff fire drills;
- provision of fire-fighting equipment;
- maintenance of the safety curtain;
- communication with the fire service;
- gangways and seating should be correctly arranged and free from obstruction;
- doors and exits should be indicated and show the method of opening;
- safe lighting arrangements;
- scenery and draperies must be non-flammable; and
- controls over smoking and overcrowding.

Outside public events, concerts and similar activities

13.65

- Health and Safety Executive (HSE) Health and Safety: Guidance Booklet.
   ○ HS(G) 195 – *The event safety guide – A guide to health, safety and welfare at music and similar events 1999.*
- Health and Safety Executive (HSE) Health and Safety: Guidance Booklet.
   ○ HS(G) 123 – *Working together on firework displays: A guide to safety for firework display organisers and operators 1999.*

Specialist information **13.66**

- Health and Safety Executive (HSE) Health and Safety: Guidance Booklet.
  - HS(G) 124 – *Giving your own firework display: how to run and fire it safely 1995.*
- Health and Safety Executive (HSE) Health and Safety: Guidance Booklet.
  - HS(G) 175 – *Fairgrounds and amusement parks: guidance on safe practice 1997.*
- Health and Safety Executive (HSE) Approved Code of Practice.
  - COP 15 – *Zoos: safety, health and welfare standards for employers and persons at work – Approved Code of Practice and Guidance Notes 1985.*
- Health and Safety Executive (HSE) Health and Safety: Guidance Booklet.
  - HS(G) 112 – *Health and safety at motor sports events: a guide for employers and organisers 1999.*
- Health and Safety Executive (HSE) Health and Safety: Guidance Booklet.
  - HS(G) 105 – *Health and safety in horse riding establishments 1993.*
- Health and Safety Executive (HSE) Health and Safety: Guidance Booklet.
  - HS(G) 154 – *Managing crowds safely 1996.*
- Health and Safety Executive (HSE) Health and Safety: Guidance Booklet.
  - L 77 – *Guidance to the licensing authority on the Adventure Activities Licensing Regulations 1996. The Activity Centres (Young Persons' Safety) Act 1995 – Guidance on regulations 1996.*

  Guide to health, safety and welfare at pop concerts and similar events.

Available from HMSO.

Offices

13.66

- *Fire Precautions (Factories, Offices, Shops and Railway Premises) Order 1989 (SI 1989 No 76).*

  Fire Precautions Act 1971. Guide to fire precautions in existing places of work that require a fire certificate. Factories, offices, shops and railway premises.

Available from HMSO.

13.67 *Specialist information*

- British Standard: *BS 5588: Fire precautions in the design and construction of buildings.*
  - Part 11:1997 Code of practice for shops, offices, industrial, storage and other similar buildings.
- London District Surveyors' Association.
  - *Guide No. 3 – Phased Evacuation.*

Premises used for public and private music, dancing etc

13.67 The issue or retention on renewal, of a licence to operate premises for public music or entertainment is subject to the applicant having complied with all fire safety requirements under the *Local Government (Miscellaneous Provisions) Act 1982.*

Similarly, in the case of premises used for private music/dancing, for example a dancing school, there must be compliance with the fire safety requirements contained within the *Private Places of Entertainment (Licensing) Act 1967.*

- *Local Government (Miscellaneous Provisions) Act 1982.*
- *Private Places of Entertainment (Licensing) Act 1967.*
- *Guide to fire precautions in existing places of entertainment and like premises.*

Available from HMSO.

- British Standard: *BS 5588 – Fire precautions in the design and construction of buildings.*
  - Part 6 – Places of assembly.

Warehouses

13.68

- *Fire Precautions (Factories, Offices, Shops and Railway Premises) Order 1989 (SI 1989 No 76).*
- British Standard: *BS 5588: Fire precautions in the design and construction of buildings.*
  - Part 11:1997 Code of practice for shops, offices, industrial, storage and other similar buildings.
- Health and Safety Executive (HSE) Health and Safety: Guidance Booklet.
  - HS(G) 76 – *Health and safety in retail and wholesale warehouses 1992.*
- Health and Safety Executive (HSE) Chemical Safety booklet.

○   HS (G) 71 – *Chemical warehousing: storage of packaged dangerous substances. 1998.*

Fire engineering

13.69   In clause 0.11 of *Approved Document B* (2000 Edition) it is suggested that:

> 'Fire safety engineering can provide an alternative approach to fire safety. It may be the only practical way to achieve a satisfactory standard of fire safety in some large and complex buildings, and in buildings containing different uses, eg airport terminals. Fire safety engineering may also be suitable for solving a problem with an aspect of the building design which otherwise follows the provisions in this document.
>
> British Standard Draft for Development *(DD) 240 Fire safety engineering in buildings* provides a framework and guidance on the design and assessment of fire safety measures in buildings. Following the discipline of DD 240 should enable designers and building control bodies to be aware of the relevant issues, the need to consider the complete fire-safety system, and to follow a disciplined analytical framework.'

This draft for development, *Fire safety engineering in buildings* was published in 1997, and is in the process of being superseded by a new British Standard Code of practice, ie British Standard *7974: 2001 Code of Practice on the Application of Fire Safety Engineering Principles to the Design of Buildings.* This Code of Practice, which has been published, is to be supported by a number of published documents (PD) as follows.

- PD0 – Guide to design framework and fire engineering procedures.
- PD1 – Initiation and development of fire within the enclosure of origin.
- PD2 – Spread of smoke and toxic gases within and beyond the enclosure of origin.
- PD3 – Structural response and fire spread beyond the enclosure of origin.
- PD4 – Detection of fire and activation of protection systems.
- PD5 – Fire Service Intervention.
- PD6 – Evacuation.
- PD7 – Probabilistic risk assessment.

It is worth assessing the general thrust of these documents.

Advantages

13.70

- fire safety measures designed with specific objectives;
- innovation in design;
- possible reduction in structural fire protection without reduction in fire safety;
- use of the most recent fire safety practices; and
- consideration of alternative fire strategies, based upon a cost benefit principle of the loss prevention measures.

Disadvantages

13.71

- need for enforcing authorities to recruit staff with the necessary skills to assess proposals;
- lack of information and data to consider acceptance of innovative proposals;
- increase in design time, negotiations and costs;
- possible restriction on subsequent use of the building; and
- conflict with the prescriptive codes – which are generally acceptable to the enforcing authorities.

The relatively new concepts outlined in the Code of Practice and in the accompanying 'published documents', will obviously be adopted in a number of new and possibly existing buildings. The general thrust of the documents will be to modify some of the existing prescriptive standards that are currently accepted.

Whether this will lead to an improvement in 'fire safety' will take many years to prove. There are undoubted advantages in being able to assess certain aspects of fire resistance during the course of a fire. The adoption of some of the techniques that are proposed will clearly indicate that by means of calculation any suggested modifications will not lessen the 'fire safety' in a particular building 'at the time it is erected'. The use of the 'probabilistic risk assessment' involving the likelihood of a fire occurring will probably increase in importance as more statistical data becomes available.

The potential downside is whether it is possible to assume what will happen in a particular building throughout its working life. A reasonable assumption for the life of most buildings is about 60 years. When the changes in working practices, use of materials and other technological advantages are considered, the 'fire load' that was assessed at the time of construction may be adversely affected. The siting of fire stations may also have an impact as fire service intervention may vary as the fire risk categories in particular parts of a brigade's area are reassessed. Some major brigades have in recent years closed ten per cent or so of their fire stations, even though the number of calls is increasing.

Employers, owners and occupiers have other responsibilities imposed upon them by legislation and the need to carry out fire risk assessments and review existing alterations to working practices, fire load etc. In many cases it would be impracticable to provide additional staircases in a building that was already constructed because an apparent reduction in standard had been accepted some years before. Where the prescriptive codes have been adopted, at the design stage, it is much more unusual for the enforcing authorities to require significant improvements to existing standards as they would have agreed to them previously.

There are going to be obvious benefits from the adoption of a 'fire engineering' approach, but enforcing authorities and employers alike will need to be satisfied that the modifications or, arguably, the relaxations that are to be accepted, in a particular case, will not adversely affect the 'fire safety' of building.

In any negotiations involving an approach where this British Standard Code of Practice, together with the published documents are likely to be used, the earliest possible contact should be made with the enforcing authorities. The enforcing authorities will need to be satisfied that all of the relevant published documents have been considered before accepting a variation to the more normally accepted standards.

With the current concerns regarding 'terrorist' activities, although they are not necessarily a 'fire safety' matter enforcing authorities may be resistant to change, eg with regard to relaxation in travel distances, fire resistance and fire loads etc.

Useful sources of information

13.72   Several government department websites can be of great assistance when searching for information regarding specific matters. The Office of the Deputy Prime Minister is extremely useful in this respect, as it contains information regarding the construction of buildings. The relevant range of Approved Documents, which support the *Building Regulations 1991 (SI 1991 No 2768)*, can, in the most part, be downloaded free of charge from their website at: http://www.odpm.gov.uk/ This department is also responsible for fire services and on the relevant section of the website are reports relating to fire statistics, fire safety and general guidance.

The Health and Safety Executive can be reached from this website or directly via http://www.hse.gov.uk/ This website contains a range of information, much of which is free and can be downloaded from the site. A particular item, which has recently been introduced and is available from the website are the *Dangerous Substances and Explosive Atmospheres Regulations 2002 (DSEAR) (SI 2002 No 2776)*.

These Regulations have been introduced progressively from 9 December 2002 and all of the Regulations will be in force from 30 June 2003. Existing premises will have to be altered and fully conform to these regulations from 2006. All new premises or where premises are altered will, after 30 June 2003 need to conform to these Regulations.

Enforcement of these Regulations is the responsibility of the following agencies.

(1) The Health and Safety Executive or local authorities depending on the allocation of premises under the *Health and Safety (Enforcing Authority) Regulations 1998 (SI 1998 No 494)*. In the main, the Health and Safety Executive will enforce the regulations at industrial premises and local authorities (Environmental Health Officers) elsewhere eg in retail premises.

(2) Fire Brigades at most premises subject to *SI 2002 No 2776* in relation to general fire precautions such as means of escape.

(3) At retail petrol filling stations in relation to storage and dispensing of petrol, liquefied petroleum gas (LPG) and any other fuel subject to *SI 2002 No 2776*, the petroleum licensing authorities.

A leaflet *Fire and Explosion – How safe is your workplace?* is available together with a number of other free leaflets on the website.

It is recommended that all fire safety managers or persons with a responsibility for fire 'visit' these websites at regular intervals as it is a relatively simple method for keeping abreast of current developments.

Health and Safety guides, Regulations and Orders are available from HMSO. General Enquiries: HMSO, St.Clements House, 2-16 Colegate, Norwich, NR3 1BQ. Tel: 01603 723011; website: www.hmso.gov.uk

14 Fire (before, during and after)

In this chapter:	
Introduction	14.1
Planning	14.9
What are 'salvage efforts'?	14.14
Review and test your plan	14.17

Introduction

14.1 A fire can be a disastrous for any business, whether it is a small corner shop or a large manufacturing plant. It is estimated that some 70 per cent of businesses involved in major fires either do not re-open or subsequently fail within three years of the fire.

Whilst an employer may have invested considerable financial resources in preventing a fire from occurring in the first place, it cannot be ignored that prevention does not always work: accidents do happen and fires will occur.

The fire risk assessments shown in **CHAPTER 7: HOW TO CARRY OUT A FIRE RISK ASSESSMENT**, that have to be carried out, will hopefully minimise the extent of any fire damage. The more attention that is given to this aspect before a fire occurs should minimise the problems that invariably happen during and after a fire.

During a fire the primary concern is, of course, for the safe evacuation of the staff and of any visitors to the premises. In this regard the fire safety routines contained in **CHAPTER 9: MANAGING FIRE SAFETY (FIRE SAFETY PROCEDURES, TRAINING AND FIRE MANUAL)** should be of considerable benefit.

An employer may consider that, in the event of a fire, the company is well insured and, all of the losses that may be incurred, are to be covered by transferring most of the risk to the insurance company. Unfortunately the owners of many businesses fail to appreciate that although the more obvious aspects of a fire loss are covered, such as the cost of rebuilding and the replacement of stock and equipment, many of the less obvious consequences of a fire are not protected. Certainly many insurance policies cover a business for consequential loss. However it is extremely difficult to quantify and anticipate all the possible losses that are likely to result from a fire.

Some of the potential losses and business interruption problems that are the consequences of a fire are discussed below.

Damaged reputation

14.2 In any business an employer will have developed a reputation with customers as a reliable supplier. When an employer is unable to continue to supply their customers the company's reputation will inevitably suffer. Many businesses form part of the local community where most of their employees and their families live. Should employees have to be laid off as a result of a fire they will invariably suffer financial hardship. Other local businesses that supply a company involved in a fire and rely on them for some of their income will also be financially affected. This 'knock on' effect in smaller communities can be considerable and will inevitably have a negative effect on a company's reputation.

Financial security

14.3 When a business is not operating normally, income will reduce and an employer may be faced with immediate cash flow problems. These problems could end up threatening the business's financial security as staff and suppliers will still need to be paid.

Loss of employees

14.4 Where an employer is unable to carry on with the business, as normal, it is quite possible that the company will need to make some, or all, of the employees redundant.

The other possibility is that employees will seek other employment, unless an employer is able to retain them on full pay. As a result, an employer may lose key employees, which will make the task of resuming normal business much more difficult.

Loss of customers and market share

14.5 No matter how sympathetic or loyal customers may be they will still need their supplies, whether it is a small item, eg a newspaper from the corner shop or a large order from a major organisation. When goods or services cannot be provided customers will obtain them from elsewhere. When the clientele have moved to another supplier, the new supplier will endeavour to retain them.

Even a small business such as a corner newspaper shop cannot be guaranteed as having a market share. Once lost, it will be extremely difficult to regain the same share of the market place.

Loss of key suppliers

14.6 Suppliers (particularly if they are supplying specialised or customised needs) can also be placed in a difficult financial situation if an employer cannot take delivery of their goods. As a result, they will urgently be seeking new customers and outlets to fill their production capacity. Once this happens an employer may find it very difficult to immediately obtain suitable supplies that are tailored to the company's specific needs when normal business activity is resumed.

An employer should prepare the business for the worst possible scenario, by pre-planning and preparing a survival plan for recovery after a fire. Irrespective of the size of the business, it can benefit from this pre-planning process.

Formulate a survival plan for recovery after a fire

14.7 History and experience have shown us that many organisations that have undergone serious fires, or similar disasters, have made the recovery process harder (or even impossible) for themselves by not planning ahead for disaster recovery.

By pre-planning an employer will considerably improve the business's chances of survival following the event. To survive, a business must be prepared for the unexpected.

As well as assisting a company to survive a fire or arson attack, the same survival plan can help a business to recover after a large-scale theft, a terrorist attack, industrial espionage, storm damage and flooding. Although an employer can never determine when, these threats will affect the business, the taking of precautionary measures can greatly reduce the consequential damage and enhance the ability to make a quick and thorough recovery.

Checklist

14.8

> A survival plan should be designed to thoroughly assess the impact of any type of business shut down. The five key steps involved in the creation of a survival plan are:
>
> - planning;

14.9 *Fire (before, during and after)*

- prevention;
- preparedness;
- response; and
- recovery.

Planning

14.9 The objective of the pre-planning process should be to systematically determine the various issues and priorities in order to develop a cost-effective and realistic strategy that is tailored to the size and nature of the business.

The disruption caused by a fire and the time during which the premises are out of use can be minimised by proper planning.

What needs to be appreciated is that there will be two distinctly different phases that must be planned for following a fire or other disaster.

Firstly, there will be the immediate, disorganised phase – which shall be termed 'Phase I'. During this time everyone will be trying to be of assistance, but in a totally unco-ordinated fashion. This is normal human behaviour and it is to be expected, as a fire is a considerable shock to both owners and employees.

During Phase I there is a very limited amount, which can be achieved. It is then followed by a period of 'makeshift-operations,' which can be quite lengthy until normal business operations can be resumed, this shall be termed 'Phase II'.

Typically, following a fire, Phase I can extend for up to a week or more while Phase II can last for several months until normal operations are restored. One of the primary objectives of pre-planning should be to keep the Phase I as short as possible in order to start the Phase II as quickly as possible. The very fact that you have prepared a survival plan will ensure that this happens.

For obvious reasons, business owners want to get back to normal as soon as possible and very often it is difficult to accept that this must be carried out in stages, if there is to be an effective recovery.

Furthermore, a survival plan is not a fixed or finalised document. It will evolve and should improve as time goes by. The plan does not have to be perfect the first time it is prepared, the most important task is that it is commenced.

The following suggestions should be considered for inclusion in the survival plan.

- The plan should be systematic and be designed to look at the common elements in all disasters, which are:
  - loss of information;
  - loss of access to information;
  - loss of access to facilities; and
  - loss of people.
- It is good practice to prepare a matrix, with the above four headings as the columns and each of your activities as a row. Then work out how an employer and employees would respond to loss of information, access to information, facilities and personnel for each of the activities.
- The survival plan should nominate a deputy in case the person normally in charge is injured in the fire or, otherwise, unavailable. The deputy should be named in the plan and delegated full authority to take charge in this situation.
- List key employees' individual responsibilities in advance and assign specific people to each task.
- Keep an accurate, up to date inventory of all goods, stock and equipment. This will be one of the first things that insurance company will require.
- Some of the more important tasks that should be treated as a priority are to notify suppliers of the situation and advise them where to deliver their goods. An employer should also advise the most important customers about the situation. An employer should inform the bank of what has happened, as their assistance may be needed in order to aid recovery.
- It is important to delegate the task of liasing with and advising employees on the current situation. Some employees may as a result of the fire and completely out of touch with the situation.
- Public relations should also be given high priority, with someone appointed to the task of liaison with the press. A positive proactive approach to the press and media will have a significant effect on a business's reputation within the community. The publicity following a fire can be of invaluable assistance as it alerts other businesses and services to the situation, some of whom may well be able to offer assistance to aid recovery.
- Make provision to protect critical paper records. Even in a fully automated computerised organisation there can be vulnerable, newly created records or documents (such as contracts, advertising, research and sales contacts) that may only exist on paper.
- Set clear priorities about business activities. After a fire, it will not be possible to return everything to normal at the same time. Decide beforehand the longest amount of time that can be allowed for each

14.10 *Fire (before, during and after)*

activity to be out of service. By following this criteria an employer will be able to advise customers accordingly.

- Because almost all companies place considerable emphasis on computer technology, a business will find it extremely difficult to survive if it loses its computer data. It is vital that all businesses have a rigid policy of making a daily backup of their computer data that should be kept *outside* of the business premises. As a business, have actual tests been made that indicate that is possible to read and restore company, computer and personal computer backup files?

- Keep copies of all vital forms and personnel records off site. This includes, eg an extra chequebook so that an employer can buy any emergency supplies that may be needed.

- Finally, ensure that key employees keep a copy of the survival plan *at home*. If an employer does not do this and the premises are destroyed by fire the plan will be destroyed as well.

Preparedness

14.10 The very fact that an employer has pre-planned and tried to anticipate the potential problems associated with a fire will mean that the prospects of survival after the event are immeasurably increased.

Prevention

14.11 The prevention of fires is dealt with in detail in **CHAPTER 10: MANAGING FIRE SAFETY (PREVENTION OF FIRE AND LIAISON WITH AUTHORITIES)**.

Recovery

14.12 Recovery from a fire or other disaster must begin immediately after the emergency occurs and should continue until all systems are back to normal.

Checklist

14.13

> The following list of suggested actions can be built into the survival plan for recovery so that the plan is appropriate:
>
> - co-ordinate activities among all personnel involved in the recovery; and

> - ascertain if the building structure is safe for personnel to enter.
>
> Continue working, in shifts if necessary, until recovery is secure and any installed protection systems have been restored.
>
> Initiate salvage operations as soon as it is safe to do so.

What are 'salvage efforts'?

14.14 Salvage efforts should include the following:

- contacting the insurance company;
- obtaining back-up equipment and supplies;
- activating any mutual-aid agreements you may have with other local companies;
- separating damaged from undamaged equipment, stock, etc;
- pumping out any standing water;
- checking electrical systems before starting up equipment;
- wiping down and covering equipment and stock;
- drying out, cleaning and testing equipment;
- retrieving building plans, equipment specifications, shop layouts;
- dehumidifying damaged areas if needed;
- maintain or re-establish security during the recovery stage, especially in critical areas;
- establish surveillance to control looting and theft, if warranted;
- set up physical access barriers as needed;
- secure software and vital records; and
- provide traffic control around the site.

Ensure all damage is documented, as recovery work proceeds and ensure that both customers and suppliers know when normal company operations will be resumed.

Documentation

14.15 A survival plan should be kept as simple as possible and must be written down. An employer should be aware that where a document is longer than 15–20 pages it is unlikely to be read or used. Indeed, there can be positive advantages in limiting the role of specific individuals to those matters that are with their own area of expertise. This means that only the fire safety

14.16 *Fire (before, during and after)*

manager and other leading members of the management team will have to concentrate on the overall implications and remedial measures that are required to resume normal trading conditions.

Some suggested items follow, to include as an appendix to the plan, which can be tailored to suit specific needs.

Checklist

14.16

> The appendix should include:
>
> - a current inventory of goods, equipment and stock;
> - availability of suitable alternative temporary accommodation;
> - a list of the telephone numbers and addresses of all key services including:
>   - ○ local authority departments;
>   - ○ architect, surveyors and structural engineers;
>   - ○ building contractors – plumbers, carpenters, electricians, heating engineers etc;
>   - ○ experts in salvage and damage control;
>   - ○ residual smoke removal experts;
>   - ○ emergency telephone numbers of the utility services – electricity, gas, water;
>   - ○ insurers, loss adjusters, loss assessors; and
>   - ○ plant hire contractors (for pumps, generators, heating equipment etc).
>
> Bearing in mind that a fire, or any other major disaster could totally destroy the business premises, provided there is no significant breach of security, key personnel should be encouraged to keep a copy of the survival plan, or that part, for which they may have specific responsibilities at home.

Review and test your plan

14.17 After completion, the survival plan needs to be reviewed with all employees on a regular basis. This does not have to be a lengthy procedure, but it will ensure that changes that have occurred in the business are taken into account and included in a revised plan.

With regard to testing, due to the practical difficulties involved, a comprehensive trial or examination of the survival plan may not always be feasible. However, any opportunity that arises, such as relocation moves, or unplanned business shutdowns, should be treated and evaluated as simulated tests of the company's survival and recovery ability.

Summary

14.18 The pre-planning process itself can be summarised in the following steps:

- provide key employees with guidelines;
- identify serious risks;
- prioritise the activities to be maintained and how to maintain them;
- assign the 'survival' team;
- take a complete inventory;
- know where to get help;
- document the plan; and
- review the plan with key employees and test it.

Finally, the underlying philosophy in an approach to survival recovery planning is that much can be achieved without undue expense. An employer can benefit greatly by preparing as much as possible beforehand. An employer should also allocate responsibilities and make management decisions now rather than wait until the incident occurs.

15 The way ahead

In this chapter:

Introduction	1.1
Possible reform of fire safety legislation	15.2
How will these proposals be taken forward, and when will they be implemented?	15.7
The current position	15.9
Alterations to some legislation	15.14
Power to make regulations	15.51
The responsible person	15.52
Application	15.53
Mitigation of the effects of a fire	15.54
Requirements of the proposed fire safety regime	15.55
Risk assessment	15.57
General duty to ensure safety	15.58
Elimination or reduction of risks from dangerous substances	15.61
Fire-fighting and fire detection	15.62
Emergency routes and exits	15.63
Maintenance	15.64
Safety assistance	15.65
Procedures for serious and imminent danger and for danger areas	15.66
Provision of information	15.67
Co-operation and co-ordination	15.68
Capabilities and training	15.69
General duties of employees at work	15.70
Guidance	15.71
Licensing	15.72

The proposals: enforcement	15.73
Crown immunity	15.74
How will the Order be enforced?	15.75
Validation of fire safety solutions – the question of public reassurance	15.76
Reassurance to the public	15.77
Charging	15.78
Analysis – re-statement of existing burdens	15.79
Necessary protection	15.80
Bain Report	15.81
Role of central and local government	15.84
Brief comment	15.85

Introduction

15.1 This is undoubtedly one of the most interesting and challenging periods concerning the methods that are to be adopted regarding the general approach to fire safety and the reform of fire safety legislation.

As indicated in paragraph **6.8** above, the appeal following the *City Logistics Ltd v the Northamptonshire County Fire Officer [2001] EWCA Civ 1216* case may well have a considerable effect upon fire safety. The decision by the Lord Justices, which restricts the fire safety measures that can be required by the enforcing authorities, will have a major impact upon the way in which fire safety legislation will be progressed.

Fire safety legislation, in this country, has generally been developed following major incidents, ie stable door legislation, this culminated in the introduction of the *Fire Precautions Act 1971 (FPA 1971)*, which, in turn was enacted in response to a number of serious fires, in particular, those at Eastwood Mill, Keighley in 1956, Hendersons' Department Store, Liverpool in 1960, and the Rose and Crown in Saffron Walden in 1969.

It is generally accepted that the present arrangements are unsatisfactory and the Office of the Deputy Prime Minister issued a long and detailed consultation document, ie '*A consultation document on the reform of fire safety legislation*' in July 2002 and some of the proposed approaches and the relevant parts of the summary of proposals are discussed later in this chapter.

The fire strike and the consequential recommendations contained in the Bains Report are also of interest, together with the Pathfinder Report that is now shown on the website of the Office of the Deputy Prime Minister.

15.2 *The way ahead*

There appears to be certain areas of conflict between some of these documents and this may well affect the final approach and lead to alterations that will need to be included in the proposed *Regulatory Reform Act* and, in due course, the relevant *Regulatory Reform (Fire Safety) Order*. The opinions of 'consultees' may also have an effect, but it is unlikely that they will materially affect the main thrust of the proposed reforms.

Fire safety engineering is one area that may well be affected in the short term. The reason being that some enforcing authorities could be concerned as to the repercussions of the Bains Report recommendations regarding the standards of fire cover in their areas. Obviously, if as a result of these proposals, there could be a delayed or reduced attendance by the fire service compared to existing arrangements the enforcing authorities may have reservations in accepting any 'fire engineering' proposals that may be significantly differ from current prescriptive codes and standards. This could be unfortunate as there have been some impressive developments in this approach to fire safety.

Possible reform of fire safety legislation

15.2 The consultation document *A consultation document on the reform of fire safety legislation* that was issued by the Government through the Office of the Deputy Prime Minister contains a suggested regime to link fire safety to risk assessment. The thrust of the consultation document is based upon the reform of fire safety legislation, by amending various Acts and Statutory Instruments to simplify, rationalise and consolidate the law with respect to fire safety in buildings in use.

The Government accepted that the provisions for fire safety are scattered among many pieces of legislation. It is sometimes inconsistent and can be difficult even for fire safety professionals to understand. For a lay person who has to comply with the legislation, it can be bewildering. The aim of the reform is to simplify, rationalise and consolidate existing legislation. It would provide for a risk based approach to general fire safety allowing more efficient, effective enforcement by the fire service and other enforcing authorities.

The proposals

15.3 It appears that the main proposals are as follows.

- So far as possible, general fire safety legislation should be reformed to create one simple fire safety regime applying to all workplaces and other non-domestic premises.

- The regime should be risk assessment-based with responsibility for the fire safety of the occupants of premises and people who might be affected by a fire resting with a defined responsible person.

- There should be no separate formal validation mechanism for higher risk premises. Fire authorities would base their inspection programmes on their assessment of the premises they considered to present the highest risk.

- There should be a duty to maintain those fire precautions required under *Building Regulations 1991 (SI 1991 No 2768)* which are for the use and protection of the fire brigade.

- There will be a new statutory duty on fire authorities to promote community fire safety, for powers of entry for the investigation of fires, and for a power to take away samples for testing.

Why are these changes needed?

15.4   Fire safety provision is scattered among many different pieces of legislation. It is sometimes inconsistent and can be difficult to understand.

Who will these proposals affect?

15.5   The proposed changes will affect: employers and virtually all those who are responsible for buildings to which the public may have access.

What will be the financial impact of the changes?

15.6   The Government consider that the financial benefits to business will be around £1.7 million from the ending of the requirement for fire certificates and £45 million to £110 million from possible savings from a reduction in the number of fires. The implementation and policy costs will total around £65 million.

How will these proposals be taken forward, and when will they be implemented?

15.7   The Government intend that the proposed changes to legislation are to be made through a *Regulatory Reform Order* under the *Regulatory Reform Act 2001*. Subject to the outcome of consultation and the Parliamentary scrutiny of the proposals, the Government propose that the changes are implemented from Spring 2004.

The Government need to be sure that an Order does not remove any necessary protection from individuals or organisations, and that it does not prevent them from exercising existing rights or freedoms that they might reasonably expect to continue to exercise. Where an Order imposes a burden, it must be desirable. It must also strike a fair balance between the public

15.8 The way ahead

interest and the interest of those who are affected by the burden being created, and the burden must be proportionate to the expected benefit.

The Government accepted that their proposals would affect employers and virtually all those who are responsible for non-domestic premises. For employers, they claim, the proposals would not impose significant additional burdens since they would recreate requirements, which already exist under the *Fire Precautions (Workplace) Regulations 1997 (SI 1997 No 1840) (as amended)*, the *Management of Health and Safety at Work Regulations 1999 (SI 1999 No 3242) (as amended)* and the *Dangerous Substances and Explosive Atmospheres Regulations 2002 (DSEAR) (SI 2002 No 2776)* (see **13.69** above). Some self-employed people and elements of the voluntary sector will be brought within the regime but many of these will already be subject to licensing requirements or the *Health and Safety at Work etc Act 1974*. Under that Act they will already be responsible for safety of people on their premises. The proposals merely clarify their specific responsibilities in respect of fire.

The removal of multiple and overlapping fire safety provisions and their replacement with a single fire safety regime should constitute the reduction of a significant burden. It is also proposed to remove the burden for the need to apply for fire certificates.

The relevant department propose to introduce the reform by means of a *Regulatory Reform Order* made under the *Regulatory Reform Act 2001*.

Regulatory Reform Order-making

15.8 Each proposal for a *Regulatory Reform Order* must satisfy a number of legal tests. The questions in the rest of this document are designed to elicit the information that the Minister will need in order to satisfy the Committees that, among other things, the proposals satisfy these tests. In particular, the *Regulatory Reform Act 2001* requires information on:

- whether any of the proposals could remove any necessary protection;

- whether any of the proposals could prevent any person from continuing to exercise any right or freedom which they might reasonably expect to continue to exercise and, if so, how they are to be enabled to continue to exercise that right or freedom;

- whether any burdens are being imposed on any person in the carrying out of an activity;

- whether any savings or increases in cost are estimated to result from the proposals and, if so, the reasons why savings or increases in cost should be expected;

- if it is practicable to make an estimate of the amount, that amount, and how it is calculated; and

- any benefits (other than savings in cost) which are expected to flow from the implementation of the proposals.

The Minister making a *Regulatory Reform Order* must be of the opinion that it does not remove any necessary protection. This means that no order can be made unless the Minister is of the opinion that it would maintain any protections that the Minster considers to be necessary. Such protection relates to the checks and balances associated with a particular regulatory regime. The protection does not have to be statutory in nature and does not have to be for the purposes originally intended by Parliament. If the Minister considers a particular protection to be no longer necessary, he or she must provide the parliamentary scrutiny committees with compelling evidence to support this view.

The current position

15.9 At present, there are two major pieces of specific fire safety legislation, the *Fire Precautions Act 1971* and the *Fire Precautions (Workplace) Regulations 1997 (SI 1997 No 1840) (as amended)*. Both apply in England and Wales and Scotland. As fire safety is a matter within the devolved competence of the Scottish Parliament, these proposals for reform will only apply in England and Wales. It will be for the Scottish Executive to consider the scope for parallel changes in Scotland. Provisions relating to fire precautions are also contained in numerous other pieces of legislation, which are not principally related to fire safety.

As mentioned in the Introduction to this chapter, fire safety legislation, up to the introduction of the Fire *Precautions Act 1971*, developed in response to a number of serious fires, in particular, at Eastwood Mill, Keighley in 1956, Hendersons' Department Store, Liverpool in 1960, and the Rose and Crown in Saffron Walden in 1969.

Fire Precautions Act 1971

15.10 Under the *Fire Precautions Act 1971*, the use of certain types of premises was designated by the Secretary of State as requiring a fire certificate. Currently there are two designating orders in force in Great Britain. One relates to hotels and boarding houses and the other to those factories, offices, shops and railway premises in which people are employed to work.

The hotels and boarding houses, which require a fire certificate are those which provide sleeping accommodation for more than six people (whether employees or guests) or if they provide sleeping accommodation for employees or guests elsewhere than on the ground or first floors of the premises.

With regard to factories, offices and shops a fire certificate is required where more than 20 people are at work at any one time or more than ten are at

15.11 *The way ahead*

work at any one time elsewhere than on the ground floor. Certificates are also required for smaller factories where significant quantities of highly flammable substances are stored.

FPA 1971 requires the occupier of designated premises to apply for a fire certificate. This will be prepared by the fire authority (in practice the local fire brigade). Before issuing a fire certificate the fire brigade will inspect the premises and satisfy themselves that:

- the means of escape in case of fire;
- the means with which the building is provided for securing that the means of escape can be safely used at all times;
- the means for fighting fire; and
- the means for giving warning in case of fire are such as may reasonably be required.

Fire Precautions (Workplace) Regulations 1997 (as amended)

15.11 *Fire Precautions (Workplace) Regulations 1997 (SI 1997 No 1840) (as amended)* were made to implement two European Directives on health and safety at work, the *European Council Framework Directive 89/391/EEC* and the *European Council Workplace Directive 89/654/EEC*. They apply to virtually all places where people are employed to work, the exceptions being construction sites, ships, mines and other areas covered by the Health and Safety Executive and other agencies. *SI 1999 No 3242*, together with those elements of the *Management of Health and Safety at Work Regulations 1999 (SI 1999 No 3242)* which are amended to impose requirements concerning general fire precautions, are correctly referred to as 'the workplace fire precautions legislation'.

This legislation requires employers to:

- carry out a fire risk assessment;
- identify the significant findings of the risk assessment;
- provide and maintain such fire precautions as are necessary to safeguard those who use the workplace; and
- provide information, instruction and training to employees about the fire precautions.

The fire precautions provided in accordance with these Regulations are intended to protect employees but must take account of other people present. Employers must take account of any duty of care they and their employees may have to other occupants of the building.

Legislative overlap

15.12 This means that in a large number of premises, two separate fire safety regimes apply based on totally different philosophies. The central aim of the *Fire Precautions Act 1971* is to ensure that, in the event of a fire, the occupants can evacuate the premises safely.

Editorial Note. This was the view expressed by the Lord Justices at the appeal in the case between *City Logistics Ltd v the Northamptonshire County Fire Officer [2001] EWCA Civ 1216*.

The *Fire Precautions (Workplace) Regulations 1997 (SI 1997 No 1840) (as amended)*, require employers to identify risks and take steps to remove or reduce them. Some premises are also subject to licensing, certification or registration regimes under which yet more fire safety requirements will be made. These include Licensing Acts, the *Theatres Act 1968*, the *Gaming Act 1968*, *Children's Homes Regulations 1991 (SI 1991 No 1506)* and many others. The Government accepts in the consultation document that this is highly confusing for businesses and as such places a burden on them.

Possible application, scope and requirements

15.13 The intention of the proposals is to remove legislative overlap and bring fire safety law into one place, the *Regulatory Reform (Fire Safety) Order*, which will be enforced, in the main, by fire authorities. The proposed Order would replace both the *Fire Precautions Act 1971* and the *Fire Precautions (Workplace) Regulations 1997 (SI 1997 No 1840) (as amended)* and as much of the remaining legislation as is practical.

This should mean that occupiers of premises designated under *FPA 1971* would no longer need to apply for a fire certificate.

The aim of the reform is to achieve clarity and to reduce the sheer volume of legislation, the Government would like, so far as is possible, to place all the requirements on the face of the Order. The Order would be based around a general duty of fire safety care with specific requirements, which will need to be met to comply with that duty.

The Government will, of course, continue to meet obligations under EU legislation. The relevant Directives in respect of fire safety are:

- *European Council Directive 89/391/EEC* on the introduction of measures to encourage improvements in the safety and health of workers at work (the Framework Directive);

- *European Council Directive 89/654/EEC* concerning the minimum safety and health requirements for the workplace (the Workplace Directive);

15.14 *The way ahead*

- *European Directive 98/24/EC* on the health and safety of workers from risks related to chemical agents at work (the Chemical Agents Directive); and

- *European Directive 99/92/EC* on minimum requirements for improving the safety and health protection of workers potentially at risk from explosive atmospheres (the Explosive Atmospheres Directive).

Alterations to some legislation

15.14   A wide range of existing legislation will need to be to amended or repealed. In amending and repealing existing legislation the Government shall not be removing any necessary protection, nor prevent any person from continuing to exercise any right or freedom, which they might reasonably expect to continue to exercise. Where the Government remove provisions relating to fire precautions they will be replaced by provisions offering equivalent protection in the new regime. The proposed Order would also create a new duty on fire authorities to promote community fire safety.

Some of the major items of legislation that will need to be completely repealed or amended, so far as aspects relating to fire safety in workplaces are concerned, would seem to include the following Acts and Regulations:

- *Building Act 1984*;
- *Building Regulations 2000 (SI 2000 No 2531); and*
- *Building (Approved Inspectors etc) Regulations 2000 (SI 2000 No 2532).*

If ongoing maintenance of fire precautions (including access for the fire service etc) is to be achieved by means of the new regime, only amendments to reflect the repeal of the *Fire Precautions Act 1971* (and revocation of the *Fire Precautions (Workplace) Regulations 1997 (SI 1997 No 1840) (as amended)*) and the introduction of the new regime should be required.

It is expected that in the *BA 1984, ss 48(4), 51B(2), 71* and *72(7)* will be repealed and the reference to 'twenty feet' in *BA 1984, s 72* be amended to '4.5 metres'.

SI 2000 No 2531, Reg 12 should be amended to reflect the new regime – as should *SI 2000 No 2532, Reg 13*.

Caravan Sites and Control of Development Act 1960

15.15   This involves a licensing regime for caravan sites. At present the local authority may impose conditions of licence including ones relating to securing proper measures for preventing and detecting fire and adequate means for fighting fire are provided and maintained. It is expected that the provision should be repealed or replaced in so far as it applies to non-domestic private accommodation.

Cinematograph (Safety) Regulations 1955. (Children) (No2) Regulation 1955, 1958, 1965 and (Amendment) Regulations 1976, 1982 and (Draft) 2001.

15.16 The safety provisions of the licensing arrangement for cinemas include fire safety and would be modified, ie the fire provisions of all cinema legislation would be removed.

However, like other licensing regimes the Government would aim to allow that a prosecution for breach of fire law would allow a cinema licence to be revoked.

In the *Cinematograph (Safety) (Amendment) Regulations 2002 (CSAR) (SI 2002 No 1903)*, proposed provisions in respect of fire safety for the disabled should not be necessary as risk assessment based precautions will take into account the needs of the disabled.

Cinematograph (Amendment) Act 1982 and Cinemas Act 1985

15.17 *Cinematograph (Amendment) Act 1982 and Cinemas Act 1985* extend earlier cinema legislation to other types of exhibition. It is expected that references to fire safety conditions should be amended to refer to compliance with the new fire safety regime.

Fire precautions Act 1971, Fire Precautions (Hotels and Boarding Houses) Order 1972, Fire Precautions (Factories, Offices, Shops and Railway Premises) Order 1989 and Fire Precautions (Application for Certificate) Regulations 1989

15.18 It is expected that the Government will repeal the *Fire precautions Act 1971*, designating Orders and Regulations.

Fire Precautions (Sub-surface Railway Stations) Regulations 1989 and Fire Precautions (Sub-surface Railway Stations) (Amendment) Regulations 1994

15.19 The Regulations, which were made following the tragic Kings Cross fire, are solely about fire safety and will fall when the *Fire Precautions Act 1971* is repealed – unless specifically cited in the new regime. The Regulations are highly prescriptive although they do allow the fire authority some discretion.

15.20 *The way ahead*

It is expected that the Regulations will be revoked in favour of the new regime.

The Fire Precautions (Workplace) Regulations 1997 and The Fire Precautions (Workplace) (Amendment) Regulations 1999

These Regulations implement European legislation. The requirements will form the basis of the new fire regime. It is expected that these Regulations will be revoked.

Fire Services Act 1947, Fire Services Act 1951 and Fire Services Act 1959

15.20 Of the *Fire Services Act 1947*, the sections of the Act dealing with fire inspectors are to be read as including a reference to the *Fire Precautions Act 1971* and the *Fire Precautions (Workplace) Regulations 1997 (SI 1997 No 1840) (as amended)*. Crown exemption is to continue for the time being and fire inspectors are to enforce for those premises.

The Government propose to amend the *Fire Services Act 1947* to reflect repeal of the *FPA 1971* and revocation of *SI 1999 No 3242* and their replacement by the *Regulatory Reform (Fire Safety) Order*.

Environment and Safety Information Act 1988

15.21 *Environment and Safety Information Act 1988* requires fire authorities to maintain a register of *Fire Precautions Act 1971, s 10* notices served (currently other than notices solely for the protection of persons at work). It is expected that the reference to *FPA 1971* in the schedule will be amended to reflect the new proposals and be extended to cover all enforcement notices served under the new fire regime.

Factories Acts 1948, 1959, 1961 and Factories Act 1961 etc (Repeals) Regulations 1976

15.22 *Factories Act 1948*, *Factories Act 1959* and *Factories Act 1961* ceased to have effect for general fire safety when the *Fire Precautions (Factories, Offices, Shops and Railway Premises) Order 1976 (SI 1976 No 2009)* was made under the *Fire Precautions Act 1971*. Where the provisions relating to means of escape etc (*FPA 1971, ss 40–51*) have not already been repealed it is proposed that this will be done insofar as those provisions relate to general fire precautions.

Fire Safety and Safety of Places of Sport Act 1987, Safety of Sports Grounds Act 1975

15.23 *Fire Safety and Safety of Places of Sport Act 1987, Sch 1, Pt I* merely amend the *Fire Precautions Act 1971* and it is proposed to revoke it. *FSSPSA 1987, Sch 1, Pt II* extends the *Safety of Sports Grounds Act 1975* to all sports grounds. *FSSPSA 1987, Sch 1, Pt III* requires safety certificates for stands at sports grounds.

The section specifically disapplies *FPA 1971* insofar as matters covered by it could be imposed by a safety certificate. However, the *Fire Precautions (Workplace) Regulations 1997 (SI 1997 No 1840) (as amended)* disapply the provisions of a safety certificate to the extent that it would require a person to contravene any provision of the workplace fire precautions legislation. Fire authorities are statutory consultees for sports ground legislation and enforce *SI 1997 No 1840*.

It is proposed that the sports ground legislation be amended so that safety certificates require compliance with the new fire regime (and contravention of fire law is to be treated as a contravention of the safety certificate). Enforcement will be by the authority enforcing the sports ground legislation. In particular it is proposed to revoke *FSSPSA 1987, s 9(1)(d), Sch I, Pt I* as they only amend *FPA 1971*.

Gaming Act 1968, Gaming (Amendment) Act 1982

15.24 A licensing provision (amended by the *Gaming (Amendment) Act 1982*) which requires premises used for gaming to be licensed. *G(A)A 1982* requires the applicant to send a copy of the application to the fire authority. The licence contains fire safety provisions and fire safety is a material factor for consideration of the grant or renewal of a licence.

It is proposed that fire safety should be removed from the conditions of licence in its current form and replaced with reference to compliance with the *Regulatory Reform (Fire Safety) Order*. It further proposed to clarify that prosecution by the fire authority should be grounds for revocation of licence etc.

Licensing Act 1961, Licensing Act 1964 and Licensing Act 1988

15.25 Licensing is the subject of separate review. *Licensing Act 1961, Licensing Act 1964 and Licensing Act 1988* require the licensing authority to be satisfied as to the safety of the premises and allow fire safety to be one of the conditions of licence. It is expected that references to safety conditions (including fire) will be altered to reflect that compliance with the new fire regime is the fire safety requirement for the grant of a licence. It is further proposed that the new fire safety order will allow that prosecution (or issue of a notice) under the new regime may be treated as if it had been taken (or served) under licensing law.

Licensing (Occasional Permissions) Act 1983

15.26   *Licensing (Occasional Permissions) Act 1983* allows 'reputable' organisations to be granted a licence to sell alcohol at functions lasting not more than 24 hours. The Justices must be satisfied that the premises will be a suitable place (ie safe). Fire is not specifically mentioned. It is proposed to amend the requirement to require compliance with the *Regulatory Reform (Fire Safety) Order*.

Licensing Act 1988

15.27   *Licensing Act 1988* amended the *Licensing Act 1964* – primarily in respect of licensing hours. LA 1988 is likely to be revoked by the Law Commission or as part of licensing reform. It seems it could be revoked as part of our reform as this would be a suitable vehicle.

Local Government (Miscellaneous Provisions) Act 1982

15.28   *Local Government (Miscellaneous Provisions) Act 1982* provides for licensing of public entertainments. LG(MP)A 1982 allows the local authority to attach conditions to the licence – which may include fire safety conditions. In addition LG(MP)A 1982 amends the *Public Health Act 1936* and the *Caravan Sites and Control of Development Act 1960* to require consultation with the fire authority. These provisions may be subject to amendment (in the amended Acts). It is proposed that references to fire safety should be amended to refer to compliance with the *Regulatory Reform (Fire Safety) Order*.

LG(MP)A 1982 also allows the fire authority to determine that 'fireman's switches' may be required for all luminous signs in their area. Such provision is also contained in some local Acts – which may be repealed in favour of national provision.

It is proposed to bring this forward as a power of fire authorities and to repeal the local Act provisions and LG(MP)A 1982 provision.

London Local Authorities Act 1991, London Local Authorities Act 1995 and London Local Authorities Act 1996

15.29   *London Local Authorities Act 1991* provides for the licensing of special treatment premises. The London Fire and Emergency Planning Authority

(LFEPA) are provided with power of entry (*LLAA 1991, s 15*). Licensing conditions (*LLAA 1991, s 6*) include public safety and, specifically, provision and maintenance of proper precautions against fire. It is proposed that the fire safety requirement should be amended to refer to compliance with the *Regulatory Reform (Fire Safety) Order*.

The LFEPA power of entry will be unnecessary as it is to be provided in the *Regulatory Reform (Fire Safety) Order* and can be repealed.

London Local Authorities Act 1995, s 16–18 provides for licensing of 'near beer' premises. A licence can be refused on the grounds that the fire precautions are not adequate. Conditions may be attached which include both public safety and provision and maintenance of proper fire precautions. It is proposed that the provision for specific fire precautions should be replaced by reference to compliance with the new regime.

London Local Authorities Act 1996 contains provisions to provide for consultation with the police and fire service in relation to applications for licences. It is not intended to remove this requirement.

Theatres Act 1968

15.30 Fire safety can form a licensing condition. It is proposed to amend the Act so that the fire safety condition would be compliant with the new fire regime.

Construction (Design and Management) Regulations 1994

15.31 *Construction (Design and Management) Regulations 1994 (SI 1994 No 3140)* make reference to information to be provided – which includes risk assessment made under *Management of Health and Safety at Work Regulations 1999 (SI 1999 No 3242)*. This by proxy will include the *Fire Precautions (Workplace) Regulations 1997 (SI 1997 No 1840)* insofar as they apply. Revocation of *SI 1997 No 1840* will break this link. Mention of the new regime will therefore need to be made in these Regulations (notably in *SI 1997 No 1840, Regs 15–19*). It is proposed to add reference to the risk assessment requirements of the *Regulatory Reform (Fire Safety) Order*.

Construction (Health, Safety and Welfare) Regulations 1996

15.32 *Construction (Health, Safety and Welfare) Regulations 1996 (SI 1996 No 1592)* contain provision (at *SI 1996 No 1592 Regs 18–21*) for emergency routes and exits, emergency procedures, fire detection and fire fighting for construction sites. The provisions, so far as they relate to fire, will probably be removed in favour of the new regime.

It is proposed that references to compliance with the *Regulatory Reform (Fire Safety) Order* should be inserted in *SI 1996 No 1592, Regs 18–21*. In addition, it is proposed that the enforcement differentiation of *SI 1996 No 1592, Reg 33* should be continued with Health and Safety Executive keeping responsibility for 'out and out' construction sites.

Control of Major Accident Hazards Regulations 1999

15.33 *Control of Major Accident Hazards Regulations 1999 (SI 1999 No 743)* provide for prevention of major accidents and prevention of harm to persons on and off the site and to the environment in the event of a major accident at sites carrying out certain industrial activities as defined in the schedule to *SI 1999 No 743*.

Major accident includes a major fire or explosion.

The person having control of an industrial activity to which *SI 1999 No 743* applies is required to notify the Health and Safety Executive of the existence of the site and to make arrangements with the local authority for an offsite emergency plan (to deal with the effects of a major accident).

It is proposed that the new regime will apply to these sites and therefore that necessary amendments be made to clarify fire authority enforcement of general fire precautions measures (note removal of *Fire Certificates (Special Premises) Regulations 1976 (SI 1976 No 2003)* at **15.34** below).

Dangerous Substances in Harbour Areas Regulations 1987

15.34 *Dangerous Substances in Harbour Areas Regulations 1987 (SI 1987 No 37), Regulation 18* imposes a duty on the owner of a berth to take all reasonable precautions in respect of fire and explosion – including at *SI 1987 No 37, Reg 18(2)(a)* means for fighting fire *SI 1987 No 37, Reg 18(2)(b)*, training in first aid fire-fighting and *SI 1987 No 37, Reg 18(2)(c)* access for the fire service. It is proposed to amend *SI 1987 No 37* to refer to compliance with the *Regulatory Reform (Fire Safety) Order*.

SI 1987 No 37, Reg 26 covers emergency plans by the harbour authority and may fall under the *Dangerous Substances and Explosive Atmospheres Regulations (DSEAR) (SI 2002 No 2776)*.

Editorial Note. This appears to be the case, in *SI 2002 No 2776*. In general terms only normal ship-board activities are exempt from the Regulations.

SI 1987 No 37,Reg 27(1) covers means of escape from the berth and means for contacting the emergency services. It is proposed that the regulation should be amended to reflect the new fire regime.

SI 1987 No 37, Reg 47(4) amends the *Fire Certificates (Special Premises) Regulations (SI 1976 No 2003)* and it will be proposed that it should be revoked.

Environment and Safety Information Act 1988

15.35 *Environment and Safety Information Act 1988* requires fire authorities to maintain a register of *Fire Precautions Act 1971, s 10* notices served (currently other than notices solely for the protection of persons at work). It is proposed that the reference to *FPA 1971* in the schedule should be amended to reflect the new regime and further proposed that the extent of the provision should be extended to cover all enforcement notices served under the new fire regime.

Fire Certificate (Special Premises) Regulations 1976

15.36 It is proposed by agreement with Health and Safety Executive to revoke the *Fire Certificate (Special Premises) Regulations 1976 (SI 1976 No 2003)*.

Health and Safety at Work etc Act 1974

15.37 The primary Health and Safety legislation. General fire safety is within its scope but not generally enforced by virtue of policy agreement. *Health and Safety at Work etc Act 1974* amends the *Fire Precautions Act 1971* (*HSWA s 75* and *Sch 8*). These provisions may be revoked (if that has not already happened). *HSWA 1974* is itself the subject of separate review by Health and Safety Executive and Commission. References to the *Fire Precautions Act 1971* are to be removed and *HSWA 1974* is to be amended to reflect the new regime. This will need to be considered further when the detail of the regime is closer to being finalised and policy agreements are reached about enforcement practice.

It is proposed to repeal *HSWA 1974, s 75* and *Sch 8* and is further proposed that *HSWA 1974* should be disapplied from application to general fire precautions in premises and other places to which the *Regulatory Reform (Fire Safety) Order* applies.

15.38 *The way ahead*

Health and Safety (Consultation with Employees) Regulations 1996

15.38   To implement European requirements the Government will need to amend the *Health and Safety (Consultation with Employees) Regulations 1996 (SI 1996 No 1513)* or otherwise link to them to provide for consultation with employees in respect of fire precautions under the new regime. It is proposed to apply the Regulations to the *Regulatory Reform (Fire Safety) Order*.

Health and Safety (Enforcing Authority) Regulations 1989

15.39   *Health and Safety (Enforcing Authority) Regulations (SI 1989 No 1903)* make local authorities responsible for enforcing certain elements of the *Health and Safety at Work etc Act 1974*, and the relevant statutory provisions. Historically fire authorities undertook some of these enforcement duties. However, the introduction of combined fire authorities has meant that only county brigades can be utilised in this way. If the Health and Safety Executive wishes fire authorities to be able to enforce, then all fire authorities should be able to do this. If not then all fire authorities should be removed from delegated authority under the regulations. It is proposed to remove fire authorities from the definition of local authority for the purposes of these Regulations. It is further proposed that reference to the *Fire Precautions Act 1971* (which defines a fire authority and so makes the Health and Safety Executive responsible for enforcement at fire authority premises) should be amended to refer to a fire authority under the *Fire Services Act 1947*.

Management of Health and Safety at Work Regulations 1999

15.40   It is proposed that references (and inclusion) of the *Fire Precautions (Workplace) Regulations 1997 (SI 1997 No 1840)* be revoked by the *Management of Health and Safety at Work Regulations 1999 (SI 1999 No 3242)*. Elements of the Regulations will be re-enacted as part of the new fire regime (risk assessment etc) to ensure continued transposition of the European Directives.

Mines and Quarries Act 1954

15.41   The provisions of the Act mainly apply to working below ground. However, *Mines and Quarries Act 1954, s 73* makes it illegal for a person to be employed in a room or confined space (which is a dangerous one) unless it has adequate means of escape in case of fire. It is proposed that the fire provisions relating to this section should be amended to require compliance with the new regime (insofar as the room is above ground).

Mines Miscellaneous Health and Safety Provisions Regulations 1995

15.42 *Mines Miscellaneous Health and Safety Provisions Regulations 1995 (SI 1995 No 2005), Reg 4(5)* requires at *(a)* that fire is included in the health and safety emergency plan. It is proposed that the provision should be amended to reflect the application of the new fire regime to surface buildings.

Railways (Safety Case) Regulations 1994

15.43 *Railways (Safety Case) Regulations 1994 (SI 1994 No 237)* requires a safety case to be prepared. The matters to be included themselves include the significant findings of risk assessments made under *Management of Health and Safety at Work Regulations 1999 (SI 1999 No 3242)* and the workplace fire precautions legislation. With the revocation of *Fire Precautions (Workplace) Regulations 1997 (SI 1997 No 1840)* it is proposed to amend these Regulations to provide a separate reference to the new regime.

Safety Representatives and Safety Committees Regulations 1977

15.44 *Safety Representatives and Safety Committees Regulations 1977 (SI 1977 No 500)* apply to health and safety. These Regulations are applied to fire safety by the *Fire Precautions (Workplace) Regulations 1997 (SI 1997 No 1840)*. It is proposed to carry application forward to the new regime in order to comply with European requirements. This may be done by amendment of the Regulations.

Children and Young Persons Act 1933

15.45 *Children and Young Persons Act 1933, ss 11* and *12* contain provisions requiring the protection of children at places of entertainment where the majority of those attending are children. The provision is general and includes safety from fire. A provision specifically for the safety of children would appear to be superfluous as the matters should be covered by risk assessment based health and safety and fire law. It is proposed that the provision should be repealed.

Children's Homes Regulations 2001, and Children's Home (Wales) Regulations 2002

15.46 *Children's Homes Regulations 2001 (SI 2001 No 3967), Reg 32* and *Children's Home (Wales) Regulations 2002 (SI 2002 No 327), Reg 31* contain fire provisions, which are to be amended to reflect the new regime.

15.47 *The way ahead*

National Health Service and Community Care Act 1990, and National Health Service and Community Care Act 1990 (Commencement No 1) Order 1990 (SI 1990 No 388)

15.47 *National Health Service and Community Care Act 1990* removed Crown exemption from the National Health Service and made minor amendment to application of the *Fire Precautions Act 1971*. It is proposed to revoke references to *FPA 1971*.

Capital Allowances Act 1990

15.48 *Capital Allowances Act 1990, ss 69* and *70* allow for expenditure on fire safety requirements in a notice under the *Fire Precautions Act 1971* or the *Fire Safety and Safety of Places of Sport Act 1987* and *Safety of Sports Grounds Act 1975* legislation to be treated as capital expenditure for tax purposes.

Subject to the views of HM Treasury, it is proposed the provisions should be revoked if they duplicate other provision or amended to reflect the new regime if tax relief will not otherwise be available.

Finance Act 1975

15.49 *Finance Act 1975, s 15* is similar to the *Capital Allowances Act 1980*. It allows tax relief to be claimed for work specified by the fire authority as required in order to gain a fire certificate or to comply with a *Fire Precautions Act 1971, s 10* notice. It is proposed that the provision should be amended to reflect the new regime.

Local Acts

15.50 Some Local Acts have been identified as containing fire safety provisions. Subject to further consideration it is proposed that these provisions should be repealed insofar as they have subsequently been overtaken by Building Regulations or will be covered by the new fire safety regime.

Power to make regulations

15.51 Circumstances might arise which could require amendments to be made to address new problems. It is intended to re-state the existing power to make Regulations, which is contained in the *Fire Precautions Act 1971*. There would be additions to the power to reflect the nature of the proposed regime. So, Regulations might deal with the carrying out of risk assessments, the principles of prevention to be applied and the general duties of employees at work. They may also impose requirements on persons other than the responsible person and make provision as to persons who are to be responsible for any contravention of the Regulations.

Any Regulations made by the Secretary of State might impose requirements:

(a) as to the provision, maintenance and keeping free from obstruction of means of escape in case of fire;

(b) as to the provision and maintenance of means for securing that any means of escape can be safely and effectively used at all material times;

(c) as to the provision and maintenance of means for fighting fire and means for giving warning in case of fire;

(d) as to the internal construction of the premises and the materials used in that construction;

(e) for prohibiting altogether the presence or use in the premises of furniture or equipment of any specified description, or prohibiting its presence or use unless specified standards or conditions are complied with;

(f) for ensuring that persons employed to work in premises receive appropriate instruction or training in what to do in case of fire;

(g) for ensuring that, in particular circumstances, specified numbers of attendants are stationed in specified parts of the premises;

(h) as to the carrying out of assessments of the risk to persons in case of fire; and

(i) as to the keeping of records of instruction or training given, or other things done in order to comply with the regulations.

Before making any Regulations, the Secretary of State would be required to consult with such persons or bodies as appear to him to be necessary. It is considered desirable that the proposed regime contains the same degree of flexibility as the existing one and that the Secretary of State should continue to be able to introduce secondary legislation to meet particular circumstances.

The responsible person

15.52 It will be necessary to ensure that the Government continue to meet their obligations under European Directives, in particular, the *Framework Directive 89/391/EEC* and *Workplace Directive 89/654/EEC*, the *Chemical Agents Directive 98/24/EC* and the *Explosive Atmospheres Directive 99/92/EC*. This means that wherever there is an employer, they will continue to be responsible for the safety of their employees. In order to achieve the broader coverage of the legislation, which the Government desire, it is proposed to extend this definition. Consequently, it is proposed that the responsible person will be:

15.52 *The way ahead*

(a) the employer in relation to any workplace which is to any extent under their control;

(b) in relation to any premises where there is no employer:

  (i) the person (whether the occupier or owner of the premises or not) who has the overall management of the premises; or

  (ii) where there is no one with overall management responsibility, the occupier of the premises; or

  (iii) where neither (i) or (ii) apply, the owner of the premises.

Inclusion of 'the owner' in the definition will mean that empty buildings would be brought within the proposed Order; for instance a new building which had received *Building Regulations 1991 (SI 1991 No 2768)* approval, but had not yet been occupied. It is not believed that this will create a significant new burden. Owners would already be subject to a duty of care in other legislation such as the *Occupiers Liability Act 1957*. They would now be under a duty to carry out a risk assessment but this would largely involve ensuring that the building was secure. It is proposed that 'owner' should be defined as in the *Fire Precautions Act 1971*.

That is, the person who receives the rackrent of the premises in question, or the person who would receive the rackrent if the premises were let at a rackrent.

Other people, such as landlords, may be in a position to exercise varying degrees of control over premises and it is proposed that they should bear a relevant degree of responsibility under the proposed Order. However, this would not detract from the primary responsibilities and duties placed on employers and other responsible persons.

Equally, people might be appointed or employed to undertake duties, which bear on the safety of the premises. The Government has in mind contractors employed to install, maintain or test fire safety equipment or systems. It would be for the responsible person to ensure that any person they employ to carry out such work is competent to do so. There are a number of ways of doing this. The responsible person might use a contractor who is certificated under a suitably accredited third party certification scheme; or they might ask for proof of qualifications or seek references. Advice about selecting competent contractors would be included in the guidance documents supporting the proposed Order.

Where it can be established that an offence has been committed under the Order (for the proposed offences) and this has been caused by the negligence, failure or deliberate misrepresentation on the part of the contractor, it is proposed that the enforcing authority should be able to take action against the contractor. Action could also be taken against the responsible person if that seemed to be justified in the circumstances of the case.

Under the proposed Order the responsible person will be responsible not only for the safety of employees, but for anyone on the premises and anyone who might be affected by a fire. This extends the burden on employers and creates a new burden on self-employed people and some people in the voluntary sector. This is covered further in the section on application below.

Application

15.53 Premises designated under the *Fire Precautions Act 1971* as requiring fire certificates include, subject to certain criteria, factories, offices, shops and railway premises where people are employed to work, and hotels and boarding houses. The *Fire Precautions (Workplace) Regulations 1997 (SI 1997 No 1840)* cover virtually all workplaces. The new fire safety regime is expected to cover virtually all workplaces and places to which the public has access, but not domestic premises. The exceptions are set out later.

The proposed Order would therefore apply to certain groups not already covered including premises used by the self-employed (but not where they work at home) and elements of the voluntary sector.

Self-employed people not working at home are already responsible, under the *Health and Safety at Work etc Act 1974*, for the safety of people on their premises. They should already be carrying out risk assessments, which should include the assessment of the risk from fire. Moreover, they are, by definition (since they do not employ anyone), likely to use smaller premises with lower levels of fire risk. In that sense, the new Order will add little or no additional requirements to existing law, but accept that some people, who have not met the requirements imposed on them by existing legislation might incur expenditure to comply with the proposed Order.

With regard to the voluntary sector, the *Management of Health and Safety at Work Regulations 1999 (SI 1999 No 3242)* already applies to charities where they are employers. The extension will therefore be to voluntary workers in premises, which are not their home. There is some uncertainty as to how far the existing law applies to these groups already. For instance, it is not clear whether volunteers working in charity shops are employees. The Government certainly consider it strikes a fair balance between the rights of the individual and of the public interest, and is proportionate in that the fire safety law should apply to such premises in the same way as it would to the shop next door, which might have employees. Another example is a body such as the Scouts Association. This body already requires all its premises to comply with *SI 1999 No 3242*. Village halls, which would frequently accommodate the Scouts and other voluntary organisations are, in many cases, subject to the fire safety requirements of the licensing law applying to places of occasional entertainment.

It is considered that the extension of fire safety law to the voluntary sector should not create significant new burdens and that the extension could

15.54 *The way ahead*

therefore be described as proportionate, as striking a fair balance between the interests of the individual and the public interest, and as helping secure the desirable overall goal of a modern risk-based general fire safety regime.

Details of the proposed application of the Order are set out in the following paragraphs. They are based on the provisions of *SI 1995 No 1840* but extend protection to non-employees.

It is proposed that the Order, except for power of entry to ascertain cause of fire shall not apply to:

(a) domestic premises;

(b) an offshore installation within the meaning of *Regulation 3* of the *Offshore Installation and Pipeline Works (Management and Administration) Regulations 1995 (SI 1995 No 738)*;

(b) a ship other than a ship which is permanently moored or is in the course of construction or repair by persons who include persons other than the master and crew of the ship;

(c) fields, woods or other land forming part of an agricultural or forestry undertaking but which is not inside a building and is situated away from the undertaking's main buildings;

(d) an aircraft, locomotive or rolling stock, trailer or semi-trailer used as a means of transport or a vehicle for which a licence is in force under the *Vehicle Excise and Registration Act 1994* or a vehicle exempted from duty under that Act; and

(e) a mine within the meaning of *section 180* of the *Mines and Quarries Act 1954*, other than any building on the surface at a mine.

These are the same exclusions as for *SI 1995 No 1840*. Other than (a) and (d), these are areas subject to alternative regimes, which the Government would not wish the proposed Order to replace.

It is proposed that the Order shall not apply to occasional work or short-term work involving:

(a) domestic service in a private household (including people such as live-in nannies); or

(b) work regulated as not being harmful, damaging or dangerous to young people in a family undertaking.

These are the same exclusions as those currently in *1999 No 3242*.

Mitigation of the effects of a fire

15.54 It is also intended that the Order should offer reasonable protection to people in the vicinity of a place who might be affected by a fire, as well as the occupants. Given that, in most circumstances, the only way to achieve this would be to prevent fire occurring in the first place or to prevent it spreading, we feel that provisions implementing such requirements would inevitably reduce the impact of fire on the environment, reduce property damage and reduce risks to firefighters.

Furthermore, the provisions of the *Dangerous Substances and Explosive Atmospheres Regulations 2002 (SI 2002 No 2776)*, on which the Health and Safety Executive consulted, addressing as they do, the potential hazards of fires in places where dangerous chemicals are stored or used, build on the existing provisions of the *Fire Precautions (Workplace) Regulations 1997 (SI 1997 No 1840)* to provide a fire safety regime which reduces the risk of serious fires. The Regulations appear to implement the safety aspects of *European Directive 98/24/EC* on the health and safety of workers from risks related to chemical agents at work (the Chemical Agents Directive) and *European Directive 99/92/EC* on minimum requirements for improving the safety and health protection of workers potentially at risk from explosive atmospheres (the Explosive Atmospheres Directive).

A dangerous substance is defined in the Regulations as:

(a) a substance or preparation which meets the criteria in the approved classification and labelling guide for classification as a substance or preparation which is explosive, oxidising, extremely flammable, highly flammable or flammable, whether or not that substance or preparation is classified under the *Chemicals (Hazard Information and Packaging for Supply) Regulations 1994 (SI 1994 No 3247)*;

(b) a substance or preparation which because of its physico-chemical or chemical properties and the way it is used or is present creates a risk, not being a substance or preparation falling within (a); or

(c) any dust, whether in the form of solid particles or fibrous materials or otherwise, which can form an explosive mixture with air or an explosive atmosphere, not being a substance or preparation falling within (a) or (b).

The Regulations place a duty on employers to carry out a risk assessment and then to apply measures to control risks and to mitigate the detrimental effects of a fire or the other harmful physical effects arising from dangerous substances. Included in this is a requirement on the employer to provide measures to avoid the propagation of fires. The Regulations will apply to the self-employed and will seek to protect other persons who may be affected. The incorporation of these provisions in our Order should not, therefore, create a new burden.

The new fire safety regime extends the principles of the Regulations by requiring the responsible person in all premises covered by the Order to take steps to mitigate the detrimental effects of a fire. Like all the specific requirements imposed by the proposed Order, this requirement will be limited to what is necessary, reasonable and practical in the circumstances. The Government believe this will allow for consistency and simplicity.

While it represents an additional duty the Government believe it desirable that people who might be affected by a fire should be protected. The additional duty is proportionate and strikes a fair balance between the rights of the individual and the public interest because it should not add significantly to existing burdens since existing fire precautions include elements, which restrict fire spread such as fire doors. The larger element of protection to people who might be affected by a serious fire would already be in place through the making of the *Dangerous Substances and Explosive Atmospheres Regulations 2002 (SI 2002 No 2776)*. In the Government's view, these provisions should help make the new regime desirable as a whole.

Requirements of the proposed fire safety regime

15.55 It is intended that the requirements of the proposed Order should reflect the requirements of the *Fire Precautions (Workplace) Regulations 1997 (SI 1997 No 1840)*, which implement the European *Workplace Directive* and *Framework Directive*. The Order must continue to implement these Directives but, in addition, will extend protection explicitly to all occupants, not just employees. Since the fire precautions put in place by employers should already take into account other people on the premises and the fact that they or their employees have a duty of care to other occupants (for example, visitors, or in the case of care homes, residents), the new regime should not impose significant additional burdens on employers.

Fire safety duties

15.56 The fire safety duties, which are proposed to be included in the Order, are set out in the following paragraphs.

Risk assessment

15.57 The responsible person would have to make a suitable and sufficient assessment of the risks to persons to which they are exposed for the purpose of identifying the measures he or she needs to take to comply with the requirements and prohibitions imposed on him or her by the Order.

Where a dangerous substance is present on the premises, the risk assessment would include consideration of all the additional relevant factors including the amount of the substance and its hazardous properties and the circumstances of the work including the work processes.

The responsible person would have to review the assessment regularly in order to keep it up to date, and if:

(a) there is reason to suspect that it is no longer valid; or

(b) there has been a significant change in the matters to which it relates (including when the premises, work processes, or organisation of the work undergoes significant changes, extensions or conversions).

Where changes to an assessment are required as a result of any such review, the responsible person would have to make them. This underlines the principle that dynamic risk assessment is an ongoing process. The responsible person could not carry out a risk assessment and then forget about it; it must be kept constantly under review.

The responsible person would not employ a young person unless he or she has made or reviewed a risk assessment taking particular account of the special factors applying to young persons set out in the *Management of Health and Safety at Work Regulations 1999 (SI 1999 No 3242)*.

The responsible person would record the significant findings of the assessment, including the measures he or she had taken in order to comply with the Order; and any group of his or her employees identified as being especially at risk if:

(a) if he or she employs five or more employees;

(b) if the premises, or people responsible for the premises, are subject to any legislation providing for licensing or certification; or

(c) if an alterations notice is in force in relation to the premises.

No new work activity involving a dangerous substance would commence unless the risk assessment had been made and the measures required by or under the Order had been implemented. It would be for the responsible person to judge what constituted a new activity. The key would be to assess whether the risk had changed and whether the risk assessment still addressed the risks adequately.

These requirements in respect of risk assessments are drawn from the *Fire Precautions (Workplace) Regulations 1997 (SI 1997 No 1840)* and the draft *Dangerous Substances and Explosive Atmospheres Regulations 2002 (SI 2002 No 2776)*. It is intended to re-state existing burdens (or burdens which will exist) which are considered to be fair and proportionate and to strike a fair balance between the interests of those affected and the public interest given the benefits in terms of safety. It is suggested that the threshold for the requirement to record the significant findings should continue to be where the employer employs five or more employees in order to maintain consistency with other health and safety legislation. This would need to be kept under review and considered again in the event of any changes to this aspect of health and safety law.

15.58 *The way ahead*

This requirement is extended to licensed premises and premises subject to an alterations notice.

General duty to ensure safety

15.58 The Government proposes that the responsible person should take such measures as are reasonable for a person in his or her position to take to ensure, so far as is reasonably practicable, that the premises, all means of access to and egress from, and any plant or substance in or on the premises is or are, in the context of general fire safety, safe.

There are already general duties in the *Health and Safety at Work etc Act 1974* for employers to ensure the health, safety and welfare of their employees and others on the premises. This includes safety from fire. The *Fire Precautions (Workplace) Regulations 1997 (SI 1997 No 1840)* served to clarify the existing duty in respect of fire though it applied only to employees. In creating a fire safety regime it will be necessary to specifically re-state the general duty in respect of fire safety. In applying the duty explicitly to non-employees this is not creating a new legal burden since this was already contained in the *HSWA 1974*. Neither should this create new practical burdens since non-employees have been protected by other legislation such as the *Fire Precautions Act 1971*.

Principles of prevention to be applied

15.59 The responsible person should implement preventive and protective measures on the basis of the following principles:

(a) avoiding risks;
(b) evaluating the risks which cannot be avoided;
(c) combating the risks at source;
(d) adapting to technical progress;
(e) replacing the dangerous by the non-dangerous or less dangerous;
(f) developing a coherent overall prevention policy, which covers technology, organisation of work and the influence of factors relating to the working environment;
(g) giving collective protective measures priority over individual protective measures; and
(h) giving appropriate instructions to employees and other persons on the premises as appropriate.

These principles are drawn from the *Fire Precautions (Workplace) Regulations 1997 (SI 1997 No 1840)*.

Fire safety arrangements

15.60 It is proposed that the responsible person should make such arrangements as are appropriate, having regard to the nature of his or her activities and the size of his or her undertaking, for the effective planning, organisation, control, monitoring and review of the preventive and protective measures. He or she should record these arrangements if:

(a) he or she employs five or more employees;

(b) the premises, or people responsible for the premises, are subject to any legislation providing for licensing or certification; or

(c) if an alterations notice is in force in relation to the premises.

'Fire safety arrangements' means all the things that the responsible person needs to do to comply with the Order, such as appointing employees to carry out various functions, establishing emergency procedures or displaying fire action notices. As mentioned at **15.58** above, it is considered that the threshold of five or more employees should be maintained for consistency with health and safety legislation.

Elimination or reduction of risks from dangerous substances

15.61 It is intended to propose that where a dangerous substance is present on the premises, the responsible person should ensure that risk related to the presence of the substance is either eliminated or reduced so far as is reasonably practicable. Wherever possible, and so far as is reasonably practicable, he or she should replace it with a substance or process which either eliminates or reduces the risk.

Where this is not reasonably practicable he or she should, so far as is reasonably practicable, apply measures consistent with the risk assessment and appropriate to the nature of the activity or operation including the following measures in order to control the risk:

(a) reduce the quantity of dangerous substances to a minimum;

(b) avoid or minimise the release of a dangerous substance;

(c) control the release of a dangerous substance at source;

(d) prevent the formation of an explosive atmosphere, including the application of appropriate ventilation, and ensure that any release of a dangerous substance which may give rise to risk is suitably collected, safely contained, removed to a safe place, or otherwise rendered safe, as appropriate;

(e) avoid:

    (i) ignition sources including electrostatic discharges; and

(ii) such other adverse conditions as could result in harmful physical effects from a dangerous substance; and

(f) segregate incompatible dangerous substances.

It is proposed that the responsible person, in all premises to which the Order would apply, should take measures to mitigate the detrimental effects of a fire. Where dangerous substances are present these will include:

(a) reducing to a minimum the number of persons exposed;

(b) ensuring measures to avoid the propagation of fires and explosions;

(c) providing explosion pressure relief arrangements;

(d) providing explosion suppression equipment;

(e) providing plant which is constructed so as to withstand the pressure likely to be produced by an explosion; and

(f) providing suitable personal protective equipment.

The responsible person should ensure that if an explosive atmosphere contains several types of flammable, combustible or flammable *and* combustible gases, vapours, mists or dusts, then the protective measures are appropriate to the greatest potential risk. He or she should also arrange for the safe handling, storage and transport of dangerous substances and waste containing dangerous substances.

Fire-fighting and fire detection

15.62 It is proposed that where necessary to ensure the safety of persons in case of fire (whether due to the features of the premises, the activity carried on there, any hazard present or any other relevant circumstances), the responsible person should ensure that:

(a) the premises are, to the extent that it is appropriate, equipped with appropriate fire-fighting equipment and with fire detectors and alarms; and

(b) any non-automatic fire-fighting equipment so provided is easily accessible, simple to use and indicated by signs.

What is 'appropriate' for the purposes of (a) above would depend on the dimensions and use of the premises, the equipment contained on the premises, the physical and chemical properties of the substances likely to be present and the maximum number of persons that may be present at any one time.

It is proposed that the responsible person should, where necessary:

(a) take measures for fire-fighting in the premises, adapted to the nature of the activities carried on there and the size of the undertaking and of the premises concerned;

(b) nominate employees to implement those measures and ensure that the number of such employees, their training and the equipment available to them are adequate, taking into account the size of, and the specific hazards involved in, the premises concerned; and

(c) arrange any necessary contacts with external emergency services, particularly as regards rescue work and fire-fighting.

These proposed requirements are drawn from the *Fire Precautions (Workplace) Regulations 1997 (SI 1997 No 1840)*. The actual purpose for which fire-fighting equipment is provided has never been clearly defined. Although the *Fire Precautions Act 1971* also contains requirements for fire-fighting equipment, we know that particular Act is primarily concerned with the safe evacuation of occupants of a building. Theoretically, someone could notice a fire in a waste paper basket, leave the room and warn other occupants to evacuate the building, leaving the fire to spread. Normally, assuming the fire precautions were adequate, everyone would escape safely. It is considered that the law should reflect the common sense view that extinguishing the fire at source must reduce the risk to life in case something goes wrong.

The proposed regime will require the implementation of measures to reduce the risk of fire and the spread of fire. This leads us to consider the purpose for which fire-fighting equipment may be required. The Government considers that the existing law is unclear in this regard and does not reflect the basis on which fire-fighting equipment is normally provided at present. Therefore it is proposed that, in addition to the earlier requirements, the proposed Order should make it clear that fire-fighting equipment should be considered as a possible means of reducing a risk of fire spreading, providing protection and for providing assistance to others, and not merely as a means of safeguarding the means of escape. It should also be considered as a possible means of mitigating the detrimental effects of a fire.

This should not impose an additional burden as it is intended as a clarification of the purposes to which fire-fighting equipment might be put and how it might be used as a particular means of addressing fire safety problems. It is believed that in most places fire-fighting equipment already exists for the purposes outlined above.

Emergency routes and exits

15.63 It is proposed that where necessary in order to ensure the safety of persons in case of fire, the responsible person should ensure that routes to emergency exits from premises and the exits themselves are kept clear at all times.

15.64 *The way ahead*

There is an intention to re-state the specific requirements in relation to emergency routes and exits, which are contained in the *Fire Precautions (Workplace) Regulations 1997 (SI 1997 No 1840)*. These include the requirements that:

(a) emergency routes and exits should lead as directly as possible to a place of safety;

(b) it should be possible for persons to evacuate the premises as quickly and as safely as possible; and

(c) the number, distribution and dimensions of emergency routes and exits should be adequate having regard to the use, equipment and dimensions of the premises and the maximum number of persons who may be present there at any one time.

Other requirements dealing with the nature of the exit doors, signage and lighting should be noted.

Editorial note. It must be remembered that when considering the requirements for emergency routes and exits to bear in mind the need to ensure full compliance with European Union legislation.

Maintenance

15.64 Where necessary, it is proposed that in order to uphold the safety of persons in case of fire, the responsible person shall ensure that the premises and any equipment provided in connection with fire-fighting, fire detection or emergency routes and exits are subject to a suitable system of maintenance and are maintained in an efficient state, in efficient working order and in good repair. This applies to facilities provided to meet the requirements of the proposed Order or any other legislation, including legislation repealed or revoked by the Order.

In the 1997 consultation document *Fire Safety Legislation for the Future*, published by the Home Office, it was stated that fire precautions should take account of the need to keep the risk to fire fighters who may have to enter a burning building to a minimum. It is believed that the measures outlined earlier for the prevention of fire and for mitigating the detrimental effects of a fire go a long way to achieving that aim. Additionally, however, the responsible person will also be required to maintain any facilities which have been provided under the Building Regulations for the use and protection of fire fighters.

This is an additional burden which – it is considered – strikes a fair balance between the interests of those affected and the public interest. At present, the law provides for the installation in new buildings of facilities specifically for the use and protection of the fire service, but little to ensure that they are fit for use when they are needed. This seems irrational and it is considered right to address this in the Order.

Editorial note. If this proposal order forms part of the Regulations this will effectively mean that any recommendation contained in British Standards and their associated Codes of Practice for maintenance are requirements. These recommendations would have to be fully implemented and suitably recorded in the Fire log book.

Safety assistance

15.65 It is proposed that the responsible person should, subject to certain exceptions, appoint one or more competent persons to assist them in undertaking the measures he or she needs to take to comply with the requirements and prohibitions imposed by the Order.

The responsible person should make arrangements for ensuring adequate co-operation between them and ensure that the number of persons appointed, the time available for them to fulfil their functions and the means at their disposal are adequate having regard to the size of the premises, the risks to which persons are exposed and the distribution of those risks throughout the premises.

If the responsible person appoints someone who is not their employee, that person should:

(a) be informed of the factors which the responsible person knows or suspects affect the safety of any other person who may be affected by the conduct of the undertaking; and

(b) have access to the information which the responsible person is required to provide under the Order.

The responsible person should also ensure that any person they appoint is given such information about any person working in the undertaking as is necessary to enable that person properly to carry out their functions.

A person would be regarded as competent where they have sufficient training and experience or knowledge and other qualities to enable them to properly assist in undertaking the measures needed to comply with the requirements of the Order.

The requirement to appoint a competent person should not apply to a self-employed employer who is not in partnership with any other person where they have sufficient training and experience or knowledge to undertake the relevant measures. Nor should it apply to individuals who are employers and who are together carrying on business in partnership where at least one of the individuals concerned has sufficient training and experience or knowledge and other qualities.

Where there is a competent person in the responsible person's employment, that person should be appointed in preference to a competent person not in their employment.

These requirements are drawn from the *Fire Precautions (Workplace) Regulations 1997 (SI 1997 No 1840)* and reflect longstanding health and safety law. In smaller businesses the employer will be able to meet the requirements of the Order himself. The purpose of these requirements is to ensure that, where necessary, the responsible person appoints someone who is competent to assist them. The Government is aware of no evidence to indicate that these requirements are not proportionate and do not strike a fair balance between the rights of those affected and the public interest. The Government believes the proposed regime, including these requirements, would be desirable as a whole.

Procedures for serious and imminent danger and for danger areas

15.66 It is proposed that the responsible person should:

(a) establish and where necessary give effect to appropriate procedures to be followed in the event of serious and imminent danger to persons;

(b) nominate a sufficient number of competent persons to implement those procedures in so far as they relate to the evacuation of persons from the premises; and

(c) ensure that no person has access to any area to which it is necessary to restrict access on grounds of safety unless the person concerned has received adequate safety instruction.

These procedures should:

- so far as is practicable, require any persons who are exposed to serious and imminent danger to be informed of the nature of the hazard and of the steps taken or to be taken to protect them from it;

- enable the persons concerned to stop work and immediately proceed to a place of safety in the event of their being exposed to serious, imminent and unavoidable danger; and

- save in exceptional cases require the persons concerned to be prevented from resuming work in any situation where there is still a serious and imminent danger.

Where a dangerous substance is present on the premises, there would be a number of additional requirements. In the event of fire arising from an accident, incident or emergency related to the presence of the dangerous substance, the responsible person should ensure that:

(a) procedures, including the provision of appropriate first-aid facilities and relevant safety drills (which shall be tested at regular intervals), have been prepared which can be put into effect when such an event occurs;

(b) information on emergency arrangements is available;

(c) suitable warning and other communication systems are established to enable an appropriate response, including remedial actions and rescue operations, to be made immediately when such an event occurs;

(d) where necessary, before any explosion conditions are reached, visual or audible warnings shall be given and persons withdrawn; and where the risk assessment indicates it is necessary, escape facilities should be provided and maintained to ensure that, in the event of danger, persons can leave endangered places promptly and safely.

It is proposed that the responsible person should ensure that the information on the emergency arrangements, procedures, warnings and other communication systems and escape facilities described above is made available to relevant accident and emergency services to ensure those services, whether internal or external to the premises, are able to prepare their own response procedures and precautionary measures.

In the event of an accident, incident or emergency related to the presence of a dangerous substance on the premises, the responsible person should ensure that:

(a) immediate steps are taken to:

   (i) mitigate the effects of the event;

   (ii) restore the situation to normal; and

   (iii) inform those persons who may be affected; and

(b) only those persons who are essential for the carrying out of repairs and other necessary work are permitted in the affected area and they are provided with:

   (i) appropriate personal protective equipment and protective clothing; and

   (ii) any necessary specialised safety equipment and plant, which shall be used until the situation is restored to normal.

These requirements, adapted to reflect the extension of protection to non-employees, are drawn from the *Fire Precautions (Workplace) Regulations 1997 (SI 1997 No 1840)* and the *Dangerous Substances and Explosive Atmospheres Regulations 2002 (SI 2002 No 2776)*.

15.67 *The way ahead*

The Government is not aware of any evidence to indicate that these requirements would not be proportionate and would not strike a fair balance between the rights of those affected and the public interest. It is believed that the proposed regime, including these requirements, would be desirable as a whole.

Provision of information

15.67 It is proposed that where the responsible person is an employer, they should provide employees and, where appropriate, other people on the premises, with comprehensible and relevant information on:

(a) the risks to them identified by the risk assessment;

(b) the preventive and protective measures;

(c) the procedures for serious and imminent danger;

(d) the identities of those persons nominated to implement fire-fighting measures and the procedures for serious and imminent danger; and

(e) the risks notified by any other responsible person sharing the premises.

The responsible person should, before employing a child, provide a parent of the child with comprehensible and relevant information on:

- the risks identified by the risk assessment;
- the preventive and protective measures; and
- the risks notified by other responsible persons.

For the purposes of this requirement, 'parent of the child' includes a person who has parental responsibility, within the meaning of *section 3* of the *Children Act 1989*, for the child.

Where a dangerous substance is present on the premises, the responsible person where they are an employer should provide employees with:

(a) the details of any such substance including:

   (i) the name of the substance and the risk which it presents;

   (ii) access to any relevant safety data sheet; and

   (iii) legislative provisions (concerning the hazardous properties of any such substance) which apply to the substance;

(b) the significant findings of the risk assessment; and

(c) suitable and sufficient information on the appropriate precautions and actions to be taken by the person in order to safeguard them and other persons.

This additional information required in respect of dangerous substances should be:

- adapted to take account of significant changes in the activity carried out or methods or work used by the responsible person; and
- provided in a manner appropriate to the risk identified by the risk assessment.

These requirements are drawn from the *Fire Precautions (Workplace) Regulations 1997 (SI 1997 No 1840)* and the *Dangerous Substances and Explosive Atmospheres Regulations 2002 (SI 2002 No 2776)*. The requirement to provide detailed information would not normally extend beyond employees. It would not, for instance, be practical to provide details of the risk assessment to anyone entering a shop. But it would be necessary to display certain information such as fire action notices and information about particular hazards.

It is considered that these requirements to be proportionate and that they strike a fair balance between the rights of those affected and the public interest. The Government believe the proposed regime, including these requirements, would be desirable as a whole.

Co-operation and co-ordination

15.68 It is proposed that where two or more responsible persons share premises (whether on a temporary or a permanent basis), each such person should:

(a) co-operate with the other responsible person concerned so far as is necessary to enable them to comply with the requirements and prohibitions imposed on them by the Order;

(b) (taking into account the nature of their activities) take all reasonable steps to co-ordinate the measures they takes to comply with the requirements and prohibitions imposed by the Order with the measures the other responsible persons are taking; and

(c) take all reasonable steps to inform the other responsible persons concerned of the risks to their employees' safety arising out of the conduct by them of the undertaking.

These requirements, adapted to reflect the definition of the 'responsible person' beyond employers, are drawn from the *Fire Precautions (Workplace) Regulations 1997 (SI 1997 No 1840)*.

It is considered that these requirements are proportionate and that they strike a fair balance between the rights of those affected and the public interest, and that the proposed regime, including these requirements, would be desirable as a whole.

15.69 *The way ahead*

Capabilities and training

15.69 It is proposed that the responsible person should ensure that employees are provided with adequate fire safety training:

(a) on their being recruited into their employment; and

(b) on their being exposed to new or increased risks because of:

  (i) a transfer or a change of responsibilities;

  (ii) the introduction of new work equipment or a change with regard to work equipment already in use;

  (iii) the introduction of new technology; or

  (iv) the introduction of a new system of work or a change concerning a system of work already in use.

This training should:

- include suitable and sufficient instruction and training on the appropriate precautions and actions to be taken by the employee in order to safeguard themselves and other persons on the premises;
- be repeated periodically where appropriate;
- be adapted to take account of any new or changed risks to the safety of the employees concerned; and
- take place during working hours.

Employees should not be charged for training. These requirements are drawn from the *Fire Precautions (Workplace) Regulations 1997 (SI 1997 No 1840)*. They are considered to be proportionate, to strike a fair balance between the rights of those affected and the public interest. This proposed regime, including these requirements, would be desirable as a whole.

General duties of employees at work

15.70 Every employee should, while at work:

(a) take reasonable care for the safety of themselves and of other persons who may be affected by their acts or omissions at work;

(b) as regards any duty or requirement imposed on their employer or any other person by the Order, co-operate with them so far as is necessary to enable performance of or compliance with that duty or requirement; and

(c) inform their employer or any other employee with specific responsibility for the safety of fellow employees:

(i) of any work situation which a person with his or her training and instruction would reasonably consider represented a serious and immediate danger to safety; and

(ii) of any matter which a person with his or her training and instruction would reasonably consider represented a shortcoming in the employer's protection arrangements for safety, insofar as that situation or matter either affects the safety of that first-mentioned employee or arises out of or in connection with their own activities at work, and has not previously been reported to their employer or to any other employee of that employer in accordance with this requirement.

These requirements are drawn from the *Fire Precautions (Workplace) Regulations 1997 (SI 1997 No 1840)* and considered to be proportionate and to strike a fair balance between the rights of those affected and the public interest. The proposed regime, including these requirements, would be desirable as a whole.

Guidance

15.71 The Government intends to publish a series of guidance documents, which will contain advice to users as to how they can meet their responsibilities under the new Order. They will need to be easy to understand but sufficiently detailed so as to be of practical use to both the responsible person and the enforcing authority. These documents will explain the principles of risk assessment and the requirements of the new law, and will give practical examples of different solutions to fire safety problems. The documents will be aimed at particular types of building and business sectors. It is intended separate guides will be published for the following premises:

- Care homes.
- Schools, boarding schools, further education establishments.
- Offices, shops.
- Factories, warehouses, superstores.
- Pubs, clubs, restaurants.
- Theatres, cinemas.
- Museums, galleries, libraries.
- Hotels, boarding houses, hostels.
- Sub-surface railway stations.

The Government will consult separately on these guidance documents and they will be submitted to the Parliamentary Scrutiny Committees as the Order passes through its scrutiny stages. The guidance should be non-statutory since the Government believes that fits well with a goal-based regime. Most guidance issued at present on fire safety matters is non-statutory, though courts will take any relevant guidance into account when deciding whether or not an offence has been committed under fire safety legislation. Any form of statutory guidance, where a failure to comply would constitute an offence, would not appear to sit well with the goal-based regime since one of the main aims is that the responsible person may choose from a number of solutions. There would be a consequent difficulty in deciding whether or not he or she had complied with the guidance.

Licensing

15.72 If the scope of the regime is to be as wide as proposed and is to meet the aim of reducing overlap with other legislation, it will be necessary to examine how the reformed law will work alongside other major regulatory regimes, in particular licensing and housing.

The benefits of the reform of fire safety legislation should be seen most clearly in respect of licensed premises. For instance, under the existing law, the operator of a hotel has to apply for a fire certificate (for which he or she has to pay the administrative costs). The precautions set out in the certificate are prescribed by the fire authority and cover all the occupants (both staff and guests). Under the *Fire Precautions (Workplace) Regulations 1997 (SI 1997 No 1840)* an employer has to carry out a risk assessment identifying fire risks, and provide the appropriate precautions to ensure the safety of employees only (while taking account of others using the premises). These regulations are also enforced by the fire authority. An employer may also have to apply for licences for drinking or gaming etc and these will contain conditions relating to fire safety. Here, the enforcing authority will be the licensing authority; they have to consult the fire authority but can impose entirely different fire safety requirements. A licensee may even have to apply to different licensing authorities for different forms of licence.

The aim of the reform has to be to make life simpler for people responsible for licensed premises without reducing the level of fire safety. For any premises there should be a single fire safety regime, and to continue to meet the United Kingdom's obligations under the *Workplace* and *Framework Directives* that regime must be based on a risk assessment prepared by the employer where there is one. Under the proposals, the employer, or other responsible person in licensed premises, will need to make a risk assessment taking account of the safety from fire of everyone on the premises and in the vicinity.

Editorial note. This approach is to be welcomed as, at the present time, there is a major overlap in control, eg building control officers, police, fire authorities and licensing officers, which can lead to confusion and differing requirements between different authorities invariably to the detriment of the applicant seeking a licence.

The Government sees no reason why that single fire safety regime could not be incorporated into the licensing arrangements. It is important to ensure the flexibility of the current arrangements whereby licensing officers can make decisions on fire safety based on their knowledge of a particular building and all the safety considerations which need to be taken into account. It is proposed to continue to make use of the fire safety expertise of many licensing officers. Effective consultation between fire authorities and licensing authorities will be crucial.

The licences should no longer contain conditions setting out specific fire safety requirements. Instead, there should be a general condition requiring compliance with the new *Fire Safety Order*. (The licence would, of course, contain conditions dealing with other aspects of safety.) The proposed new law will be enforced by the fire authority, but it is expected that the fire authority – before agreeing the risk assessment – will take into account other linked risks that may be drawn to their attention by the licensing authority as part of a statutory consultation process. It will be necessary for the risk assessment to be recorded regardless of the number of employees. The fire authority would also be required to draw the attention of the licensing authority to any relevant information. Additionally, the fire authority would be required to consult the licensing authority before taking any enforcement action (though, as now, failure to do so would not invalidate the enforcement action). Fire authorities would still be able to prohibit or restrict the use of premises if circumstances so demanded without consultation. Any action taken by the fire authority in respect of a breach resulting in a prosecution under fire legislation should automatically count as grounds for endorsement, revocation or refusal, whichever is appropriate, and enforcement action should be taken into account on re-application for a licence.

Because any breach of the fire safety law would constitute a breach of the licence conditions, the licensing authority would be able to take action to address deficiencies in fire precautions by enforcing the licensing law. Where a new problem arose requiring a new fire safety solution, the licensing officer would be able to advise the licensee as to whether, in their opinion, the new solution proposed would still meet the requirements of the fire safety law and thereby meet the conditions of that particular licence. Licensing officers would thus be able to take decisions in circumstances where it might not be practical to consult a fire safety officer. This would normally apply to short term solutions to counter immediate difficulties. Long term solutions involving significant changes to the risk assessment would normally need to be approved by the fire authority.

In line with the general proposal that where different enforcing authorities have responsibility for a premises, they should consult each other. The licensing authority would be required to consult the fire authority on receipt of an application for a licence, and before taking any enforcement action in respect of the condition requiring compliance with the Fire Safety Order. This would not prevent the licensing authority from acting immediately to close premises or prevent an event from taking place if the risks were so serious as to demand such action. Again, this would be carried out using their powers under licensing laws.

15.73 *The way ahead*

The proposals: enforcement

15.73 Who will enforce the new Fire Safety Order?

The Government believe that, in the majority of cases, the fire authority should be the primary enforcing authority for the proposed Order. There are, however, some areas where it would be more appropriate for other bodies to take the lead, eg nuclear installations and ships under construction or repair for which the Health and Safety Executive would be the appropriate enforcing authority. It is expected the Ministry of Defence's fire service to be the enforcing authority for places used by the armed forces of the Crown, visiting forces and international headquarters or defence organisations designated for the purposes of the *International Headquarters and Defence Organisations Act 1964*.

In the case of premises subject to safety certification under the *Safety of Sports Grounds Act 1975* and the *Fire Safety and Safety of Places of Sport Act 1987*, it is proposed that the local authority responsible for the issue of the safety certificate be the enforcing authority. This work is carried out by a committee on which the fire authority sits.

Crown immunity

15.74 Both the *Fire Precautions Act 1971* and the *Fire Precautions (Workplace) Regulations 1997 (SI 1997/1840)* apply, to a limited extent, to premises either occupied by the Crown or owned by the Crown but not occupied by it. The Crown is immune from prosecution under fire safety legislation. The Government is considering its position on the question of Crown immunity. It is believed that – whatever conclusion is reached – there needs to be consistency of application to the Crown across health and safety law and fire safety legislation. For that reason there has been no attempt to address the question of Crown immunity separately in this reform.

The intention is to maintain the status quo in respect of fire safety legislation. The Government could, however, allow for flexibility in the future by identifying parts of the proposed Order as subordinate, which would allow subsequent amendment by either negative or affirmative resolution order.

In the meantime, the Government proposes that the Crown Premises Inspection Group of HM Fire Service Inspectorate would enforce the proposed Order in relation to:

(i) premises owned or occupied by the Crown;

(ii) the Houses of Parliament;

(iii) the United Kingdom Atomic Energy Authority.

It would be possible under the proposed Order for the Secretary of State to authorise other persons or bodies to enforce the requirements if they consider it appropriate. This reproduces an existing provision in the *Fire Precautions Act 1971*.

How will the Order be enforced?

15.75 Enforcing authorities will be under a duty to enforce the new Fire Safety Order. But the new regime is based on the principle that employers and people responsible for activities giving rise to risk have the responsibility for the fire safety of their premises and the people who use them. This means enforcing authorities will need to enforce the new legislation in a way which reflects that principle.

Where the enforcing authority are not satisfied that the fire precautions are adequate, rather than stipulate exactly what the responsible person must do, it will be for the enforcing authority to advise them where the law has not been complied with, why they are of that opinion and, where necessary, require them to take action. This is in line with existing health and safety law.

The Government expects enforcing authorities to enforce the law in accordance with the Cabinet Office's *Enforcement Concordat*. It is also open to Ministers, under *section 9* of the *Regulatory Reform Act 2001*, to issue a code of practice to be followed by enforcing officers. Failure to comply with such a code could be taken into account by a court in determining whether a person has failed to comply with any provision of the proposed Order. It is expected that most minor breaches could be dealt with through informal action such as verbal or written advice without the need to serve a formal notice. Where the breach of the law is more serious, or where informal action has not been effective, the enforcing authority might then issue a formal enforcement notice, which explains how the law has been breached. The authority might (though it would not be obliged to) add a schedule to the notice setting out means by which the contravention could be rectified. The schedule would acknowledge that the responsible person may choose a reasonable alternative method of achieving compliance.

Validation of fire safety solutions – the question of public reassurance

15.76 In *Fire Safety Legislation for the Future* it was envisaged that there might be some form of formal validation procedure for high risk premises. It was felt that something of this nature might be required to provide the public with the kind of reassurance currently offered by the certification process. One of the problems with such an approach is that it gives rise to the question of how one defines 'high risk'. The Government believes that it is not practicable to incorporate a definition in the law that would be sufficiently comprehensive and robust as to capture all those premises, which might give rise to concern.

15.77 *The way ahead*

It is considered that one of the fundamental advantages of the reform is that fire authorities will be freed to apply their resources to the inspection of premises which they consider, on the basis of their local knowledge, and with the assistance of guidance issued by the Secretary of State, to present the most risk. The Government is not proposing that the plans should be approved by the Secretary of State.

To reinforce this approach it is proposed that there should be a duty on the enforcing authority to institute, develop and maintain an enforcement programme which would include details of how the authority might determine the frequency with which it will inspect premises to which the Order applies in order to monitor and encourage compliance with the law. In developing such a programme the authority would have to take into account all the information available about the use and associated fire risks in the premises in its area.

Such a requirement would make it sufficiently clear that inspection is an integral element of the enforcement process while leaving the actual decision as to the frequency of inspection at the discretion of the enforcing authority.

The Government believes that the requirement to prepare an enforcement programme would form part of the systems of protections under the proposed new risk-based fire safety system, and as such is proportionate. It is not considered that it would affect any rights or freedoms that people or the fire authority could reasonably expect to continue to exercise. In the view of the Government, it should strike a fair balance between the interest of those affected and the public interest.

Reassurance to the public

15.77 The Government has considered very carefully the question of public reassurance and the implications of removing the requirement for fire certificates. It is important to consider what it is about the existing arrangements which gives the public their sense of assurance. It is believed that this assurance comes from enforcement by the fire service (on the knowledge that the fire service has inspected the premises, examined the fire precautions and declared themselves satisfied). The fire service will, of course, continue to do this.

The fact that the responsible person will be responsible for carrying out the risk assessment does not free them to provide fire precautions to a lower standard than existed before. The new regime is goal-based and, as explained, the law will set out the goals very clearly as specific requirements. The responsible person can determine how the goals may be achieved, but it will be for the enforcing authority, in enforcing the law, to assess whether they have done so. Effective enforcement is therefore key to the whole process and this underlines the importance of the authority being seen to have an enforcement programme, which targets its inspections on the basis of risk.

Charging

15.78 In *Fire Safety Legislation for the Future* it was considered whether there might be circumstances where the fire authority should be able to charge for providing advice. At present, the advice given on request under *section 1(1)(f)* of the *Fire Services Act 1947* has to be provided free of charge. Many fire authorities favour this on the grounds that a charge would deter people from seeking advice, to the detriment of fire safety. Charging would also seem to be inconsistent with the risk assessment-based approach and the principles of the *Enforcement Concordat*.

On the other hand, businesses may on occasion turn to fire authorities for specialist advice, because it is free, rather than pay fees to consultants. In some cases, fire authorities find themselves allocating fire safety professionals to work full time on a major development with no legal power to charge for this service. The Government had difficulty in arriving at a clear distinction between the giving of advice and the fulfilment of a statutory duty as a consultee. In any case, both would remain statutory duties. On balance, therefore, it was concluded that fire authorities should not be able to charge for any services related to fire safety.

Analysis – re-statement of existing burdens

15.79 Many of the requirements contained in the proposed Order re-state existing burdens contained in provisions of the workplace fire precautions legislation. These are the provisions relating to:

(a) risk assessment;

(b) the principles of prevention to be applied;

(c) fire safety arrangements;

(d) fire fighting and fire detection;

(e) emergency routes and exits;

(f) safety assistance;

(g) procedures for serious and imminent danger and for danger areas;

(h) provision of information;

(j) co-operation and co-ordination;

(k) persons working host employers' or self-employed persons' undertakings;

(l) capabilities and training; and

(n) general duties of employees at work;

All these requirements reflect longstanding health and safety law and, in most cases, implement European Union legislation. The Government is not aware of any evidence to indicate that they are not proportionate to the safety benefits they bring or that they do not strike a fair balance between the burden they impose on employers and the self-employed and the people they are intended to protect.

Necessary protection

15.80 Existing fire safety law provides a level of protection from fire to occupants of premises, which are put to certain uses. There is nothing in the proposals for reform, which would justify any reduction in the levels of protection already achieved by fire certification, safety certification or other existing legislative provision.

Protection should be extended to persons who might be affected by a fire.

Health and safety law already offers protection to people who might be affected by accidents in the workplace. The Government think there is a desirable benefit in protecting anyone who might be affected by a fire, not just occupants of buildings. It is believed that the *Dangerous Substances and Explosive Atmospheres Regulations 2002 (SI 2002 No 2776)* will go some way to providing such protection since they seek to mitigate the effects of fires involving the most dangerous materials. It is proposed to extend the duty on the responsible person to mitigate the detrimental effects of a fire to all premises covered by the proposed Order. As fire precautions in most buildings already contain elements aimed at stopping the spread of fire, this should not be a significant new burden and could be considered proportionate.

There is to be a new duty to maintain those fire precautions and facilities required under building regulations and fire safety legislation, including those which are for the use and protection of the fire brigade.

In these circumstances the Government considers the imposition of this new burden to be reasonable and proportionate to the problem being addressed. Any measure, which enables the fire service to fight fires more efficiently and safely, must strike a fair balance between those affected and the public. At present, the law provides for the installation in new buildings of facilities specifically for the use and protection of the fire service, but nothing to ensure that they are fit for use when they are needed. It is desirable that this be rectified.

Bain Report

15.81 A detailed report entitled *The Future of the Fire Service: Reducing Risk, Saving Lives – The Independent Review of the Fire Service (December 2002)* was produced by Professor Sir George Bain, and Professor Michael Lyons and Sir Anthony Young (who were the other members of his enquiry team). This enquiry was instituted following the pay claim submitted by the Fire Brigades Union early in 2002.

List of recommendations

15.82 Some of the proposals of the Bain Report are listed below with paragraph numbers relating to that report, where appropriate.

Risk and community fire safety

15.83 The work on risk-based fire cover should be taken forward through a series of incremental steps as follows:

(a) The Government should give fire authorities the power to deploy resources differently from the present requirements.

(b) The Government should instruct each fire authority to develop a Risk Management Plan that will save more lives and provide better value for money.

(c) Fire authorities should be required to consult their communities and key stakeholders in the preparation of their plans.

(d) Chief officers should be empowered to implement their authority's plan. (Paragraph 5.12.)

(e) The Office of the Deputy Prime Minister should issue the necessary guidance to implement a risk-based approach to fire cover as a matter of urgency. (Paragraph 5.19.)

(f) The Office of the Deputy Prime Minister should amend or remove *section 19* of the *Fire Services Act 1947* as soon as possible. (Paragraph 5.20.)

(g) The Government should legislate to put the Fire Service on a new statutory basis. (Paragraph 5.23.)

(h) The Government should commit itself to submit an annual report to Parliament on the Fire Service. (Paragraph 5.23.)

(j) The Government should put in hand the work necessary to produce new options for the Fire Standard Spending Assessment (SSA) linked to the role of the future Fire Service and its restated objectives so that a new formula can be introduced for 2006/07 and earlier if possible. (Paragraph 5.25.)

Role of central and local government

15.84 With regard to the role of central and local government, the Bain Report suggests the following.

(a) A new body should replace the Central Fire Brigades Advisory Council (CFBAC) for England and Wales. (Paragraph 6.8.)

15.84 *The way ahead*

(b) The existing local government performance management framework should be used to set national priorities for the Fire Service, more specifically:

(c) National priorities for the Fire Service should reflect what it could contribute in the context of the Shared Priorities agreed with local government.

(d) The national Public Service Agreement (PSA) or Service Delivery Agreement (SDA) should articulate more clearly what national government expects of all fire authorities.

(e) In the light of forthcoming best value and performance improvement guidance, fire authorities should consider how Best Value reviews can help to address whether existing services are the most efficient and effective means of meeting the needs of users and the wider community.

(f) The Government should discuss with the Audit Commission and the Accounts Commission an assessment process for fire authorities, building on lessons from the Comprehensive Performance Assessment (CPA) process. (Paragraph 6.12.)

(g) As a matter of priority, the Government should establish a strategic-level, high capability co-ordination infrastructure to deal with New Dimension work. (Paragraph 6.16.)

(h) All fire authorities which retain separate control rooms should be required to demonstrate to the Audit Commission and the Accounts Commission that their retention is likely to be cost effective against national performance standards. (Paragraph 6.17.)

Collaboration and co-operation should include:

- introducing common training standards and reducing training and other facilities duplicated within or across brigades;
- making more use of the facilities of local colleges of further education;
- co-ordinating procurement, including timetables, sharing and using best practice;
- developing operational policies and strategies to deal with the New Dimension;
- sharing experience in rolling out the reform agenda, particularly in human resources;
- sharing best practice in management; and
- developing local strategic partnerships. (Paragraph 6.21.)

Additionally, the report recommends brigades should investigate the potential for developing First Responder Partnerships (paragraph 6.26) and the Local Government Association and the Convention of Scottish Local Authorities should take steps to develop the contribution of elected members on fire authorities and to ensure that they give stronger leadership in the future (paragraph 6.28).

Note The remaining recommendations mainly deal with Implementation Recruitment and Pay etc. The full report is available on the Internet at http://www.irfs.org.uk/docs/future/

Brief comment

15.85 The *Bain Report* is, of course, one of many reports that have been made following enquiries into the Fire Service, and with the previous reports only partial implementation has subsequently followed. It is probable that the recommendations that have been made in the Report will, if accepted, only be implemented in part. The Fire Brigades Union did not make a contribution, although they were in the list of Consultees on the Consultation document on the reform of fire safety legislation. Therefore, it is reasonable to assume that they will have views on that document, which could affect the recommendations made in the *Bain Report*.

Officers from Her Majesty's Fire Inspectorate inspect all Fire Brigades on an annual basis to ensure that adequate standards are being maintained. A report is produced and a number of these annual reports are available on the website of the Office of the Deputy Prime Minister at http://www.odpm.gov.uk These reports are usually favourable in their content and the areas for disagreement between the Fire Service employers and the Fire Brigades Union immediately become apparent.

Should the Government decide to amend or remove *section 19* of the *Fire Services Act 1947*, which could allow Fire Authorities to modify the existing standards, there may be a possible reduction in the number of fire stations, appliances and/or personnel. This will have an impact upon the way in which individual fires can be fought. At present the number of fire stations, appliances and personnel cannot, normally, be reduced without prior approval from the Office of the Deputy Prime Minister.

Should reductions occur it might well place even more emphasis upon the need for fire risk assessments to be properly conducted. Any reduction in standard could also affect fire safety in a building being linked, where Fire Service intervention is one of the parameters, to a Fire Engineering solution.

The current average cost of a fire in public sector premises is £44,300.00, and in commercial premises the average cost is £63,600.00. This could rise unless considerable care is taken in the alterations to fire cover.

Obviously there is always scope for the alteration and improvement in standards and it must be hoped that all parties involved will approach any negotiations in a sensible way.

Appendix   Sources of further information

Introduction

A.1   There is a wide range of information relating to Fire Safety. In this Appendix some of the more important 'current' sources of information are explained and listed. A number of the British Standards or other Codes of Practice may have been superseded. However, they are still listed here as they are 'recognised' as the appropriate document in the Approved Documents that support the *Building Regulations 2000 (SI 2000 No 2531)*.

Means of escape in case of fire

A.2   This is the most important issue in Fire Safety. Employees or persons visiting premises must be able to escape quickly and safely from a building. The most important sources of information concerning escape are the British Standards 5588 series of Codes of Practice together with *Approved Document B* (2000 Edition).

The relevant documents in this series that relate to workplaces are as follows.

British Standard 5588: Fire precautions in the design, construction and use of buildings

- Part 0: 1996 Guide to fire safety codes of practice for particular premises/applications.
- Part 1: 1990 Code of Practice for residential buildings (amended September 1993).
  Note: This Code of Practice has only limited references to workplaces.
- Part 4: 1998 Code of Practice for smoke control using pressure differentials.
- Part 5: 1998 Code of Practice for fire-fighting stairs and lifts (amended June 1962 and March 1999).
- Part 6: 1991 Code of Practice for places of assembly (amended December 1998 and August 1999).
- Part 7: 1997 Code of Practice for the incorporation of atria in buildings (amended August 1999).
- Part 8: 1998 Code of Practice for means of escape for disabled people.

- Part 9: 1999 Code of Practice for ventilation and air conditioning ductwork.
- Part 10: 1991 Code of Practice for shopping complexes.
- Part 11: 1997 Code of Practice for shops, offices, industrial, storage and other buildings (amended January 1998).

Building Regulations Approved Document B (2000 Edition) and Approved Document B – Fire safety (2002 Amendments): B1 Means of warning and escape

Fire alarms and automatic fire detection systems

A.3   In the event of an outbreak of fire the operation of the fire alarm systems either by the operation of a manual call point or the detection of fire by automatic fire detection systems are the earliest warning that most people in a building will receive. There are a considerable number of Codes of Practice available covering different aspects of these systems, the majority of the most important are as follows.

British Standard 5839: Fire detection and alarm systems for buildings

- Part 1: 2002 Code of Practice for system design, installation, commissioning and maintenance.
- Part 2: 1983 Specification for manual call points.
- Part 3: 1988 Specification for automatic release mechanisms for certain fire protection equipment.
- Part 5: 1988 Specification for optical beam and smoke detectors.
- Part 6: 1995 Code of Practice for the design and installation of fire detection and alarm systems in dwellings.
- Part 8: 1998 Code of Practice for the Design, Installation and Servicing of Voice Alarm Systems.
- PD 6531: 1997 Queries and interpretations on BS 5839: Parts 1 and 4 (as amended).

British Standard 5445: Components of automatic fire detection systems

- Part 5: 1977 Heat sensitive detectors – point detectors containing a static element.
- Part 7: 1984 Specification for point-type smoke detectors using scattered light, transmitted light or ionisation.

A.3 *Sources of further information*

- Part 8: 1984 Specification for high temperature heat detectors.
- Part 9: 1984 Methods of test of sensitivity to fire.

British Standard EN 54: Fire Detection and Fire Alarm Systems

BS EN 54–1:1996	Introduction
BS EN 54–2:1998	Control and indicating equipment
BS EN 54–3:2001	Fire alarm devices. Sounders
BS EN 54–4:1998	Power supply equipment
BS EN 54–5:2001	Heat detectors. Point detectors
BS EN 54–7:2001	Smoke detectors. Point detectors using scattered light, transmitted light or ionization
BS EN 54–10:2002	Flame detectors. Point detectors
BS EN 54–11:2001	Manual call points

Note: Standards in preparation

pREN 54–13	Systems requirements
pREN 54–15	Point type multi-sensor fire detectors incorporating a smoke sensor (using scattered light, transmitted light or ionization) on combination with a heat sensor
pREN 54–16	Components for voice alarm systems
pREN 54–17	Short-circuit isolators
pREN 54–18	Input/output devices
pREN 54–20	Aspirating smoke detectors
pREN 54–21	Alarm transmission routing devices/fault warning routing devices

- Line type heat detectors
- Alarm devices other than audible ie light emitting beacons
- prEN 14604 Self contained smoke alarms
- Radio interconnected components/devices

Technical Reports/Specifications in preparation

CEN/TS 54–14	Guidelines for planning, design, installation, commissioning, use and maintenance of fire detection and alarm systems
CEN/TS 14568	Interpretation of specific clauses of EN54–2:1997

British Standard 7807: 1995

- Code of Practice for design, installation and servicing of integrated systems incorporating fire detection and alarm systems and/or other security systems for buildings other than dwellings.

Fire extinguishing installations and equipment on premises

A.4 There a variety of fire extinguishing installations and equipment provided in buildings and the suitable sources of information are as follows.

British Standard 5306: Fire extinguishing installations and equipment on premises

- Part 0: 1986 Guide for the selection of installed systems and other fire equipment.
- Part 1: 1976 Hydrant systems, hose reels and foam inlets.
- Part 2: 1990 Specification for sprinkler systems.
- Part 3: 2000 Maintenance of portable fire extinguishers.
- Part 4: 2001 Specification for carbon dioxide systems.
- Part 5.1: 1992 Halon systems. Specification for Halon 1301 total flooding systems.
- Part 5.2: 1984 Halon 1211 total flooding systems.
- Part 6.1: 1988 Foam systems. Specification for low expansion foam systems.
- Part 6.2: 1989 Foam systems. Specification for medium and high expansion foam systems.
- Part 7: 1988 Specification for powder systems.
- Part 8: 2000 Selection and installation of portable fire extinguishers. Code of Practice.

Fire tests on building materials and structures

A.5 This British Standard outlines a wide range of measures relating to the fire testing of materials and structures to ensure that they are suitable.

A.6 *Sources of further information*

British Standard 476: Fire tests on building materials and structures

- Part 3: 1975 External fire exposure roof test.
- Part 4: 1970 Non-combustibility test for materials.
- Part 6: 1989 Method of test for fire propagation for products.
- Part 7: 1997 Method of test to determine the classification of the surface spread of flame of products.
- Part 10: 1983 Guide to the principles and application of fire testing.
- Part 11: 1982 Method for assessing the heat emission from building materials.
- Part 12: 1991 Method of test for ignitability of products by direct flame impingement.
- Part 13: 1987 Method of measuring the ignitability of products subjected to thermal irradiance.
- Part 15: 1993 Method for measuring the rate of heat release of products.
- Part 20: 1987 Method for determination of the fire resistance of elements of construction (general principles).
- Part 21: 1987 Methods for determination of the fire resistance of loadbearing elements of construction.
- Part 22: 1987 Methods for determination of the fire resistance of non-loadbearing elements of construction.
- Part 23: 1987 Methods for determination of the contribution of components to the fire resistance of a structure.
- Part 24: 1987 Method for determination of the fire resistance of ventilation ducts.
- Part 31.1: 1983 Methods for measuring smoke penetration through doorsets and shutter assemblies. Method of measurement under ambient temperature conditions.
- Part 32: 1989 Guide to full scale fire tests within buildings.
- Part 33: 1993 Full-scale room test for surface products.

Other British Standards and Codes of Practice relating to fire

A.6   There are many other British Standards and their associated Codes of Practice which relate to fire. The following list is not exhaustive. Several of these documents have been the main guidance for several years and, as mentioned in **16.1** above, they are still considered to be the current guidance.

British Standard 1635: 1990 Recommendations for graphic symbols and abbreviations for fire protection drawings

British Standard 2782: Methods of testing plastics

British Standard 4422: Glossary of terms associated with fire

- Part 1: 1987 General terms and phenomena of fire.
- Part 2: 1990 Structural fire protection.
- Part 3: 1990 Fire detection and alarm.
- Part 4: 1994 Fire extinguishing equipment.
- Part 5: 1989 Smoke control.
- Part 6: 1988 Evacuation and means of escape.
- Part 7: 1988 Explosion detection and suppression means.
- Part 9: 1990 Marine terms.

British Standard 5255: 1989 Specification for thermoplastics waste pipes and fittings

British Standard 5266: Emergency lighting

- Part 1: 1999 Code of Practice for the emergency lighting of premises other than cinemas and certain other specified premises used for entertainment.
- Part 2: 1998 Code of Practice for electrical low mounted way guidance systems for emergency use.
- Part 3: 1981 Specification for small power relays (electromagnetic) for emergency lighting applications up to and including 32 A.
- Part 4: 1999 Code of Practice for design, installation, maintenance and use of optical fibre systems.
- Part 5: 1999 Specification for component parts of optical fibre systems.
- Part 6: 1999 Code of Practice for non-electrical low mounted way guidance systems for emergency use. Photoluminescent systems.
- Part 7: 1999 Emergency lighting.

A.6   *Sources of further information*

British Standard 5268: Part 4: Structural use of timber

- Part 4.1: 1978 Fire resistance of timber structures. Recommendations for calculating fire resistance of timber members.

- Part 4.2: 1990 Fire resistance of timber structures. Recommendations for calculating fire resistance of timber stud walls and joisted floor constructions.

British Standard 5395: Stairs, ladders and walkways

- Part 1: 2000 Code of Practice for the design, construction and maintenance of straight stairs and winders.

- Part 2: 1984 Code of Practice for the design of helical and spiral stairs.

- Part 3: 1985 Code of Practice for the design of industrial type stairs, permanent ladders and walkways.

British Standard 5438: 1976

- Methods of test for flammability of vertically oriented textile fabrics and fabric assemblies subjected to a small igniting flame.

British Standard 5438: 1989

- Methods of test for flammability of textile fabrics when subjected to a small igniting flame applied to the face or bottom edge of vertically oriented specimens.

British Standard 5499: Fire safety signs, notices and graphic symbols

- Part 1: 2002 Graphical symbols and signs. Safety signs, including fire safety signs. Specification for geometric shapes, colours and layout.

- Part 2: 1986 Fire safety signs, notices and graphic symbols. Specification for self-luminous fire safety signs.

- Part 3: 1990 Fire safety signs, notices and graphic symbols. Specification for internally-illuminated fire safety signs.

- Part 4: 2000 Safety signs, including fire safety signs. Code of practice for escape route signing.

- Part 5: 2002 Graphical symbols and signs. Safety signs, including fire safety signs. Signs with specific safety meanings.

- Part 6: 2002 Graphical symbols and signs. Safety signs, including fire safety signs. Creation and design of graphical symbols for use in safety signs. Requirements.

- Part 11: 2002 Graphical symbols and signs. Safety signs, including fire safety signs. Water safety signs.
- EP 250 BSI Electronic Book. Graphical signs and symbols CD-ROM.

British Standard 5628: Code of Practice for use of masonry

- Part 1: 1992 Structural use of unreinforced masonry.
- Part 2: 2000 Structural use of reinforced and pre-stressed masonry.
- Part 3: 2001 Materials and components, design and workmanship.

British Standard 5720: 1979 Code of Practice for mechanical ventilation and air conditioning in buildings

British Standard 5906:1980 Code of Practice for storage and on-site treatment of solid waste from buildings

British Standard 5950: Structural use of steelwork in building

- Part 8: 1990 Code of Practice for fire resistant design.

British Standard 6266: 2002 Code of Practice for fire protection for electronic data processing installations

British Standard 6336: 1998 Guide to the development of fire tests, the presentation of test data and the role of tests in hazard assessment

A.6 *Sources of further information*

British Standard 6387: 1994 Specification for performance requirements for cables required to maintain circuit integrity under fire conditions

British Standard 6651: 1999 Code of Practice for protection structures against lightning

British Standard 7157: 1989

- Method of test for ignitability of fabrics used in the construction of large tented structures.

British Standard 7346: Components for smoke and heat control systems

- Part 1: 1990 Specification for natural smoke and heat exhaust ventilators.
- Part 2: 1990 Specification for powered smoke and heat ventilators.
- Part 3: 1990 Specification for smoke curtains.

British Standard 7974: 2001 Code of Practice on the Application of Fire Safety Engineering Principles to the Design of Buildings

- PD0: 2002: Guide to design framework and fire engineering procedures.
- PD1: Initiation and development of fire within the enclosure of origin.
- PD2: 2002: Spread of smoke & toxic gases within & beyond the enclosure of origin.
- PD3: Structural response and fire spread beyond the enclosure of origin.
- PD4: Detection of fire and activation of protection systems.
- PD5: 2002: Fire Service intervention.
- PD6: Evacuation.
- PD7: Probabilistic risk assessment.

Note: A number of these 'published documents' (PD) are still in the course of preparation. (At the time of going to print only PD0, PD2 and PD5 are published.)

British Standard 8110: Structural use of concrete

- Part 1: 1985 Code of Practice for design and construction.
- Part 2: 1985 Code of Practice for special circumstances.
- Part 3: 1985 Design charts for singly reinforced beams, doubly reinforced beams and rectangular columns.

British Standard 8202: Coatings for fire protection of building elements

- Part 1: 1995 Code of Practice for the selection and installation of sprayed mineral coatings.
- Part 2: 1992 Code of Practice for the use of intumescent coating systems to metallic substrates for providing fire resistance.

British Standard 8214: 1990 Code of Practice for fire door assemblies with non-metallic leaves

British Standard BS 8313: 1997 Code of Practice for accommodation of building services in ducts

CP 1007: 1955

- Maintained lighting for cinemas.

Building Research Establishment Codes

- Design methodologies for smoke and heat exhaust ventilation. BR 368, BRE 1999.
- (Revision of design principles for smoke ventilation in enclosed shopping centres. BR 186, BRE, 1990.)
- BR 258 – Design approaches for smoke control in atrium buildings.
- BR 1999 – Assessing the fire performance of external cladding systems: a test method, BRE Fire Note 9.
- BR 187 – External fire spread: Building separation and boundary distances, BRE, 1991.
- BR 135 – Fire performance of external thermal insulation for walls of multi-storey buildings, BRE 1988.
- Heat radiation from fires and building separation. Fire Research Technical Paper 5, HMSO, 1963.

A.7 *Sources of further information*

- BR 128 – Guidelines for the construction of fire-resisting structural elements, BRE, 1988.
- Increasing the fire resistance of existing timber floors, BRE Digest 208, 1988.

Loss Prevention Council

- The LPC design guide for the fire protection of buildings 2000.

Healthcare Technical Memoranda (HTM's) and FIRECODE

- HTM 81 – Fire precautions in new hospitals.

Local Acts (England and Wales)

Many local Acts allow the enforcing authorities to impose higher 'Fire Safety' standards in specified buildings than is currently outlined in the *Building Regulations 2000 (SI 2000 No 2531)*, Approved Document B (2000 Edition).

Other sources

A.7

- Fire protection for structural steel in buildings (Second edition – revised). ASFP/SCI/FTSG, 1992 (available from the ASFP, Association House, 235 Ash Road, Aldershot, Hants GU12 4DD and the Steel Construction Institute, Silwood Park, Ascot, Berks SL5 7QN).
- Code of Practice. Hardware essential to the optimum performance of fire-resisting timber doorsets. Association of Builders Hardware Manufacturers, 1993. (Available from the ABHM, 42 Heath Street, Tamworth, Staffs B79 7JH.)
- Design, construction, specification and fire management of insulated envelopes for temperature controlled environments. The International Association of Cold Storage Contractors (European Division), 1999. (Available from the IACSC, Downmill Road, Bracknell, Berks RG12 1GH.)
- The Chartered Institution of Building Services Engineers.
- The CIBSE Guide 'Fire Engineering' 1998.
- Technical Memoranda 'Relationships for Smoke Control Calculations 1995'. (Available from CIBSE, 222 Balham High Road, Balham, London SW12 9BS.)

Websites

A.8 There is a considerable amount of information available on the internet and some of the websites where information relating to Fire Safety matters can be obtained are as follows.

- The Office of the Deputy Prime Minister website covers both building control legislation as well as fire service information: http://www.odpm.gov.uk/

- The Health and Safety Executive website contains a wide range of information relating to all aspects of safety and risk assessment: http://www.hse.gov.uk/

- The Health and Safety Bookfinder website provides a catalogue of all HSE's publications and multimedia products available from HSE Books: www.hsebooks.co.uk

- Her Majesty's Stationery Office website is available at http://www.hmso.gov.uk/

- The simplest and easiest route to find both national and local government information and departments is at http://www.ukonline.gov.uk/

  Use of this site allows access to Building Control Departments in some cases, with facilities to obtain application documents and respond online.

- To request further information on any British Standards Institution (BSI) products and services or to purchase standards, publications or merchandise, contact BSI, 389 Chiswick High Road, LONDON W4 4AL (Tel: 020 8996 9000/Fax: 020 8996 7400) or www.bsi-global.com

- The Chartered Institution of Building Services Engineers website is available at www.cibse.org

- The British Automatic Sprinkler Association is available at http://www.basa.org.uk/

- The British Fire Protection Systems Association (BFPSA) website is available at www.bfpsa.org.uk

- The Fire Protection Association website is available at http://www.thefpa.co.uk/

Table of Cases

R v F Howe & Son (Engineers) Limited [1999] 2 All ER 249   6.8
City Logistics Ltd v the Northamptonshire County Fire Officer
    [2001] EWCA Civ 1216   6.8
R v Associated Octel Co Ltd [1996] 4 All ER 846   6.10
City Logistics Ltd v the Northamptonshire County Fire Officer
    [2001] EWCA Civ 1216   15.1
City Logistics Ltd v the Northamptonshire County Fire Officer
    [2001] EWCA Civ 1216   15.12

Table of Statutes

Activity Centres (Young Persons' Safety) Act 1995	13.66
Building Act 1984	15.14
s 47	3.20
s 48(4)	15.14
s 51B(2)	15.14
s 71	15.14
s 72	15.14
(7)	15.14
Capital Allowances Act 1990	15.48
s 69	15.48
s 70	15.48
Children Act 1989	
s 3	15.67
Children and Young Persons Act 1933	15.45
s 11	15.45
s 12	15.45
Cinemas Act 1985	15.17
Cinematograph (Amendment) Act 1982	15.17
Caravan Sites and Control of Development Act 1960	15.15, 15.28
Capital Allowances Act 1980	15.49
Disability Discrimination Act 1995	4.1, 4.2, 4.3, 4.10, 4.13, 4.14
Part I	4.1
II	4.1
III	4.4
Employers' Liability (Compulsory Insurance) Act 1969	6.40
Employment Act 1988	2.5
s 25	2.5
Employment and Training Act 1973.	2.5
s 2	2.5
Environment and Safety Information Act 1988	15.21, 15.35
Explosives Act 1875	2.39
Factories Acts 1948	15.22
Factories Acts 1959	15.22
Factories Act 1961	2.1, 15.22
Finance Act 1975	15.48
s 15	15.48
Fire Precautions Act 1971	2.1, 2.2, 2.4, 2.6, 2.9, 2.26, 2.30, 2.36, 2.38, 2.40, 3.17, 3.28, 3.33, 5.2, 6.8, 8.1, 9.1, 9.4, 10.24, 10.25, 11.4, 11.16, 11.23, 11.43, 11.61, 13.44, 13.53, 13.57, 13.58, 13.62, 13.67, 15.1, 15.9, 15.10, 15.12, 15.13, 15.14, 15.18, 15.19, 15.20, 15.21, 15.22, 15.23, 15.35, 15.37, 15.39, 15.47, 15.48, 15.51, 15.52, 15.53, 15.58, 15.62, 15.74
s 1	2.5
(2)	2.25, 3.17
(4)	2.23
s 2	2.25, 3.17
s 3	2.25, 2.31, 3.17
s 4	2.31

Fire Precautions Act 1971 *contd*

s 5(2)(a)(b)	2.13
(2A)(a)-(c)	2.11
(3)(a)-(d)	2.16
(4)(a)(b)	2.16
s 5A	2.16, 2.29, 2.31
(1)	2.29
(2)(a)(b)	2.29
(3)	2.29
(4)(a)(b)	2.29
(5)	2.29
(6)(a)(b)	2.29
(7)(8)	2.29
(9)(a)(b)	2.29
s 5B	2.29
(1)-(5)	2.29
s 6(1)(b)-(e)	2.21, 2.23
(2)	2.23
(5)	2.21, 2.23, 2.29
(6)	2.23
s 7(1)(2)	2.29
(4)	2.29
s 8(1)	2.23
(2)(a)-(c)	2.21
(3)	2.21
(4)(a)(b)	2.21
(4)(b)	2.29
(5)	2.29
(a)(b)	2.23
(6)(7)	2.29
(a)-(c)	2.23
(9)	2.23, 2.29
s 8A	2.29
(1)	2.29
(2)(a)(i)-(ii)	2.29
(b)(c)	2.29
(3)(a)(b)	2.29
s 9	2.23, 2.29
(1)(a)-(g)	2.29
(2)(a)(b)	2.29
(3)(4)	2.29
(5)(a)(b)	2.29
s 9A	2.26, 2.27
s 9B	2.27
s 9D	2.27, 3.20
(1)(a)-(c)	2.27
s 9E(1)-(3)	2.27
s 9F	2.27
(1)	2.27
(2)(a)(b)	2.27
s 10	3.16, 3.17, 15.21, 15.35, 15.49
ss 10-10B	3.18
s 10(1)(a)(b)	2.25, 3.17
(a)	2.31
(2)(3)	2.25, 3.17
(4)(a)-(c)	2.25, 3.17
(5)-(7)	2.25, 3.17

Table of Statutes

Fire Precautions Act 1971 – *contd*	
s 10A	2.25
(1)-(3)	2.25
s 10B	2.25
(1)(2)	2.25
(3)(a)(b)	2.25
s 12	2.23, 2.31
s 17	3.30
s 18	2.31
(1)	3.13
s 19	2.29, 2.30, 2.31, 2.33
(1)(a)-(d)	2.31
(b)	2.33
(2)(a)-(e)	2.31
(3)(a)-(d)	2.31
(4)(5)	2.31
(6)(a)(b)	2.31
s 20	2.33
s 22	2.32
(1)(a)-(d)	2.33
(2)(a)(b)	2.33
s 23(1)(2)	2.35
s 40(2)(a)(b)	3.26
ss 40-51	15.22
Fire Safety and Safety of Places of Sport Act 1987	2.1, 3.20, 13.63, 15.23, 15.48, 15.73
s 9(1)(d),	15.23
Sch 1, Part I	15.23
II	15.23
III	15.23
Fire Services Act 1947	10.24, 15.20, 15.39, 15.85
s 1(1)(f)	15.78
s 19	15.83, 15.85
s 24	3.13
s 33(1)	3.13
Fire Services Act 1951	15.20
Fire Services Act 1959	15.20
Gaming Act 1968	13.54, 15.24
Gaming (Amendment) Act 1982	15.24
Health and Safety at Work etc Act 1974	2.15, 3.30, 6.1, 6.6, 6.10, 6.23, 6.30, 6.31, 6.32, 6.33, 6.37, 6.46, 6.49, 6.51, 9.3, 13.16, 15.7, 15.37, 15.39, 15.53, 15.58
s 1(2)	6.5
s 2	6.32, 9.3
(1)	3.30, 6.32, 9.3
(2)	6.32, 9.3
(3)	6.32, 6.46
ss 2-6	6.31
s 3	3.30
(1)	6.8, 6.10, 13.16
s 5(2A)	9.4
s 15	6.2
ss 21, 22	6.31
s 33	6.31
s 42	6.31
s 53	6.21
s 75	15.35
Sch 8	15.35

Table of Statutes

Health and Safety at Work etc Act 1974 – *contd*	
Part 1	2.15
International Headquarters and Defence Organisations Act 1964	15.73
Licensing Act 1961	15.25
Licensing Act 1964	15.25, 15.27
Licensing Act 1988	15.25, 15.27
Licensing (Occasional Permissions) Act 1983	15.26
Local Government (Miscellaneous Provisions) Act 1982	13.67, 15.28
London Local Authorities Act 1991	15.29
s 6	15.29
s 15	15.29
London Local Authorities Act 1995	15.29
ss 16-18	15.29
London Local Authorities Act 1996	15.29
Merchant Shipping Act 1995	3.25
s 313(1)	3.25
Mines and Quarries Act 1952	2.39
s 73	15.41
s 180	15.53
National Health Service and Community Care Act 1990	15.47
Nuclear Installations Act 1965	
s 1	2.39
s 2	2.39
Occupiers Liability Act 1957	15.52
Offices, Shops and Railway Premises Act 1963	2.1
Petroleum (Consolidation) Act 1928	13.52
Police (Health and Safety) Act 1997	6.23
Places of Sport Act 1987	15.48
Planning and Compensation Act 1991	10.31, 13.18
Planning (Listed Buildings and Conservation Areas) Act 1990	13.3
Private Places of Entertainment (Licensing) Act 1967	13.67
Public Health Act 1936	15.28
Safety of Sports Grounds Act 1975	2.1, 3.20, 15.23, 15.48, 15.73
Single European Act 1987	6.2
Art 138	6.2
Theatres Act 1968	15.12, 15.30
s 12(1)	13.64
Town and Country Planning Act 1990	10.31, 13.18
s 55	10.32, 13.18
Vehicle Excise and Registration Act 1994	3.26, 15.53

Table of Statutory Instruments

Asbestos (Licensing) Regulations 1983
 (SI 1983 No 1649)   6.28
Asbestos (Prohibitions) Regulations 1992
 (SI 1992 No 3067)   6.28
Asbestos (Prohibitions) (Amendment) Regulations 1999
 (SI 1999 No 2373)   6.28
Building (Approved Inspectors etc) Regulations 2000
 (SI 2000 No 2532)   15.14
Building Regulations 1991
 (SI 1991 No 2768)   2.40, 3.32, 10.28, 10.29, 10.32, 10.33, 11.1A, 11.4, 11.7, 11.13,
  11.14, 11.18, 11.50, 13.6, 13.17, 13.18, 13.72, 15.3, 15.52
Building Regulations 2000
 (SI 2000 No 2531)   15.14, A.2, A.6
 Reg 12   15.14
 Reg 13   15.14
Building Regulations (Amendment) (No 2) Regulations 1999
 (SI 1999 No 3410)   3.32
Building Regulations (Northern Ireland) 1994
 (SI 1994 No 243)   3.32, 11.1A, 11.50
Building Standards (Scotland) Regulations 1991
 (SI 1991 No 158)   3.32, 11.1A, 11.50
Carriage of Dangerous Goods (Amendment) Regulations 1999
 (SI 1999 No 303)   6.28
Chemicals (Hazard Information and Packaging for Supply) Regulations 1994
 (SI 1994 No 3247)   15.54
Chemicals (Hazard Information and Packaging for Supply) (Amendment)
  Regulations 1999
 (SI 1999 No 3165)   6.28
Children's Homes Regulations 1991
 (SI 1991 No 1506)   13.48, 13.51
Children's Homes Regulations 2001
 (SI 2001 No 3967)   15.46
Children's Home (Wales) Regulations 2002
 (SI 2002 No 327)   15.46
Civil Procedure Rules 1998
 (SI 1998 No 3132)   6.40
Cinematograph (Amendment) Regulations 1982
 (SI 1982 No 1856)   15.16
Cinematograph (Children) (No 2) Regulations 1955
 (SI 1955 No 1909)   15.16
Cinematograph (Safety) Regulations 1955
 (SI 1955 No 1129)   13.49, 15.16
 Reg 2   13.49
 Reg 5   13.49
 Reg 6   13.49
 Reg 24   13.49
Cinematograph (Safety) (Amendment) Regulations 2002
 (SI 2002 No 1903)   15.16

Confined Spaces Regulations 1997
  (SI 1997 No 1713) 6.11, 6.47, 13.52
Construction (Design and Management) Regulations 1994
  (SI 1994 No 3140) 6.10, 6.36, 13.15, 13.32, 13.33, 13.34, 13.35, 13.52, 15.31
  Reg 14 13.36, 13.37
  Reg 19 13.38
Construction (Health, Safety and Welfare) Regulations 1996
  (SI 1996 No 1592) 3.25, 6.13, 6.36, 13.15, 13.52, 15.32
  Reg 20(2) 6.21
Control of Asbestos at Work Regulations 1987
  (SI 1987 No 2115) 6.28
Control of Industrial Major Accidents Hazard Regulations 1984
  (SI 1984 No 1902) 6.15
Control of Lead at Work Regulations 1998
  (SI 1998 No 543) 6.14
Control of Major Accident Hazards Regulations 1999
  (SI 1999 No 743) 6.15, 15.33
Control of Substances Hazardous to Health Regulations 1988
  (SI 1988 No 1657) 6.9
Control of Substances Hazardous to Health Regulations 1999
  (SI 1999 No 437) 6.16, 6.36
Dangerous Substances and Explosive Atmospheres Regulations 2002
  (SI 2002 No 2776) 6.35, 13.72, 15.7, 15.34, 15.54, 15.57, 15.66, 15.67, 15.80
Dangerous Substances in Harbour Areas Regulations 1987
  (SI 1987 No 37) 15.34
  Reg 18(2)(a) 15.34
  Reg 18(2)(c) 15.34
  Reg 26 15.34
  Reg 27(1) 15.34
  Reg 47(4) 15.34
Disability Discrimination Codes of Practice (Education) (Appointed Day) Order 2002
  (SI 2002 No 2216) 4.14
Diving at Work Regulations 1997
  (SI 1997 No 2776) 6.17
Electricity at Work Regulations 1989
  (SI 1989 No 635) 6.22, 6.36, 13.52
Equipment and Protective Systems Intended for Use in Potentially Explosive Atmospheres 1996 Regulations
  (SI 1996 192) 6.36
Factories Act 1961 etc (Repeals) Regulations 1976
  (SI 1976 No 2004) 15.22
Fire Certificates (Special Premises) Regulations 1976
  (SI 1976 No 2003) 2.1, 2.38, 2.39, 3.29, 15.33, 15.34, 15.36
  Reg 2(1)(a) 2.38
Fire Precautions (Application for Certificate) Regulations 1989
  (SI 1989 No 77) 15.18
Fire Precautions (Factories, Offices, Shops and Railway Premises) Order 1976
  (SI 1976 No 2009) 15.22
Fire Precautions (Factories, Offices, Shops and Railway Premises) Order 1989
  (SI 1989 No 76) 2.1, 2.5, 13.53, 13.58, 13.62, 13.66, 13.68, 15.18
Fire Precautions (Hotels and Boarding Houses) Order 1972
  (SI 1972 No 238) 2.1, 2.5, 13.57, 15.18
Fire Precautions (Sub-surface Railway Stations) Regulations 1989
  (SI 1989 No 1401) 13.58, 15.19
Fire Precautions (Sub-surface Railway Stations) (Amendment) Regulations 1994
  (SI 1994 No 2184) 15.19

Fire Precautions (Workplace) Regulations 1997
   (SI 1997 No 1840)   2.1, 2.2, 3.1, 3.2, 3.3, 3.4, 3.5, 3.10, 3.19, 3.28, 6.18, 7.1, 7.2,
      7.10, 7.23, 7.24, 8.1, 8.2, 9.5, 9.25, 10.3, 10.24, 11.58, 12.1,
      13.44, 15.7, 15.9, 15.11, 15.12, 15.13, 15.14, 15.19, 15.23, 15.31,
      15.43, 15.44, 15.53, 15.54, 15.55, 15.57, 15.58, 15.62, 15.63,
      15.65, 15.66, 15.67, 15.68, 15.69, 15.70, 15.72, 15.74

Part II	3.2, 7.1, 7.20
Part III	6.21
Reg 2	3.4
Reg 3	3.4
(5)	3.26
Reg 4	7.23, 7.25
(2)	9.5
Reg 5	7.20
Reg 6	7.27
Reg 11	3.14, 3.15
Reg 12	3.18
Reg 13	3.14, 3.20
(5)	3.19
Reg 14	3.22
Reg 15–19	15.31
Reg 15	3.24

Fire Precautions (Workplace) (Amendment) Regulations 1999
   (SI 1999 No 1877)   3.1, 3.2, 3.3, 3.19, 8.1, 12.1, 15.19
Health and Safety (Consultation with Employees) Regulations 1996
   (SI 1996 No 1513)   6.36, 15.38
Health and Safety (Display Screen Equipment) Regulations 1992
   (SI 1992 No 2792)   6.4, 6.34
Health and Safety (Enforcing Authority) Regulations 1989
   (SI 1989 No 1903)   15.39
Health and Safety (Enforcing Authority) Regulations 1998
   (SI 1998 No 494)   6.35, 13.72
Health and Safety (First-Aid) Regulations 1981
   (SI 1981 No 917)   6.36
Health and Safety (Safety Signs and Signals) Regulations 1996
   (SI 1996 No 341)   3.31, 6.36, 11.20, 11.21, 11.22, 11.23, 11.42, 11.69, 13.52
Health and Safety (Training for Employment) Regulations 1990
   (SI 1990 No 1380)   6.36
Health and Safety (Young Persons) Regulations 1997
   (SI 1997 No 135)   6.21
Highly Flammable Liquids and Liquefied Petroleum Gases Regulations 1972
   (SI 1972 No 917)   13.52
Ionising Radiation Regulations 1985
   (SI 1985 No 1333)   2.39, 6.19
Ionising Radiations Regulations 1999
   (SI 1999 No 3232)   6.19
Level Crossings Regulations 1997
   (SI 1997 No 487)   6.25
Lifting Operations and Lifting Equipment Regulations 1998
   (SI 1998 No 2307)   6.20, 6.47
National Health Service and Community Care Act 1990 (Commencement No 1)
      Order 1990
   (SI 1990 No 38)   15.47
Noise at Work Regulations 1989
   (SI 1989 No 1790)   6.22, 6.36
Nursing Homes and Mental Nursing Homes Regulations 1984
   (SI 1984 No 1578)   13.59

Management of Health and Safety at Work Regulations 1992
(SI 1992 No 2051)   6.4, 6.21, 6.34, 9.32
Reg 3   6.9
Reg 4   6.44
Management of Health and Safety at Work Regulations 1999
(SI 1999 No 3242)   3.2, 3.7, 6.4, 6.21, 6.34, 6.37, 6.47, 7.1, 7.3, 7.8, 8.2, 10.3,
                            11.60, 11.61, 13.38, 13.40, 15.7, 15.11, 15.31, 15.40, 15.43, 15.53,
                                                                                                                                        15.57
Reg 3   3.7, 6.47
Reg 4   6.21
Reg 5   3.7, 6.47
   (1)   6.47
      (b)   6.21
      (c)   6.21
   (2)   6.47
   (2)(a)   6.21
Reg 7(1)   6.21, 6.47
   (8)   6.21, 6.47
Reg 8   3.7
Reg 9   6.21
Reg 10   3.8, 3.9, 3.13, 7.8
Reg 11   3.11
Reg 21   6.21
Reg 22   6.37
Reg 24   6.21
Reg 25   6.21
Reg 27   6.21
Management of Health and Safety at Work (Amendment) Regulations 1994
(SI 1994 2865)   6.21
Manual Handling Operations Regulations 1992
(SI 1992 2793)   6.4, 6.30, 6.34, 11.42
Mines Miscellaneous Health and Safety Provisions Regulations 1995
(SI 1995 No 2005)   15.42
Offshore Electricity and Noise Regulations 1997
(SI 1997 No 1993)   6.22, 6.36
Offshore Installations and Pipeline Works (First-Aid) Regulations 1989
(SI 1989 No 1671)   6.21
Offshore Installations and Pipelines Work (Management and Administration)
Regulations 1995
(SI 1995 No 738)   3.25
Reg 3   15.53
Police (Health and Safety) Regulations 1999
(SI 1999 No 860)   6.23
Personal Protective Equipment at Work Regulations 1992
(SI 1992 No 2966)   6.4, 6.16, 6.34
Provision and Use of Work Equipment Regulations 1992
(SI 1992 No 2932)   6.4, 6.24, 6.34
Provision and Use of Work Equipment Regulations 1998
(SI 1998 No 2306)   6.4, 6.24, 6.47
Reg 6   6.24
Reg 25–30   6.24
Reg 31–35   6.24
Railways (Safety Case) Regulations 1994
(SI 1994 No 237)   15.43
Railway Safety (Miscellaneous Provisions) Regulations 1997
(SI 1997 No 553)   6.25

Table of Statutory Instruments

Reporting of Injuries, Diseases and Dangerous Occurrences Regulations 1995
(SI 1995 No 3163)   6.36, 13.38
Residential Care Homes Regulations 1984
(SI 1984 No 1345)   13.59
Safety Representatives and Safety Committees Regulations 1977
(SI 1977 No 500)   6.36, 15.44
Standards for School Premises Regulations 1972
(SI 1972 No 2051)   13.60
Town and Country Planning (Control of Advertisements) Regulations 1992
(SI 1992 No 666)   10.38
Town and Country Planning (General Permitted Development) Order 1995
(SI 1995 No 418)   10.33
Transport of Dangerous Goods (Safety Advisers) Regulations 1999
(SI 1999 No 257)   6.26
Work in Compressed Air Regulations 1996
(SI 1996 No 1656)   13.52
Workplace (Health, Safety and Welfare) Regulations 1992
(SI 1992 No 3004)   6.4, 6.34
Working Time Regulations 1998
(SI 1998 No 1883)   6.27

Table of European Legislation

Council Directive 89/654/EEC	15.11, 15.13, 15.52
Council Directive 89/391/EEC	6.2, 15.11, 15.13, 15.52
Council Directive 92/57/EEC	13.34
Council Directive 92/58/EEC	11.20
Council Directive 93/104/EC	6.27
Council Directive 94/33/EC	6.27
Council Directive 98/24/EC	15.13, 15.52, 15.54
Council Directive 99/92/EC	15.13, 15.52, 15.54

Index

A

Accident Prevention Advisory	
Unit (APAU)	6.44, 6.45
Accidents	
costs of	6.45
fatal	6.36
Major Accident Prevention	
Policy	6.15
Advertisements	
planning applications	10.38
Air supported structures	3.32, 13.45
Alarms *see* **Fire alarms**	
Appeals	
enforcement notice	3.21, 3.22
fire certificate exemption	2.29
improvement notice	2.26, 2.27
prohibition notice	2.24, 2.25
Arson	
costs of	1.2
definition	1.5
generally	1.2, 10.10
increase in	1.2
prevention	1.4, 1.5
reasons for	1.3
Arson Prevention Bureau	1.2, 1.5
Asbestos	6.28
Assembly points	7.22, 11.65
Association of British Insurers	1.2, 1.5
Atrium buildings	13.46
Automatic fire detection systems	
annual inspection and test	12.9
beam detectors	11.26
British Standard	11.27, 11.30, A.3
buildings, protecting	
Type L1	11.30
Type L2	11.30
Type L3	11.30
Type M	11.30
Type P1	11.30
Type P2	11.30
control panel	11.26
daily attention by user	12.5
flame detectors	11.26
generally	7.25, 11.26
heat detectors	11.26

Automatic fire detection systems – *contd*	
life, protecting	
Type L1 – Life 1	11.28
Type L2 – Life 2	11.28
Type L3 – Life 3	11.28
Type M – Manual System	11.28
maintenance and testing	7.26, 12.3
service routines and record entries	12.11
monthly attention by user	12.7
property, protecting	
Type P1 – Property 1	11.29
Type P2 – Property 2	11.29
quarterly inspection and test	12.8
smoke detectors	11.26
types of protection	11.27
weekly attention by user	12.6
wiring and cabling	11.26
wiring check	12.10
Automatic sprinklers *see* **Sprinkler systems**	

B

Bain Report	15.81–15.85
Beam detectors	11.26
see also **Automatic fire detection systems**	
Bed and breakfast accommodation *see* **Hotels and boarding houses**	
Bingo halls	13.54
Blankets	
fire blankets	11.44, 11.69
Boarding houses *see* **Hotels and boarding houses**	
Bodies corporate	
corporate killing, proposed offence of	6.7
offences by	2.34, 2.35
Bomb alerts and threats	
bags, searching	8.43
being alert and on guard	8.45
evacuation	
premature	8.26
procedure	8.30, 8.48
generally	8.2, 8.22, 8.35–8.39
glass	8.42

387

Index

Bomb alerts and threats – *contd*	
high-explosive bomb	8.36
Home Office guidance	8.33, 8.34
incendiary bomb	8.38
key holders	8.46
plan of action	8.23
postal bombs	8.39
precautions	8.40–8.49
preventive measures	8.31
re-entering property	8.49
search	8.27–8.29, 8.48
bags, of	8.43
security measures	8.32, 8.41
telephone warnings	8.24, 8.47
checklist	8.25, 8.50
tidiness	8.44
vehicle bomb	8.37
visitors	8.43

British Standards	
automatic fire detection systems	11.27, 11.30, A.3
emergency lighting	A.6
fire alarms	12.37, A.3
fire detection and alarm systems	11.27, 11.30
fire engineering	5.2, 13.69
fire extinguishers	11.32, 11.39, 11.40, 11.41, A.4
fire tests on building materials and structures	A.5
generally	5.1, 5.2, A.1–A.6
Kitemark	5.1, 5.3
management systems	6.47
means of escape	A.2
pressurisation systems	12.31
relationship with international and European standards	5.3
risk assessments	7.2
sprinkler systems	11.46, 12.24

British Standards Institution (BSI)	5.1–5.4

Building control	
building notice application	10.29
inspection stage	10.29
meaning	10.28
operation of	10.29
permitted development	10.33
plan stage	10.29
structural alterations and extensions	13.17
see also Planning	

Building Regulations	3.32, 10.28
Approved Documents	3.32, 5.2
doors	3.32
emergency exits	11.13
emergency lighting	11.18

Building Regulations – *contd*	
fire alarms	3.32
means of escape	3.32, 11.13
sprinkler systems	11.50

C

Caravan sites	13.47, 15.15, 15.28
Casinos	13.54

Causes of fire	
arson	
costs of	1.2
definition	1.5
generally	1.2, 10.10
increase in	1.2, 1.5
prevention	1.4, 1.5
reasons for	1.3
best practice guidelines	10.21
cigarettes and matches *see* smoking *below*	
cooking	10.16
cooking appliances	1.1
electrical appliances	1.1, 10.12
electrical distribution	1.1
fuel	10.7–10.8
generally	1.1, 10.4
heating appliances	8.14, 10.14
hot processes	10.15, 10.18
'hot work'	1.7
housekeeping standards	10.20
ignition sources	10.9
lighting equipment	10.12
naked flames	10.13
office equipment	10.19
smoking	1.1, 8.15, 10.2, 10.3, 10.11, 10.21
static electricity	10.17
'triangle of fire'	10.6

CE mark	5.3

Children	
protection of	15.45
Children's homes	13.48, 15.46

Cigarettes and matches *see* Smoking	
Cinemas	13.49, 15.16, 15.17
disability discrimination	4.8
City Logistics case	6.8, 15.1
Cleanliness and tidiness	8.8
see also Housekeeping	

Committee for Electrotechnical Standardisation (CENELEC)	5.3
Community homes	13.51
Compartmentation	11.6

Computers	
computer suites	13.50

Index

Computers – *contd*
  display screen equipment 6.34
  records, protecting 8.21
Consequences of fire
  customers, loss of 14.5
  damaged reputation 14.2
  employees, loss of 14.4
  financial security 14.3
  generally 14.1
  key suppliers, loss of 14.6
  market share, loss of 14.5
  mitigation of 15.54
Conservation areas
  planning applications 10.38
Construction
  fire risk during 1.7
  see also Planning; Structural alterations and extensions
Construction (Design and Management) Regulations 1994
  application of 13.34
  client, duties of 6.36
  contractors' role 6.10, 6.36, 13.38
  exclusions 13.35
  generally 6.36, 13.32, 15.31
  health and safety file 6.36
  health and safety plan 6.36
  objectives 13.33
  planning supervisor 6.36, 13.36
  principal contractor 6.36, 13.15, 13.37
Construction sites 13.52
Consultation with employees 6.36, 15.38
Contributory negligence 6.42
Cooking
  cause of fire, as 10.16
Corporate killing
  proposed offence of 6.7
Costs
  accidents, of 6.45
  arson, of 1.2
Crown immunity 15.20, 15.47, 15.74

D

Damage control/salvage 8.16, 9.34
Damages
  assessment of 6.41
Dancing
  premises used for 13.67
Dangerous goods
  transport of 6.26
Defences
  contributory negligence 6.42
  due diligence 2.36–2.37, 6.48

Defences – *contd*
  injuries not reasonable
    foreseeable 6.42
  voluntary assumption of risk 6.42
Deluge and re-cycling
  installations 11.48E
Deregulation 6.5
Designated premises
  fire certificate 2.1, 2.2, 15.10
Detection systems *see*
  Automatic fire detection systems
Disability Discrimination Act 1995
  access to goods, facilities,
    services and property 4.1, 4.2
  alterations to premises 4.10
  auxiliary aids and services,
    meaning 4.8
  definition of 'disability' 4.1
  employers, requirements on 4.7
  generally 4.1
  good practice checklist 4.12, 4.13
  less favourable treatment 4.1
  meaning 4.6
  means of escape 4.11, 8.20
  publications 4.14
  reasonable adjustments 4.1, 4.7
  rights under 4.3
  service providers
    'direct to the public' 4.4
    exemptions 4.4
    generally 4.2
    legal responsibilities 4.5
  special aids, provision of 4.9
Display screen equipment 6.34
Doors
  Building Regulations 3.32
  emergency exits 11.11, 11.13, 11.14
  exit signs 7.19, 11.20, 11.72
  fire doors 11.7, 11.67, 12.37
    automatic releases 11.8, 12.37
    inspection 12.37, 12.38
    generally 7.17, 7.20, 15.63
    inner rooms 11.15, 11.16
    locking 11.14, 11.16
    maintenance 11.16, 11.67
    number and width 7.18, 11.14
    self-closing 12.37
    sliding doors 11.16
    storey exits 7.17
    types 11.67
Dry risers 11.55, 11.70
Due diligence defence 2.36, 2.37, 6.48

389

E

Electrical equipment
  faulty   10.12
Electrical fires
  fire extinguishers   11.32
Electricity
  static   10.17
Emergency lighting
  British Standard   A.6
  Building Regulations   11.18
  generally   7.19, 11.17, 11.18
  inspection   7.26, 11.73
    daily attention by user   12.16
    frequency of   7.26, 12.15
    log for   12.19
    monthly attention by user   12.17
    subsequent tests   12.18
  maintenance   11.73
  siting   11.19
Emergency plan   7.21, 7.22
Emergency procedures
  evacuation see Evacuation
  generally   11.1E
  large premises   11.59
  staff training   11.57–11.60
Emergency signs   11.20, 11.21–11.23
Employees
  consultation with   6.36, 15.38
  fire safety awareness   8.6, 9.1
  health and safety duties   6.32
  information for   3.8, 3.9, 7.8
  loss of   14.4
  teamwork   8.6, 9.1
  training
    emergency procedures   11.57–11.60
    for employment   6.36
    fire safety see Fire safety training
  voluntary   15.53
  work experience programmes   6.36
Employers' liability insurance   6.40
Enforcement notices   3.19, 3.20
  appeal rights   3.21, 3.22
  extension of   3.19, 3.20
  offence   3.23, 3.24
  withdrawal   3.19, 3.20
Environment and Safety Information Act 1988   15.21, 15.35
Equipment
  faulty electrical and lighting equipment   10.12
  fire-fighting see Fire-fighting equipment
  first aid   6.36
  lifting equipment   6.20

Equipment – contd
  office equipment   10.19
  provision and use   6.24
Escape see Means of escape
European Committee for Standardisation (CEN)   5.3
European Telecommunications Standard Institute (ETSI)   5.3
Evacuation
  assembly points   7.22, 11.65
  bomb alerts and threats   8.30, 8.48
    premature evacuation   8.26
  disabled persons   8.20
  drills   9.10, 11.61
  members of the public   8.20
  policy   8.20
  times   7.17
  visitors   8.20
Exits
  emergency exits   11.11, 11.13, 11.14
  generally   7.17, 7.20, 15.63
  number and width   7.18, 11.14
  signs   7.19, 11.14, 11.20
  storey exits   7.17
  see also Doors
Extensions see Structural alterations and extensions
Extinguishers see Fire extinguishers
Eyesight tests   6.34

F

Factories   2.1, 13.53, 15.18, 15.22
  fire certificate   2.4, 2.5, 2.18, 15.10
False alarm
  prevention during testing   12.4
Falsification of documents   2.32, 2.33
Fatal accidents   6.36
Faulty electrical and lighting equipment   10.12
Filling stations   13.55
Fire
  causes of see Causes of fire
  consequences of see Consequences of fire
  numbers of fires   10.2
  recovery after see Survival plan
Fire action notices   7.22, 11.62–11.65
Fire alarms
  British Standards   12.37, A.3
  Building Regulations   3.32
  generally   11.24, 11.25
  maintenance   7.26, 11.71, 12.3–12.11
    annual inspection and test   12.9
    daily attention by user   12.5
    monthly attention by user   12.7

Index

Fire alarms – *contd*
  maintenance – *contd*
    prevention of false alarms   12.4
    quarterly inspection and test   12.8
    service routines and record
      entries   12.11
    testing   12.3, 12.4, 12.8
    weekly attention by user   12.6
    wiring check   12.10
Fire authority
  duties and functions   10.24, 15.39
  inspection powers   2.22, 2.23, 2.30, 2.31, 3.13
Fire awareness   8.1, 9.1
Fire blankets   11.44, 11.69
Fire certificate
  application for
    generally   2.2, 2.7, 15.10
    interim duties   2.7, 2.9–2.11
    occupier, by   2.8
    owner, by   2.8
    stages of   2.12–2.17
    types of premises   2.8
  boarding house   2.4, 2.5, 15.10
  bodies corporate, offences by   2.34, 2.35
  change of conditions, where   2.21, 2.22, 2.23
  City Logistics case   6.8
  consultation with fire authority   10.25
  contents   2.18
  definition   2.6
  designated premises   2.1, 2.2, 15.10
  due diligence defence   2.36, 2.37
  duties under   2.20, 10.25
  employees
    meaning   2.5
    number of   2.4
  exempt premises   2.1, 2.5, 2.28, 2.29
  existing certificate   2.19–2.21
  factories   2.4, 2.5, 2.18, 15.10
  falsification   2.32, 2.33
  generally   2.1
  Health and Safety Executive   2.1
  hotel   2.4, 2.5, 15.10
  inspection   2.6, 15.10
  fire authority inspectors' powers   2.30, 2.31
  fire authority powers   2.22, 2.23
  interim duties   2.7, 2.9–2.11
  issue of   2.6, 2.15, 2.16
  offences
    bodies corporate, by   2.34, 2.35
    falsification of documents   2.32, 2.33
  offices   2.4, 2.5, 15.10
  railway premises   2.4, 2.5
  refusal   2.15, 2.16

Fire certificate – *contd*
  requirement for   2.3, 2.4
  shops   2.4, 2.5, 15.10
  special premises   2.1, 2.38, 2.39, 3.29, 15.36
  staff training, and   9.4
  tax relief, and   15.48, 15.49
Fire dampers
  inspection   12.37, 12.38
Fire defence equipment *see* **Equipment**
Fire detection systems *see* **Automatic fire detection systems**
Fire doors *see* **Doors**
Fire drills   9.10, 11.61
  see also **Evacuation**
Fire engineering
  advantages   13.70
  British Standard   5.2, 13.69
  disadvantages   13.71
  generally   5.2, 13.69
Fire extinguishers
  A triple F (AFFF)   11.39
  BCF   11.37
  British Standards   11.32, 11.39, 11.40, 11.41, A.4
  carbon dioxide   11.38
  Class A fires   11.31, 11.32
  Class B fires   11.31, 11.32
  Class C fires   11.31
  Class D fires   11.31
  colour coding   11.40
  dry powder   11.36
  electrical fires   11.32
  fire blankets   11.44, 11.69
  foam   11.35
  legal note   11.43
  maintenance   11.69, 12.12
  portable
    classes of fire   11.32
    colour coding   11.40
    servicing   11.69, 12.12
  signs indicating   11.42
  siting of   11.42
  standards   11.32, 11.39, 11.40, 11.41, A.4
  training in use of   9.6, 11.43
  types of fire   11.31
  types of   11.33
  water   11.34, 11.42
Fire marshals
  generally   8.5, 9.16
  role   8.5, 9.17, 9.18
  training module   9.19, 9.20
Fire Precautions Act 1971
  amendments   2.1

Index

Fire Precautions Act 1971 – *contd*
application, *City Logistics* case 6.8, 15.1
Crown immunity 15.74
designated premises 2.1, 2.2
designating orders 2.1, 2.2
due diligence defence 2.36, 2.37
falsification of documents 2.32, 2.33
fire certificate *see* Fire certificate
Fire Precautions (Workplace)
 Regulations 1997 compared 2.2
generally 2.1, 9.4, 15.10
improvement notice 2.1, 2.26, 2.27
inspection powers
 fire authority 2.22, 2.23
 fire authority inspectors 2.30, 2.31
means of escape 2.1
means of fighting fire 2.1
nature of 2.1
occupiers/owners, duties on 2.1, 3.28
offences by bodies corporate 2.34, 2.35
prohibition notice 2.1, 2.24, 2.25, 3.16–3.18
publications 2.40
special premises 2.38, 2.39
staff training 9.4

Fire Precautions (Workplace) Regulations 1997 and 1999
application of 3.3
co-operation and co-ordination 3.10, 15.69
Crown immunity 15.74
emergency routes and exits 7.20, 15.63
employers, duties on 2.2, 3.28
enforcement 3.12
enforcement notices 3.19, 3.20
 appeal rights 3.21, 3.22
 extension of 3.19, 3.20
 offence 3.23, 3.24
 withdrawal 3.19, 3.20
European approach 3.2
excepted workplace 3.25, 3.26
fire detection 7.25
Fire Precautions Act 1971
 compared 2.2
fire risk assessment *see* Fire risk assessment
fire-fighting equipment 7.25, 15.62
 maintenance 7.27
generally 2.1, 3.1, 6.18, 9.5, 15.11
offences 3.14
prohibition notice 3.16–3.18
risk assessment 2.1
scope 3.5
self-compliance concept 3.1
serious cases
 offences 3.14–3.15
 prohibition notices 3.16–3.18

Fire Precautions (Workplace) Regulations 1997 and 1999 – *contd*
training 15.69
workplace
 definition 3.4
 persons in control 3.4

Fire risk assessment
actions following 3.8
adequacy of fire precautions 7.7
appropriate to workplace 7.3
categories of fire risk 7.11–7.14
checklist 7.28–7.30
co-operation and co-ordination 3.10–3.11
concept of 7.2
emergency lighting 7.19
emergency plan 7.21
escape routes
 emergency lighting 7.19
 emergency plan 7.21
 evacuation times 7.17
 exit signs 7.19
 general principle 7.16–7.22
 length of 7.17
 number and width of exits 7.18
 Regulations, and 7.20
evacuation times 7.17
evaluation of risks 7.7
exit signs 7.19
fire detection system 7.7
fire warning system 7.7
five steps approach 7.4–7.9
generally 3.2, 3.6, 3.7, 7.1, 7.3, 10.3, 14.1
hazard
 definition 7.2
 identification 7.2, 7.5
high fire risk 7.12
information for employees 3.8, 3.9, 15.67
location of people at significant risk 7.6
low fire risk 7.14
maintenance and testing of fire safety equipment 7.26
means of escape 7.7, 7.10
means of fighting fire 7.23–7.25
fire safety training 7.24
new Fire Safety Order proposals 15.57
normal fire risk 7.13
recording of findings and action taken 3.8, 7.8
review of 7.9
risk, definition 7.2

Index

Fire safety legislation,
  proposals for reform *see*
  New Fire Safety Order
  proposals
Fire safety manager 9.12–9.15
  general responsibilities 9.14
  job specification 9.13
  role 9.12
Fire safety training
  assistance with 9.35
  continuity training 9.31
  contractors briefing 9.29
  damage control/salvage 9.34
  fire extinguishers, use of 9.6, 11.43
  fire marshals 8.5, 9.16–9.20
  fire safety manager 9.12–9.15
  fire teams 9.33
  Fire Training Manual 9.9, 9.11
  general staff training 9.25–9.27
  generally 7.24, 8.18, 9.1, 15.69
  legal responsibilities 9.2–9.6
  level of training 9.8
  manual *see* Fire Training Manual
  practice of systems 9.10
  record of training 9.32
  refresher training 9.31
  security staff 9.21–9.24, 11.74
  staff induction briefing 9.28
  trainers 9.30
  training policy 9.7–9.10
Fire service 10.24
  Bain Report 15.81–15.85
  see also Fire Authority
Fire shutters
  inspection 12.37, 12.38
Fire teams 9.33
Fire tests on building materials
  and structures
  British Standard A.5
Fire Training Manual 9.9, 9.11
Fire warning 11.1C, 11.24
Fire-fighting equipment
  extinguishers *see* Fire
    extinguishers
  fire blankets 11.44, 11.69
  'first aid' 7.23, 7.26, 8.5, 11.1D
  fixed *see* Fixed fire suppression
    systems
  generally 7.23, 8.17, 11.1D, 11.70
  hose reels 7.26, 11.42, 11.45, 11.69
  maintenance and testing 7.26, 12.12, 12.13
  portable 11.69, 12.12, 12.13
  sprinklers *see* Sprinkler systems
First aid 6.36
'First aid' fire-fighting
  equipment 7.23, 7.26, 8.5, 11.1D

Fixed fire suppression systems
  design and installation 11.53
  dry risers 11.55, 11.70
  foam inlets 11.54
  generally 11.1D, 11.51
  maintenance 11.52, 11.70
  wet risers 11.56
Flame detectors 11.26
  see also Automatic fire detection
    systems
Flammable liquids 8.12
Foam inlets 11.54
Fuel
  cause of fire, as 10.7, 10.8
  liquefied petroleum gas (LPG) 6.35, 8.9

G

Gaming houses 13.54, 15.24
Garages 13.55
Glass 8.42, 11.67
Guesthouses *see* Hotels and
  boarding houses

H

Harbours 15.34
Hazard
  definition 7.2
  identification 7.2, 7.5
  major accident hazard 6.15
  substances hazardous to health 6.16
Health and safety
  absolute duties 6.33
  absolute requirements 6.33
  advice, sources of 6.53
  arrangements 3.7, 6.46
  asbestos 6.28
  civil law
    civil procedure 6.40
    contributory negligence 6.42
    damages 6.41
    defences 6.42
    duty of care 6.38
    generally 6.37, 6.43
    injuries not reasonably
      foreseeable 6.42
    time limit for action 6.40
    vicarious liability 6.39
    voluntary assumption of risk 6.42
  confined space 6.12
  consultation with employees 6.36, 15.38
  contractors, employers'
    responsibility as to 6.10
  contributory negligence 6.42

393

Index

Health and safety – *contd*
corporate killing, proposed offence of	6.7
costs of accidents	6.45
criminal law	6.30–6.32
fines	6.30
offences	6.30
damages	6.41
dangerous goods, transport of	6.26
defences	
civil law	6.42
due diligence	6.48
deregulation, and	6.5
designers, duties of	6.32
disabled persons	6.50
display screen equipment	6.34
driving at work	6.17
duties of all people	6.52
duty of care	6.38
employees	
consultation with	6.36, 15.38
duties of	6.32
employers	
contractors, responsibilities as to	6.10
duties of	6.32
equipment, provision and use	6.24
European Union	
implementation of directives	6.3
role of	6.2
eyesight tests	6.34
fatal accidents	6.36
file	6.36
first aid	6.36
future trends	6.8
generally	6.1
hearing protection	6.36
importers, duties of	6.32
improvement notice, failure to comply	6.31
injuries not reasonably foreseeable	6.42
interference and misuse	6.32
ionising radiations	6.19
Labour government	6.6
lead	6.14
level crossings	6.25
levels of duty	6.33
lifting equipment	6.20
major accident hazards	6.15
manslaughter	6.7
manual handling operations	6.34
manufacturers, duties of	6.32
'minded to' procedure	6.5, 6.6
noise	6.22, 6.36
offences	6.30
high profile cases	6.8
penalties	6.30, 6.31

Health and safety – *contd*
offshore electricity	6.22
penalties	6.30, 6.31
personal protective equipment	6.16
plan	6.36
police	6.23
policies	6.46
practicable measures	6.33
premises	6.32
prohibition notice, failure to comply	6.31
public liability	6.49
railways	6.25
reasonably practicable measures	6.32, 6.33
reporting of injuries, diseases and dangerous occurrences	6.36
risk assessments	3.2, 3.7, 6.9, 6.10, 6.33, 6.47
safety advisers, transport of dangerous goods	6.26
safety committee	6.36
safety representatives	6.36
'six pack'	6.4, 6.9, 6.34
statutory duties	6.30
substances hazardous to health	6.16
suppliers, duties of	6.32
supply chain regulation	6.10
training for employment	6.36
unauthorised entry	6.51
vicarious liability	6.39
visitors	6.50
voluntary assumption of risk	6.42
work experience programmes	6.36
working time	6.27
workplace, fire safety in	3.30
young persons	6.50

Health and Safety at Work Act 1974 2.1, 3.30, 6.1, 6.5, 6.6, 9.3, 15.37
absolute requirements	6.33
breach	6.31
duties under	
all people	6.52
employees	6.32
employers	6.32, 13.16
levels of	6.33
manufacturers, suppliers etc	6.32
not to charge	6.32
premises	6.32
public liability	6.49
unauthorised entry	6.51
enforcement	15.39
health and safety policy	6.46
interference and misuse	6.32
practicable measures	6.33
public liability	6.49

Index

Health and Safety at Work Act 1974
– *contd*
reasonably practicable measures 6.33
Health and Safety Commission 6.3, 6.5
Health and Safety Executive
fire certificates 2.1
generally 6.5, 10.39–10.41, 13.72
ionising radiations 6.19
publications 6.44, 6.45, 6.53
role of 10.40
transport of dangerous goods 6.26
Hearing protection 6.36
Heat detectors 11.26
see also **Automatic fire detection systems**
Heating appliances
cause of fire, as 8.14, 10.14
Historic buildings
generally 13.2
listed buildings 13.3
consent 13.4
grants for 13.5
planning applications 10.38
means of escape 13.9
proprietary systems 13.8
risk management 13.10
self-help 13.11,–13.12
special problems 13.6
upgrading structural elements 13.7
Hose reels 7.26, 11.42, 11.45
maintenance 11.69
Hospitals 13.56
Hot processes
cause of fire, as 10.15, 10.18
'hot work' permit system 1.7, 13.26–13.30
Hotels and boarding houses 2.1, 13.57, 15.18
disability discrimination 4.7, 4.8
fire certificate 2.4, 2.5, 15.10
Housekeeping 8.1, 8.8, 8.44, 10.20
Hydrants 11.70

I

Ignition sources 10.9
Improvement notice 2.1, 2.26, 2.27
failure to comply 2.26, 2.27, 6.31
Incendiary bomb 8.38
see also **Bomb alerts and threats**
Information for employees 3.8, 3.9, 7.8
Inner rooms 11.15, 11.16

Insurance 1.5, 8.2, 11.46, 11.75, 14.1
employers' liability insurance 6.40
International Electrotechnical Commission (IEC) 5.3
International Organisation for Standardisation (ISO) 5.3

L

Lead 6.14
Legislation, proposals for reform *see* **New Fire Safety Order proposals**
Liability
public 6.49
vicarious 6.39
Liaison with authorities
fire service 10.24
generally 10.22
Health and Safety Executive 10.39–10.41
local authorities 10.27–10.38
see also **Building control**
reasons for 10.23
Licensing of premises 15.25–15.30
new Fire Safety Order proposals 15.72
Lighting
emergency lighting
British Standard A.6
Building Regulations 11.18
daily attention by user 12.16
generally 7.19, 11.17, 11.18
inspection 7.26, 11.73, 12.15
log for inspections 12.19
monthly attention by user 12.17
siting 11.19
subsequent tests 12.18
faulty equipment 10.12
generally 8.14
rooflights 3.32
Liquefied petroleum gas (LPG) 6.35, 8.9, 13.31
Liquids
flammable 8.12
Listed buildings *see* **Historic buildings**
Local authorities
building control *see* **Building control**
licensing of premises 15.28, 15.29
London
licensing of premises 15.29
Loss control
definition 8.19

Index

M

Machinery	8.13
see also Equipment	

Maintenance
automatic fire detection systems	7.26, 12.3, 12.11
buildings	8.11
doors	11.16, 11.67
emergency lighting	7.26, 11.73
fire alarms	7.26, 11.71, 12.3–12.11
fire extinguishers	11.69, 12.12
fire safety equipment	7.26, 7.27
fixed fire suppression systems	11.52, 11.70
hose reels	11.69
means of escape	11.67, 11.68
new Fire Safety Order proposals	15.64
planned maintenance programmes	11.66–11.73
pressurisation systems	12.31 12.36
repair services	12.3
service records	12.3, 12.11, 12.13
sprinkler systems	12.24–12.30
stand-by generator	12.21–12.23

Major Accident Prevention Policy (MAPP)	6.15
Management cycle	6.47

Management of Health and Safety at Work Regulations 1999
co-operation and co-ordination	3.11
costs of accidents	6.45
danger areas, procedures for	3.7
enforcement	3.13
evacuation procedure	11.61
generally	6.21, 6.34, 6.44, 15.40
information for employees	3.9, 7.8
offences	3.15
risk assessment	3.2, 3.7, 6.9, 6.47, 7.1, 10.3
record of findings	7.8
see also Fire risk assessment	
serious cases	3.15
serious and imminent danger, procedures for	3.7
voluntary workers	15.53

Manslaughter	6.7, 6.30
corporate offence	6.7

Masonry
| British Standard | A.6 |

Means of escape
automatic fire door releases	11.8
British Standards	A.2
Building Regulations	3.32, 11.13
compartmentation	11.6

Means of escape – contd
confined spaces	15.41
disabled persons	4.11
emergency exits	11.11, 11.13, 11.14
emergency lighting	11.17, 11.18
siting	11.19
emergency signs	11.20, 11.21–11.23
escape routes	7.16–7.20, 11.11, 11.12
emergency lighting	7.19
evacuation times	7.17
exit signs	7.19, 11.20
length of	7.17
number and width of exits	7.18
protected route	11.4
travel distances	11.4
exits	7.17, 7.20, 15.63
emergency exits	11.11, 11.13, 11.14
number and width	7.18
signs	7.19, 11.20
storey exits	7.17
fire doors	11.7
automatic releases	11.8
generally	2.1, 7.7, 11.2
historic buildings	13.9
Home Office guidance	7.10
inner rooms	11.15, 11.16
integral	11.67
maintenance	11.67, 11.68
measures to facilitate	11.1B, 11.10–11.23
mines	15.41
planned maintenance	11.68
principles	11.2–11.4
risk assessment	7.7
staircases	11.9, 13.9
structural protection	11.1A, 11.5–11.9, 11.67
travel distances	11.4
unsatisfactory	11.3

Mines	15.41, 15.42

Music
| premises used for | 13.67 |

N

Naked flames
| cause of fire, as | 10.13 |

National Health Service
| Crown immunity | 15.47 |

Negligence
	6.37
contributory	6.42
time limit for action	6.40
vicarious liability	6.39

New Fire Safety Order proposals
alterations to current legislation	15.14
application	15.53

Index

New Fire Safety Order proposals – *contd*
capabilities and training	15.69
charging	15.78
co-operation and co-ordination	15.68
Crown immunity	15.20, 15.74
danger areas, procedures for	15.66
dangerous substances, risks from	15.61
emergency routes and exits	15.63
employees, general duties of	15.70
enforcement	15.73, 15.75
EU legislation, and	15.13
exceptions	15.53
fire detection	15.62
fire safety arrangements	15.60
fire safety duties	15.56
fire safety regime requirements	15.55–15.80
fire-fighting	15.62
general duty to ensure safety	15.58
generally	15.2–15.8
guidance	15.71
information, provision of	15.67
licensing	15.72
maintenance	15.64
mitigation of effects of fire	15.54
necessary protection	15.80
order-making procedure	15.8
possible application, scope and requirements	15.13
principles of prevention	15.59
public reassurance	15.77
re-statement of existing burdens	15.79
responsible person	15.52, 15.65
risk assessment	15.57
serious and imminent danger, procedures for	15.66
validation of fire safety solutions	15.76
New systems	12.2

Notices
fire action notices	7.22, 11.62–11.65
Nursing homes	13.59

O

Offences
bodies corporate, by	2.34, 2.35
corporate killing, proposed offence of	6.7
enforcement notices	3.23, 3.24
falsification of documents	2.32, 2.33
Fire Precautions (Workplace) Regulations 1997	3.14
serious cases	3.14, 3.15
health and safety	
high profile cases	6.8
penalties	6.30, 6.31

Offences – *contd*
improvement notice	2.26, 2.27, 6.31
manslaughter	6.7, 6.30
prohibition notice	2.24, 2.25, 6.31
serious cases	3.14–3.18

Office equipment
cause of fire, as	10.19
Offices	13.66, 15.18
fire certificate	2.4, 2.5, 15.10
Outside public events	13.65

P

Planning
permitted development	10.33
planning department	10.30
planning permission	10.32, 13.18
advertisements	10.38
application forms	10.37
conservation areas	10.38
development	13.18
full application	10.36
internal alterations	13.18
listed building applications	10.38, 13.4
making an application	10.37
outline application	10.35
types of application	10.34–10.36
town and country planning	10.31

see also Building control

Postal bombs	8.39

see also **Bomb alerts and threats**

Pressurisation systems
acceptance test	12.32
British Standard	12.31
frequency of test	12.35
generally	12.31
information to be made available	12.33, 12.34
log for routine test	12.36
Preventative measures	11.1F, 11.66
Prohibition notice	2.1, 2.24, 2.25, 3.16–3.18
failure to comply	2.24, 2.25, 6.31
Public address systems	11.25
Public events, outside	13.65
Public liability	6.49
Public and private music, dancing etc, premises used for	13.67

R

Radiation
ionising radiations	6.19

Index

Railway premises 6.25, 13.58, 15.18, 15.19, 15.43
fire certificate 2.4, 2.5
Re-cycling installations 11.48E
Recovery after fire *see* Survival plan
Refurbishment
  fire risk during 1.7
Regulations
  power to make 15.51
Regulatory Reform Order proposals *see* New Fire Safety Order proposals
Residential and nursing homes 13.59
Responsible person 15.52, 15.65
Risk
  definition 7.2
  monitoring 8.19
  transfer 8.19
  voluntary assumption of 6.42
Risk assessment
  concept of 7.2
  definition 7.2, 8.19
  fire risk *see* Fire risk assessment
  five steps approach 7.4
  health and safety 3.2, 3.7, 6.9, 6.10, 6.33, 6.47
  management cycle 6.47
Risk management
  generally 8.19
  historic buildings 13.10
  loss control 8.19
  risk monitoring 8.19
  risk transfer 8.19
Rooflights 3.32

S

Safety signs and signals *see* Signs and signals
Salvage efforts 9.34, 14.14
Schools 13.60
Security
  arson, preventing 1.4, 1.5
Security staff
  fire safety role 9.22
  generally 9.21, 11.74
  training 9.23, 9.24, 11.74
Servicing *see* Maintenance
Shopping centres 13.61
Shops 13.62, 15.18
  fire certificate 2.4, 2.5, 15.10
Signs and signals
  British Standard A.6
  emergency signs 11.20, 11.21–11.23
  European Union Directive 11.20
  exit signs 7.19, 11.20, 11.72

Signs and signals – *contd*
  fire action notices 7.22, 11.62–11.65
  fire extinguishers 11.42
  fireman's switches 15.28
  generally 3.31, 6.36, 11.72
  maintenance of 3.31
  pictograms 3.31, 11.20
Smoke control systems
  acceptance test 12.40
  generally 12.39
  pressure differentials *see* Pressurisation systems
  routine testing 12.41
Smoke detectors 11.26
  see also Automatic fire detection systems
Smoking
  cause of fire, as 1.1, 8.15, 10.2, 10.3, 10.11, 10.21
Special premises
  fire certificate 2.38, 2.39, 3.29
Sports facilities 2.1, 13.63, 15.23
  capital allowances 15.48
Sprinkler systems
  acceptance tests 12.29
  alternate systems 11.48B
  automatic pump starting test 12.28
  British Standard 11.46, 12.24
  Building Regulations 11.50
  commissioning and acceptance tests 12.25
  completion certificate and documents 12.29A
  deluge and re-cycling installations 11.48E
  diesel engine restarting test 12.28
  dry pipe 11.48C
  generally 11.46, 11.47
  initial testing 12.27
  installation pipework 12.26
  lead acid plante batteries 12.28
  legislation 11.50
  log for routine examinations, inspections and test 12.30
  maintenance 12.24–12.30
  pre-action 11.48D
  standards 11.46, 11.49
  types of 11.48
  water motor alarm test 12.28
  weekly routine 12.28
  wet pipe 11.48A
Staff *see* Employees
Staircases 11.9, 13.9
Stand-by generator
  inspection 12.21–12.23
Static electricity
  cause of fire, as 10.17

Index

Statistics 10.3
Storage 8.10
Structural alterations and extensions
building control 13.17
code of practice 13.20–13.24
contractors
advising of fire safety policy 13.25–13.30
dealing with 13.19
potential hazards 13.31
principal contractor 13.33, 13.37, 13.42
role 13.38
designing out fire 13.24
emergency procedures 13.23
fire safety plan 13.21
generally 13.14
health and safety plan
checklist 13.40
generally 13.33, 13.39, 13.40
health and safety file 13.43
preparation by planning supervisor 13.41
principal contractor 13.33, 13.42
'hot work' permits 13.26–13.30
legal responsibility 13.16
liquefied petroleum gas, use of 13.31
planning control 13.18
planning supervisor 13.33, 13.36, 13.41
principal contractor 13.33, 13.42
Regulations 13.15
site fire safety co-ordinator, role of 13.22
Survival plan
documentation 14.15, 14.16
generally 9.34, 14.7, 14.8
planning 14.9–14.13
preparedness 14.10
prevention 14.11
recovery 14.12
review and test of 14.17
salvage efforts 9.34, 14.14

T

Tax relief
fire certificate compliance, and 15.48, 15.49

Terrorism *see* **Bomb alerts and threats**
Theatres 13.64, 15.30
disability discrimination 4.8
Timber, structural use
British Standard A.6
Town and country planning *see* **Planning**
Training
emergency procedures 11.57–11.60
for employment 6.36
fire safety *see* Fire safety training
'Triangle of fire' 10.6

V

Vehicle access 3.32
Vehicle bomb 8.37
see also **Bomb alerts and threats**
Vicarious liability 6.39
Visitors
bomb threats 8.43
evacuation of 8.20
health and safety 6.50
Voluntary assumption of risk 6.42

W

Warehouses 13.68
Warning
fire warning 11.1C, 11.24
Websites A.8
Wet risers 11.56
Wiring check 12.10
Work experience programmes 6.36
Workplace
definition 3.4
excepted 3.25, 3.26
fire risk assessment *see* Fire risk assessment
fire safety policy 8.2
persons in control 3.4, 3.28
publications 3.33
Workshops 13.55

Y

Young persons
protection of 15.45

A complete range of reference works for Health and Safety Professionals

Order Details

Title ISBN		Product Code	Unit Price	Total
New for 2003				
Workplace Accident Handbook	0 7545 2023 4	WAH	£60.00	
Fire Safety Training Manual	0 7545 2183 4	FST	£60.00	
Stress Management Handbook/Managing Stress in the Workplace	0 7545 1269 X	MSIW	£50.00	
Risk Assessment Workbook - Retail	0 7545 1890 6	RAWR	£60.00	
Risk Assessment Workbook - Education	0 7545 1891 4	RAWE	£60.00	
Risk Assessment Workbook - Leisure	0 7545 1892 2	RAWL	£60.00	
Risk Assessment Workbook - Health	0 7545 1887 6	RAWH	£60.00	
Risk Assessment Workbook - Office	0 7545 1889 2	RAWO	£60.00	
Risk Assessment Workbook - Manufacturing	0 7545 1888 4	RAWM	£60.00	
Looseleafs				
Health and Safety at Work*	0 7545 0815 3	HSWMW	£99.00	
Environmental Law and Procedures Management*	0 7545 0811 0	ENLMW	£95.00	
Handbooks				
Guide to Managing Employee Health	0 7545 1886 8	MGEH02	£59.50	
Managing Violence in the Workplace	0 7545 1967 8	MVW	£65.00	
Corporate Killing: Manager's Guide to Legal Compliance	0 7545 1066 2	CKNL01	£50.00	
Health and Safety at Work Handbook 15th Edition	0 7545 1751 9	HSW15	£70.00	
Office Health and Safety 3rd edition	0 7545 1267 3	OHS3	£72.50	
Practical Risk Assessment 4th edition	0 7545 2247 4	PRAH4	£50.00	
Online				
www.hsedirect.com			2 Users £444.15	
			one user £265.25	
			Total amount due	

● Confirm Delivery Details

mail _____
ame _____
b Title _____
ompany _____
ddress _____
_____
stcode _____ Tel _____

● Make Payment Choice

exisNexis Butterworths Tolley account, no. _____
 by credit card as follows: ☐ Mastercard ☐ Amex ☐ Visa
y credit card no. is _____
nd its expiry date is _____
r with the attached cheque for £ _____
(made payable to **LexisNexis Butterworths Tolley**)
y VAT registration no. is _____
☐ Please supply the above product(s) with a 21 day money back guarantee (EU countries only)
gnature/Date _____

+We are pleased to be able offer you these titles on an annual subscription basis. Simply tick the appropriate box below.

☐ Please send me the above title(s). I understand that I will be sent all future editions automatically on publication, with a 10% discount. I am at liberty to cancel this arrangement at any time, or return the book if it does not meet with my requirements.

☐ Please send me the above title(s). I do not wish to participate in the subscription scheme, however, please advise me when a new edition becomes available.

❹ Return Your Order

 Freepost
Paul Holliday Marketing Department Butterworths Tolley
FREEPOST 6983, London WC2A 1BR

 Phone
Please phone Customer Services on: +44 (0)20 8662 2000

 Fax
Please fax your order to: +44 (0)20 8662 2012

 Online Bookshop
Visit our Online Catalogue and Bookshop at www.lexisnexis.co.uk to order securely online.

Bookshop
You can also order copies via our specialist bookshop which stocks a wide range of our titles. It is based at 35 Chancery Lane, London WC2A 1EL, nearest tubes are Chancery Lane and Temple.
www.lexisnexis.co.uk

☐ If you do NOT wish to be kept informed by mail of other LexisNexis products and services, please tick here.

☐ If you do NOT wish your mailing details to be passed on to companies approved by LexisNexis, to keep you informed of their products and services, please tick here.

* Pay As You Go (PAYG)
I understand that I will continue to receive, until countermanded, future updating material issued in connection with this work and will be invoiced for these issues when I receive them.

21 Day Money Back Guarantee

The book(s)/product(s) mentioned on this order form, are available in EU countries only with 21 days money back guarantee. This allows you to try it/them out and assess its/their usefulness before buying. If you should decide not to purchase the book(s)/product(s), simply return it/them to us in good condition and your invoice will be cancelled and any payment made refunded. All orders placed with us are subject to our standard terms of trade, available on request and with the invoice. The book(s)/product(s) detailed on this order form is/are intended to be (a) general reference guide(s) to the law and cannot be (a) substitute(s) for legal advice.

www.lexisnexis.co.uk

Thank you for your order, prices are subject to change without notice.

35 Chancery Lane London WC2A 1EL A division of Reed Elsevier (UK) Ltd
Registered office 25 Victoria Street London SW1H 0EX
Registered in England number 2746621 VAT Registered No. GB 730 8595 20

A complete range of reference works for Employment Professionals

Order Details

Title ISBN		Product Code	Unit Price	Total
Books				
Managing Absence and Leave	0 7545 1953 8	IRSMAL	£89.00	
Managing Diversity in the Workplace	0 7545 1955 4	IRSMDW	£89.00	
Managing Employee Representatives	0 7545 1954 6	IRSMER	£89.00	
Employment Tribunals	0 7545 1488 9	TETH	£49.95	
Managing Dismissals	0 7545 1255 X	IMDF	£35.00	
Managing Fixed Term Part Time Workers	0 7545 1662 8	MFTPTW	£35.00	
Managing Business Transfers	0 7545 1661 X	MBTTMO	£37.00	
Looseleafs*				
Employment and Personnel Procedures	0 7545 0833 1	EPLMW	£95.00	
Termination of Employment	0 7545 0824 2	TOEMW	£95.00	

Total amount due

Confirm Delivery Details

ail _____

ne _____

Title _____

npany _____

ress _____

_____

_____

code _____ Tel _____

Make Payment Choice

sNexis Butterworths Tolley account, no. _____

y credit card as follows: ☐ Mastercard ☐ Amex ☐ Visa

credit card no. is _____

its expiry date is _____

vith the attached cheque for £ _____
made payable to **LexisNexis Butterworths Tolley**)

VAT registration no. is _____

Please supply the above product(s) with a 21 day money back guarantee (EU countries only)

nature/Date _____

We are pleased to be able offer you these titles on an annual ubscription basis. Simply tick the appropriate box below.

ase send me the above title(s). I understand that I will be sent all future editions automatically on ublication, with a 10% discount. I am at liberty to cancel this arrangement at any time, or return the ok if it does not meet with my requirements.

ase send me the above title(s). I do not wish to participate in the subscription scheme, however, ase advise me when a new edition becomes available.

❹ Return Your Order

Freepost
Nick King Marketing Department Butterworths Tolley
FREEPOST 6983, London WC2A 1BR

Phone
Please phone Customer Services on: +44 (0)20 8662 2000

Fax
Please fax your order to: +44 (0)20 8662 2012

Online Bookshop
Visit our Online Catalogue and Bookshop at
www.lexisnexis.co.uk to order securely online.

Bookshop
You can also order copies via our specialist bookshop which stocks a wide range of our titles. It is based at 35 Chancery Lane, London WC2A 1EL, nearest tubes are Chancery Lane and Temple.
www.lexisnexis.co.uk

☐ If you do NOT wish to be kept informed by mail of other LexisNexis products and services, please tick here.

☐ If you do NOT wish your mailing details to be passed on to companies approved by LexisNexis, to keep you Informed of their products and services, please tick here.

* Pay As You Go (PAYG)
I understand that I will continue to receive, until countermanded, future updating material issued in connection with this work and will be invoiced for these issues when I receive them.

21 Day Money Back Guarantee
The book(s)/product(s) mentioned on this order form, are available in EU countries only with 21 days money back guarantee. This allows you to try it/them out and assess its/their usefulness before buying. If you should decide not to purchase the book(s) /product(s), simply return it/them to us in good condition and your invoice will be cancelled and any payment made refunded All orders placed with us are subject to our standard terms of trade, available on request and with the invoice. The book(s)/product(s) detailed on this order form is/are intended to be (a) general reference guide(s) to the law and cannot be (a) substitute(s) for legal advice.

www.lexisnexis.co.uk

Thank you for your order, prices are subject to change without notice.

35 Chancery Lane London WC2A 1EL   A division of Reed Elsevier (UK) Ltd
Registered office   25 Victoria Street   London SW1H 0EX
Registered in England number 2746621 VAT Registered No. GB 730 8595 20